The Asian Tsunami and Post-Disaster Aid

Sunita Reddy
Editor

The Asian Tsunami and Post-Disaster Aid

Springer

Editor
Sunita Reddy
Centre of Social Medicine and Community Health and Adjunct Faculty at the Special Centre for Disaster Research
Jawaharlal Nehru University
New Delhi
India

ISBN 978-981-13-0181-0 ISBN 978-981-13-0182-7 (eBook)
https://doi.org/10.1007/978-981-13-0182-7

Library of Congress Control Number: 2018938637

Printed on acid-free paper

This Springer imprint is published by the registered company Springer Nature Singapore Pte Ltd.
The registered company address is: 152 Beach Road, #21-01/04 Gateway East, Singapore 189721, Singapore

Foreword

Disaster management in India has gone through a paradigm shift. To prevent disasters and to better prepare ourselves to manage disasters, we have adopted some of the best practices: used technological innovations and adopted novel ways of rescuing our people. We have understood the importance of investment in disaster preparedness, especially for strengthening early-warning systems, and made appropriate investments for improving the same. Efforts are being made to make best use of the expertise available in space science, meteorology, earth sciences, and seismology not only in our country but worldwide. Apart from prevention and preparedness, relief and mitigation are equally important.

The Tsunami catastrophe that struck Asia in December 2004 was one of the worst natural disasters in modern history. There was an overwhelming response by the people of India to help those who were affected by the tsunami, and the entire post-disaster situation was managed with national resources. In fact, India even managed to offer some assistance to Indonesia, Sri Lanka, and the Maldives.

The book "**The Asian Tsunami and Post-Disaster Aid**" delves into issues relating to managing aid in relief, recovery, reconstruction, and rehabilitation. Many crucial concerns related to aid, such as issues of distribution, accountability, transparency, dependency, and appropriateness have been analyzed. Its implications on the various facets of the lives of the affected communities have also been brought out. The book also explores issues ranging from shelter and livelihood, to the emergence of ethnic, caste-based, tribe–non-tribe conflicts in the post-disaster scenario, arising out of aid distribution. The role of agencies attempting relief and reconstruction, coordination between civil society and defence establishments, NGOs, religious groups, and others has also been analyzed. The book poses important questions with regard to the utilization of aid in post-tsunami regions in order to evaluate the impact on affected and non-affected communities.

I am confident that the meaningful analyses and research that have been presented in the book will help in framing strategies and evolving systems to effectively manage aid in relief, recovery, and rehabilitation in the post-disaster scenario.

New Delhi, India
June 2017

Kiren Rijiju
Minister of State for Home Affairs
Government of India

Acknowledgements

The journey of this book began with a display of Nature's destructive strength—The Asian Tsunami of 2004—a strength no human can challenge or conquer. It ended in the beautiful, almost heavenly locales of Mashobra in 2016 and Zurich in 2017, where the last few lines were penned down. This book is a culmination of an international conference *'Re-Examining Disaster, Recovery and Reconstruction: Social Science Perspectives on the Tsunami'*, which was organized on 14–15 January 2008, by the Center of Social Medicine and Community Health, in collaboration with the National Institute of Disaster Management (NIDM). As this book is primarily about long-term recovery and rehabilitation, documenting the decade-long experiences required time to assimilate, absorb, and reflect. The task which was especially time-consuming was analyzing what went wrong and documenting the lessons learnt after the disaster.

The international conference had brought together 40 social scientists, from over a dozen countries and a number of disciplines, to discuss social science perspectives on the Tsunami. I am thankful to Dr. Dhar Chakraborti, the then Executive Director, NIDM, for his support and collaboration. I would like to express my sincere appreciation to Lt. General Janak Raj Bhardwaj, the then member of NDMA, Professor Paul Greenough, late Professor B. B. Bhattacharya, former Vice-Chancellor of JNU, Mr. Dunu Roy, Director of Hazard Center, and other scholars, for the stimulating deliberations in this conference. Thanks to all my students who volunteered to be part of the seminar, especially Dr. Ranvir Singh, who was responsible for the beautiful and creative design of the seminar invitations and brochures, and for the technical support provided by Golak, and to Shishir for digging out the relevant literature for this book.

I would also like to thank all the institutions which supported this international conference; Jawaharlal Nehru University (JNU), Indian Council of Social Science Research (ICSSR), India Sri Lanka Foundation (ISLF), United Nation's International Strategy for Disaster Reduction (UNISDR), and the European Union (EU).

Each scholar who contributed to this book has been very kind and generous enough to revise and re-write his/her paper. Thanks for reposing faith in me to get this book published. It was 2017, by the time the book was completed and ready to be published. The final push to publish was the dawn of the realization that it was ‘now or never’. It was also my moral responsibility to ensure that this work saw the light of day, in order to share the cross-cultural experiences of the tsunami-affected communities as well as the lessons learnt during the disaster recovery and long-term rehabilitation. In addition, an important task was to bring out the challenges faced by the governments and the humanitarian agencies in responding to the Asian Tsunami, especially during reconstruction and rehabilitation to pave the way for a better understanding of disaster mitigation in future. On behalf of all the scholars, I express my heartfelt gratitude to the respondents in the field who shared their hopes, aspirations and painful experiences of facing disaster.

This book would not have been possible without the support and encouragement of various scholars, friends, well-wishers and institutions. I would like to convey my heartfelt gratitude to my mentor and colleague Dr. Imrana Qadeer, who encouraged me to follow my interest in disaster studies way back in 2004 when I joined JNU. My sincere thanks to my colleagues who took time to read through the drafts, especially Prof. Sanghmitra Acharya, Prof. P. C. Joshi, Prof. Santosh Kumar, Mr. Kamal Kishore, Prof. Vinod C. Menon and Prof. Madha Suresh, for giving their valuable comments. A note of appreciation to Shefali and Amrisha for their meticulous proofreading of chapters.

I would also like to thank few of my friends who have always stood by me and encouraged me. A special mention to Dr. Hari Mohan, Dr. Praveen Verma, and Aditi. My gratitude to Pushpinder and Gurlad Singh for providing me with a quiet and peaceful place to write down the final pages of this volume. Not to forget my appreciation the assistance of Kaushal and Bhagat Singh in everyday chores.

I am also glad to have an encouraging colleague in Prof. Amita Singh, with whom I realized a vision for a Special Centre for Disaster Research (SCDR) in JNU, which will set standards of research on disasters in India. I would like especially acknowledge the honourable Union Minister of State for Home Affairs, Shri Kiren Rijiju, as well as Prof. Sudhir Kumar Sopory, former Vice-Chancellor of JNU, who took a keen interest in setting up this Centre and have been inspiring all of us to contribute to disaster research.

This journey wouldn’t have been possible without the support of my family. My children, Shefali and Utkarsh, who understood well my passion for work and encouraged me to complete this long-pending task. They promised to be responsible and careful and gave me the confidence to get away and work peacefully. Thanks to my husband Neeraj for his support and being a good critic. I derive strength from the unconditional love and support of none other than my mom and dad, my brother Prem, and my sister-in law, Jo, who have always been there for me.

Sunita Reddy

Contents

The Asian Tsunami and Post-Disaster Aid: Critical Perspectives 1
Sunita Reddy
1 Contextualizing Post-disaster Aid 3
2 Asian Tsunami Aid: Colossal and Generous Funding 5
3 Positive Changes After Disasters 7
4 Disaster Aid in Post Asian Tsunami: Issues and Concerns 9
5 The Scope and Structure of the Volume 11
5.1 Section 1: Aid, Conflict and Differential Distribution 11
5.2 Section 2: Build Back Better? Relief Materials, Livelihoods, Shelters, and Discontentment 13
5.3 Section 3: Coping Strategies, Systems, and Aid Effectiveness . . . 16
6 Comparative Experiences and Lessons Learnt 18
7 Positive Global Development in Humanitarianism 20
8 Suggestions and Recommendations 21
References 22

Part I Aid, Conflict and Differential Distribution

An Ethno-political Accounting of the Sri Lankan Tsunami Impact and Recovery 27
Randall Kuhn
1 Introduction 27
2 Humanitarian Assistance and Conflict 29
3 Population and Civil Conflict in Coastal Sri Lanka 30
4 The Tsunami and Its Victims 33

5 Down the Rabbit Hole: Development Assistance Database in Practice 35
5.1 Aid by Sector 40
5.2 Aid by Funding Agency 42
5.3 Aid Allocations and Disbursements Over Time 43
6 Conclusions 47
References 48

Learning from Nature: Responding to Natural Disaster and the Need for a Multi-faceted Approach in Law 51
V. T. Thamilmaran
1 Introduction 51
2 The Immediate Response to the Tsunami: A Short-Lived Bubble of Spirited Humanitarianism 52
3 The Tsunami and the Minorities Affected by the Civil War 53
4 The Politics of the Response of the State 55
5 Issues of Good Governance 60
6 Conclusion 62
References 63

Social Interfaces in Disaster Situations: Analyzing Rehabilitation and Recovery Processes Among the Fisherfolk of Tamil Nadu After the Tsunami in India 65
Sunil D. Santha
1 Introduction 65
2 The Progression of Vulnerability: The Pre-impact Phase 67
2.1 Root Causes 67
2.2 Dynamic Pressures 68
2.3 Unsafe Conditions 70
3 Social Interfaces in Tsunami Rehabilitation and Recovery in Tamil Nadu 72
3.1 Strengthening Access to Resources: Customary Property Rights and the Deregulation of the Coastal Environment 72
3.2 Rebuilding Livelihoods: Ecological Insecurity and Resource Use Conflicts 73
3.3 Providing Relief: Defining the Affected, Identities and Conflicts 74
3.4 Designing Rehabilitation and Mainstreaming Development 75
4 Discussion: Disaster Recovery, or Re-creating Vulnerabilities? 76
References 77

Disaster, Development, and Water: The Reconstruction and (Re)Fabrication of Hazardous Waterscapes in Post-tsunami India 79
Luke Juran
1 Introduction: The Historical and Theoretical Context 79
2 Methods 80
3 Setting the Stage 82
4 Water Quality Woes in Post-tsunami India 84
5 Coping with (Poor) Water Quality in Post-tsunami India 93
6 Conclusions 97
References 98

An Applied Anthropological Perspective on Localizing Post-disaster Aid: Lessons from Post-tsunami Aceh, Indonesia 101
Patrick Daly
1 Introduction 101
2 Localizing Aid to Support Recovery and Reconstruction 103
3 The 2004 Tsunami and Reconstruction in Aceh 104
4 Community Infrastructure and Recovery 106
4.1 Cultural Spaces and Places 106
4.2 Localized Leadership and Decision-Making 108
4.3 Formal and Customary Legal Systems 110
4.4 Community Mobilization and Volunteerism 112
5 Discussion 114
References 115

Land, Natural Resources and Sustainable Development in Aceh, Indonesia: Confronting Inequalities Through Post-tsunami and Post-conflict Recovery 121
Jane Dunlop
1 Introduction 121
2 A Background to Land and Natural Resource Rights in Aceh: Exploitation, Conflict and Natural Disaster 122
3 GAM's Fight for an Independent Aceh 122
4 The Impact of the Tsunami 124
5 The Memorandum of Understanding (MoU) 125
6 Post-tsunami and Post-conflict—Rehabilitation and Reconstruction 125
6.1 Land Tenure Security 126
6.2 Women's Rights to Land 128
6.3 Landless Survivors of the Tsunami 130
6.4 Grievances Over Land 132
7 Economic Recovery and Long-Term Development 133
8 Conclusion 137
References 138

Part II Build Back Better? Relief Materials, Livelihoods, Shelters and Discontentment

The Sustainability of Humanitarian Aid: The Nicobar Islands as a Case of 'Complex Disaster' 143
Simron Jit Singh, Marina Fischer-Kowalski and Willi Haas
1 Introduction 143
2 The Pre-tsunami Socioecological System of the Nicobar Islands 144
3 The Tsunami and the Humanitarian Response in the Nicobar Islands 148
3.1 The Role of Aid Organizations 150
3.2 The Impact of State Interventions 152
4 What Is a 'Complex Disaster'? 154
5 Relief for the Nicobar Islands: A Case of Complex Disaster? 158
References 163

The Tsunami and Its Aftermath: Resilience and Rupture of the Social Fabric Among the Nicobarese 167
Mohammed A. Abid
1 Introduction 167
2 The Island of Katchal 168
3 Initial Response of the Community 171
4 Aid and Assistance by the State 172
5 Role of Non-state Players 175
6 Housing and Habitat 175
7 Social Change 177
8 Economy and Livelihoods 179
9 Political Reorganization 180
10 The Issue of Land Rights 182
11 Public Infrastructure 184
12 Conclusion 185
References 186

New Andamans: The Aftermath of Tsunami in the Andaman and Nicobar Islands 187
Sunita Reddy
1 Introduction 187
2 Aid, Interventions, and the Gap Between the Bureaucratic and Emergent Norm 188
3 Andaman and Nicobar Islands 190
3.1 Geographical Location 190
3.2 Historical Significance of the Islands 191
3.3 'Settler' Population and 'Illegal Settlements' 192
3.4 Indigenous Tribes of the Andaman and Nicobar Islands 192

3.5 Depletion of Ecological Treasures and a Disaster in the Making 193
3.6 Overcrowding of the Islands and Encroachment of Land 194
4 Rehabilitation Process 195
4.1 Impact of the Tsunami 195
4.2 Challenges in Rehabilitation 196
4.3 Relief Measures 197
4.4 Aid Agencies 198
4.5 Debates Over Development in the Islands After the Tsunami 199
4.6 Livelihood Rehabilitation 201
4.7 Linkages of Shelters with Livelihoods 202
4.8 'Tsunami Houses' 204
4.9 Loss of Neighbourhood and Networks 206
5 Discussion 206
References 208

Post-disaster Housing: Lessons Learnt from the 2004 Tsunami of Sri Lanka 211
Nisha Arunatilake
1 Introduction 211
2 Government's 2005 Policy on Housing 213
2.1 The Two-Pronged Process 213
2.2 Donor-Driven Relocation 215
2.3 Owner-Driven Rebuilding 215
3 The Government's Revised Policy on Housing 216
4 Methodology 217
5 Results 218
5.1 Characteristics of the Households 218
5.2 Issues Relating to the Tsunami Housing Programme: Owner-Driven Rebuilding 219
5.3 Donor-Driven Relocation Programme 221
5.4 Progress and Perceptions of Process One and a Half Years on 223
5.5 Donor Coordination 228
5.6 Perceptions of the Change in Housing Programme 229
6 Conclusions 230
6.1 Coverage and Access to Assistance 231
6.2 Speed of Receiving Assistance 231
6.3 Welfare Effects 231
6.4 Coordination and Grievance Resolution 232
References 232

Part III Coping Strategies, Systems and Aid Effectiveness

Aid, Religion and Recovery in Post-tsunami Thailand 235
Monica Lindberg Falk
1 Introduction 235
2 The Impact of the Tsunami on Thailand 237
3 Humanitarian Aid 238
3.1 The Thai Government's Relief Efforts 238
3.2 National Foundations and Relief Work 239
3.3 International Cooperation 241
3.4 International Relief Work 241
3.5 Relocation and Housing Issues 242
3.6 Livelihood and Compensation 242
4 Religion in Post-disaster Recovery 243
5 Buddhist Temples in Post-disaster Phang Nga 244
5.1 Temples and Aid Distribution 246
6 Survivors' Experiences of Post-tsunami Aid 247
7 Conclusion 249
References 251

Memory Interventions: Gujarat and Sri Lanka 253
Edward Simpson and Malathi de Alwis
1 Introduction 253
2 A Note on Natural Disaster and Social Process 255
3 The Gujarat Earthquake 256
4 The Tsunami and Sri Lanka 261
5 A Comparative Conclusion 265
References 268

Conflicting Paradigms and the Danger Discourse: Re-thinking Indian Disaster Management Framework in the Post-tsunami Era 271
Pravin Kushwaha
1 Paradigms of Disasters, Development, and Vulnerability 271
2 Unpacking the 'Paradigm Shift' in Indian Disaster Management Framework 275
3 Vulnerability and Post-tsunami Recovery in India 280
4 Aid Politics and Re-thinking Indian 'Paradigm Shift' 284
5 Conclusion 286
References 286

Tsunami Recovery Impact Assessment and Monitoring System (TRIAMS): Methodological Approach for Recovery 289
Lakshman Dissanayake
1 Introduction 289
2 Objective 290

3 Methodology 291
4 Selection of the Sample 291
5 Methods of Consultation 293
5.1 Consultation of Affected Communities 293
5.2 Consultation Process for Livelihood Sector 293
5.3 Consultation Process for Women 293
5.4 Consultation Process for Disabled People 294
5.5 Consultation Process for Children 294
5.6 Consultation Process for Elderly 294
6 Application of Formative Dialogue Research Methodology 294
7 Approach 295
8 User-Friendly Format and Maps 296
9 Use of Statistics to Examine the Internal Consistency of Information 296
10 Median Values for Further Analysis 297
11 Special Note on the Four Areas of Recovery 302
References 302

Tsunami Aid and Its Effectiveness 305
P. G. Dhar Chakrabarti
1 Introduction 305
2 Unprecedented Humanitarian Assistance 306
3 Spatial and Sectoral Distribution of Aid 309
4 Effectiveness of Tsunami Aid 313
5 Linking Relief, Rehabilitation, and Development 318
6 Lessons Learnt from Tsunami Aid 320
7 Conclusion 326
References 329

About the Editor

Sunita Reddy is Associate Professor at the Centre of Social Medicine and Community Health, School of Social Sciences, and also Adjunct Faculty in Special Centre for Disaster Research (SCDR) Jawaharlal Nehru University (JNU), Delhi. She is an anthropologist specializing in medical anthropology, and teaches public health in JNU. She has researched on various issues related to health, violence against women and children, migrant women workers in construction and domestic sectors, medical tourism, surrogacy, and has published various research papers in national and international journals. She has delivered lectures in many Indian and foreign universities over the past two decades. She has also researched on disaster issues from a social science perspective, and has published a book, *Clash of Waves: Post-Tsunami Relief and Rehabilitation in Andaman and Nicobar Islands* (2013). She has been in the core group on the formulation of guidelines on disasters in India, and is also a core group member in starting the SCDR in JNU. She is co-editor of *Marginalization in Globalizing Delhi: Land, Livelihoods and Health* (Springer 2017). She is a founder member of 'Anthropos India Foundation', a trust that promotes visual and action anthropology (www.anthroposindiafoundation.com). She is active in sports, and co-curricular activities, and is involved in social service through a community-based organization called 'SATAT'.

Abbreviations

ADB	Asian Development Bank
ANI	Andaman and Nicobar Islands
ANIIDCO	Andaman and Nicobar Islands Integrated Development Corporation
ANPATR	Andaman and Nicobar Protection of Aboriginal Tribes Regulation
APWD	Andaman Public Works Department
BPN	Badan Pertanahan Nasional
BRA	Badan Reintegrasi-Damai Aceh
BRR	Badan Rehabilitasi dan Rekonstruksi
CBOs	Community Based Organizations
CBPFs	Country Based Pooled Funds
CDA	Community-Driven Adjudication
CERF	Central Emergency Response Fund
CFA	Cease Fire Agreement
CHS	Core Humanitarian Standard
CoHA	Cessation of Hostilities Agreement
CRF	Calamity Relief Fund
CRZ	Coastal Regulation Zone
CTC	Chairman Tribal Council
CUCEC	Colombo University Community Extension Centre
DAC	Development Assistance Committee
DAD	Development Assistance Database
DAP	Dollars per Affected Population
DAT	Damage Assessment Team
DAU	Dana Alokasi Umum
DDMA	District Disaster Management Authority
DEC	Domestic Energy Consumption
DMA	Disaster Management Act
DMC	Disaster Management Centre
DMC	Domestic Material Consumption
DNRF	National Disaster Response Force

DRR	Disaster Risk Reduction
DS	Divisional Secretary
DSD	Divisional Secretariat Division
DVD	Digital Video Disc
ECHO	European Community Humanitarian Office
ECLAC	Economic Commission for Latin America and the Caribbean
ECOSAN	Ecological Sanitation
FBOs	Faith Based Organizations
FCA	The Forgotten Crises Assessment
FD	Fixed Deposit
FEPs	Forcibly Evicted Peoples
FFI	Flora and Fauna International
FGD	Focus Group Discussion
FTS	Financial Tracking System
GA	Government Agent
GAM	Gerakan Aceh Merdeka
GH	Government Hospital
GHD	Good Humanitarian Donorship
GN	Grama Niladhari
GND	Grama Niladhari Division
GOI	Government of India
GoI	Government of Indonesia
GoSL	Government of Sri Lanka
GSW	Grama Sevaka Wasama
HIC	Humanitarian Information Centre
HPC	High Power Committee
IDLO	International Development Law Organisation
IDP	Internally Displaced People
IFC	International Finance Corporation
ILO	The International Labour Organisation
IMF	International Monetary Fund
IMFL	Indian Manufactured Foreign Liquor
INGO	International Non-Governmental Organization
IPS	Institute of Policy Studies
JBIC	Japan Bank for International Cooperation
JHU	Jathika Hela Urumaya
JVP	Janatha Vimukthi Perumana
KPA	Komite Peralihan Aceh
LARC	Legal Awareness Raising Campaign
LBH	Lembaga Bantuan Hukum
LoGA	Law on Governing Aceh
LPG	Liquid Petroleum Gas
LRRD	Linking Relief, Rehabilitation, and Development
LTTE	Liberation Tigers of Tamil Eelam
MCLE	Mediation and Community Legal Empowerment

MDF	Multi Donor Fund
MEFA	Material and Energy Flows
MoU	Memorandum of Understanding
MSF	Médecins sans Frontières
NDMA	National Disaster Management Authority
NDMC	National Disaster Management Centre
NDMP	National Disaster Management Plan
NGO	Non-Governmental Organization
OCHA	Office for Coordination of Humanitarian Response
PAR	Pressure and Release
PDNA	Post Disaster Need Assessment
PoU	Point of Use
PRI	Panchayati Raj Institutions
P-TOMS	Post-Tsunami Operational Management Structure
PVTGs	Particularly Vulnerable Tribal Groups
PWD	Public Works Department
RADA	Reconstruction and Development Agency
RALAS	The Reconstruction of Aceh Land Administration Systems
RO	Reverse Osmosis
RRR	Relief, Rehabilitation, and Reconstruction
SBY	Susilo Bambang Yudhoyono
SC	Save the Children
SDMA	State Disaster Management Authority
SHG	Self Help Group
SLFP	Sri Lanka Freedom Party
SLR	Sri Lankan Rupees
SRO	Special Relief Officer
TAFREN	The Task Force for Rebuilding the Nation
TEC	The Tsunami Evaluation Coalition
TISS	Tata Institute of Social Sciences
TNA	Tamil National Alliance
TNI	Tenaga Negara Indonesia
TRC	Total Residual Chlorine
TRIAMS	Tsunami Recovery Impact Assessment and Monitoring System
TRO	Tamil Rehabilitation Organisation
TV	Television
TWA	Tin, Wood, and Asbestos
TWAD	Tamil Nadu Water Supply and Drainage Board
UDA	Urban Development Authority
UN	United Nations
UNDP	United National Development Programme
UNFCCC	United Nations Framework Convention on Climate Change
UNICEF	The United Nations Children's Fund
UNISDR	United Nations International Strategy for Disaster Reduction
UNORC	UN Office of the Recovery Coordinator

UNP	United National Party
USD	US Dollars
UT	Union Territory
VRC	Village Rehabilitation Committee
WATSAN	Water and Sanitation
WB	World Bank
WHS	World Humanitarian Summit

List of Figures

An Ethno-political Accounting of the Sri Lankan Tsunami Impact and Recovery

Fig. 1 Tsunami deaths by district, National Disaster Management Centre 31
Fig. 2 Average donor commitment per adjusted affected population, by month of project initiation and province 33
Fig. 3 Average donor disbursements per adjusted population, by month of project initiation and province 46

Learning from Nature: Responding to Natural Disaster and the Need for a Multi-faceted Approach in Law

Fig. 1 Regional Breakdown of progress in housing construction made through the donor driven scheme as on April 2006 (percentage of needs being met or in the process of being met) 56

Social Interfaces in Disaster Situations: Analyzing Rehabilitation and Recovery Processes Among the Fisherfolk of Tamil Nadu After the Tsunami in India

Fig. 1 The PAR model: progression of vulnerability among the traditional fisherfolk in South India 71

Disaster, Development, and Water: The Reconstruction and (Re)Fabrication of Hazardous Waterscapes in Post-tsunami India

Fig. 1 The 14 study villages in Nagapattinam District, Tamil Nadu, and Karaikal District, Puducherry 81

Fig. 2 These taps in New Nambiyar Nagar are surrounded by standing water. Some taps are only accessible by jamming rubber hoses into the pipe, which inevitably fall into the water from time to time. Other taps are completely submerged, with hoses barely reaching out from the floodwater. 88
Fig. 3 Floods and waterlogging, primarily during the monsoon, represent hazards to water quality, not to mention hygiene, mobility, and quality of life. Here, standing water is visible in New Nambiyar Nagar (**a**), Saveriyarkovil (**b**), and Andana Pettai (**c**), while a septic tank is surrounded by water in New Nambiyar Nagar (**d**). 89
Fig. 4 Drainage ditches are purposely blocked by cement in Akkaraipettai (**a**), sand in Samanthanpettai (**b**), and solid waste in Mandapathur (**c**), while water is obtained from a tap hovering above an open drainage ditch in Kilinjilmedu (**d**) 89
Fig. 5 Taps have been lowered to subsurface levels in order to access water in Saveriyarkovil (**a**) and Akkaraipettai (**b**). In Kizhakasakudimedu (**c**), a young girl fills a small container with water and subsequently pours it into a larger container near surface level 91
Fig. 6 Access at this subsurface tap in Kizhakasakudimedu has been obstructed by monsoon rains, rendering quality poor and access impossible 92
Fig. 7 Storage of water is a fixture of post-tsunami village life. Here, water is stored in an unused bathroom (**a**), unused kitchen (**b**), and storage tank provided by non-profits (**c**) 93

The Sustainability of Humanitarian Aid: The Nicobar Islands as a Case of 'Complex Disaster'

Fig. 1 Location of the Nicobar Islands. 145
Fig. 2 Comparing material and energy flows (MEFA) before and after the tsunami 161

The Tsunami and Its Aftermath: Resilience and Rupture of the Social Fabric Among the Nicobarese

Fig. 1 Family details of the Saul family: survivors out of a family of 95 members 182

New Andamans: The Aftermath of Tsunami in the Andaman and Nicobar Islands

Fig. 1 Map of Andaman and Nicobar Islands 191

Post-disaster Housing: Lessons Learnt from the 2004 Tsunami of Sri Lanka

Fig. 1 Perception on location, house quality, and access to services, relocated 228

Memory Interventions: Gujarat and Sri Lanka

Fig. 1 'Tiger Hill', Indraprastha, Gujarat 258
Fig. 2 The preserved 'Ruins' of Adhoi, Gujarat. 260
Fig. 3 Section of mural depicting the train being swept away, from the 'Fernandopulle Memorial', Peraliya, Southern Province, Sri Lanka 262
Fig. 4 Counter-memorial at Peraliya, Southern Province, Sri Lanka. 263
Fig. 5 Navalady memorial, Batticaloa, Eastern Province, Sri Lanka. 264

Tsunami Aid and Its Effectiveness

Fig. 1 International humanitarian response to Indian Ocean Tsunami. 306
Fig. 2 Highest and lowest responses to UN coordinated appeals (2004–2013) 308
Fig. 3 Distribution of Tsunami aid to affected countries 310
Fig. 4 Tsunami aid for relief and reconstruction 311
Fig. 5 Sectoral allocation of Tsunami aid by major donors 312
Fig. 6 The UN cluster system 322
Fig. 7 Disaster-related international financing 1980–2010 325

List of Tables

An Ethno-political Accounting of the Sri Lankan Tsunami Impact and Recovery

Table 1 Tsunami impacts according to death toll, homes destroyed, and affected population, by district 34
Table 2 Donor assistance database description and exclusions from analysis 36
Table 3 Donor project allocations and aid commitment by province 38
Table 4 Donor aid disbursement by province 39
Table 5 Donor aid disbursement per affected population, by sector and province 40
Table 6 Donor aid disbursement per affected population, by funding agency and province 42
Table 7 Donor aid disbursement per affected population, by funding agency and time 44

Learning from Nature: Responding to Natural Disaster and the Need for a Multi-faceted Approach in Law

Table 1 Summary of assessed reconstruction needs and funding commitments 55
Table 2 Total amount of aid received in Sri Lanka 61

An Applied Anthropological Perspective on Localizing Post-disaster Aid: Lessons from Post-tsunami Aceh, Indonesia

Table 1 Estimates of tsunami damage in Aceh and Nias 104
Table 2 Source of funding allocated to specific projects 104
Table 3 Summary of funding allocations by sectors 105

The Sustainability of Humanitarian Aid: The Nicobar Islands as a Case of 'Complex Disaster'

Table 1 Medium term consequences of disasters and complex disasters 159

The Tsunami and Its Aftermath: Resilience and Rupture of the Social Fabric Among the Nicobarese

Table 1 Demographic impact of the disaster on the Nicobarese 168
Table 2 Nicobarese population in Katchal 169
Table 3 Damage to plantation crops in Katchal 170
Table 4 Changes in land cover features in Katchal due to the tsunami 170
Table 5 Tribal council of Katchal as on 26 December 2004 170
Table 6 Types of ex gratia payments 172
Table 7 Temporary Shelter sites in Katchal which later became Permanent Settlements 176
Table 8 Reorganized tribal council of Katchal Island in 2006–7 181

New Andamans: The Aftermath of Tsunami in the Andaman and Nicobar Islands

Table 1 Loss of property: private and public 196

Post-disaster Housing: Lessons Learnt from the 2004 Tsunami of Sri Lanka

Table 1 Housing reconstruction assistance policy 214
Table 2 Housing reconstruction: process flow and managing agency 214
Table 3 Housing situation as in July 2006 (by eligibility under 2005 housing programme) 226
Table 4 Perceptions on rebuilding of houses by source of funds 227
Table 5 Quality of houses: relocated households 227

Tsunami Recovery Impact Assessment and Monitoring System (TRIAMS): Methodological Approach for Recovery

Table 1 Distribution of the sample size for the FGDs 292
Table 2 Gamma statistic for selected variables, Galle District 297
Table 3 Median values for variables, Southern Districts, Sri Lanka 297
Table 4 Level of recovery: FGD for tsunami-affected women's sector, Kahawagammedda, GN Division, Bentota DS Division, Galle District 298

Tsunami Aid and Its Effectiveness

Table 1 Contributions to the United Nations for humanitarian assistance to the victims of the Indian Ocean Tsunami (in USD) 309

The Asian Tsunami and Post-Disaster Aid: Critical Perspectives

Sunita Reddy

The Asian Tsunami of 2004 drew worldwide attention due to the sheer scale of destruction: massive loss of lives, livelihoods, and infrastructure. It attracted unprecedented international attention, assistance, and aid. While there has been a lot written on the tsunami, mostly by natural scientists, engineers, disaster management, and policy specialists, yet a sustained body of scholarly work in the social sciences on long-term rehabilitation has not been developed. The existing perspectives from different disciplines are understandably influenced by the enormous loss of property and material devastation and the efforts to manage and mitigate disasters. However, the fact remains that disasters are social phenomenons and there are acute socio-economic and political dimensions to the devastation, having long-term implications, especially on recovery. Such social disruptions call for a social science perspective—against a regional and political backdrop—to address a range of issues, including the primary issue of displacement that results from disaster and subsequent development.

The tsunami in 2004 witnessed two major tidal waves, which ushered the massive destruction. However, the aid that followed was 'the third wave' (Silva 2009) and its sheer magnitude was hitherto unseen. Due to the sheer magnitude of the aid, it has also been referred to as the 'second tsunami'. The enormous humanitarian response was due to various other reasons: the disaster's uniqueness; high mortality rate; foreign tourists affected; media coverage of the large-scale death and destruction. The overwhelming humanitarian response was also because of the wide geographical reach of the disaster, which affected 14 countries.

While aid is necessary for long-term recovery from a major disaster, managing it appropriately and effectively is a challenge. Managing aid for relief, reconstruction,

S. Reddy (✉)
Centre of Social Medicine and Community Health and Adjunct Faculty at the Special Centre for Disaster Research, Jawaharlal Nehru University, New Delhi, India
e-mail: sunitareddyjnu@gmail.com

S. Reddy (ed.), *The Asian Tsunami and Post-Disaster Aid*,
https://doi.org/10.1007/978-981-13-0182-7_1

and rehabilitation raises a lot of issues and concerns like its appropriateness, as well as distribution, accountability, and transparency. Aid sometimes undermines community resilience and leads to increased dependency and irreversible changes for either good or bad.

In this volume, aid is considered to be a generic category, which includes assistance, in both cash and kind, by the national/federal/state Governments, Non-Governmental Organizations (NGOs), humanitarian agencies (national and international), as well as Faith Based Organizations (FBOs) and Community Based Organizations (CBOs). There is no demarcation or mention of the source of aid, but in general, the volume attempts to capture the post-disaster aid and its consequences, both positive and negative, for the communities.

This book on disaster rehabilitation is based on empirical and longitudinal research by the scholars from the countries affected, as well as others, with an insider and outsider perspective, drawn by or already studying the subject prior the tsunami. They shared lessons learnt and possible solutions for managing aid in long-term reconstruction and rehabilitation. It includes work on aid and its implications for various facets of life in the affected communities. The book dwells upon issues of conflicts within and between the communities; some inherent and some manifesting post-disaster. It also focuses on intra-community conflicts that arise out of aid distribution. Thus, the book examines the accountability, conduct, effectiveness, delivery, politics, and ethics of 'Aid'.

This book covers major geographical areas affected by tsunami: Tamil Nadu, Puducherry, and the Andaman & Nicobar Islands in India; Banda Aceh in Indonesia; Sri Lanka; and, Thailand. The scope of the volume is to re-examine the current paradigms of relief, recovery, and rehabilitation, guiding the global humanitarian response. The papers in this volume explore issues ranging from shelters, livelihoods, emergence, and re-emergence of ethnic, caste, and tribe–non-tribe conflicts in the post-disaster scenario. Most of the papers in this volume are based on qualitative research and are ethnographic studies. With a theoretical underpinning and methodological rigor, these longitudinal research findings on the long-term rehabilitation post-tsunami raise many concerns and lessons to be learnt, with cross-cultural experience from people's perspective. The papers have tried to recover the 'emic' perspectives, through individual and collective narratives about the process of rehabilitation, their expectations and desires, and how they have both benefited and lost in the process. This book has an academic rigor, is unbiased, value-neutral, and critically looks at the accepted and taken-for-granted concepts in the disaster discourse.

What distinguishes this volume, however, is that we intend to problematize concepts that are normally taken for granted, including the key concepts of disaster, recovery, and reconstruction. We must bear in mind that agencies attempting relief and reconstruction may do so with certain theories and assumptions, leading to a particular trajectory of recovery and rehabilitation. The possible incongruence felt between the needs and expectations of the affected, the kind of imagined and actual practice of relief by agents, which includes the state, NGOs, civil society, and religious groups, is something that needs to be systematically addressed and re-examined. It is this incongruence that has led to questions about whose needs are best served by the aid: those of donor agencies, local governments and their bureaucracy, construction sector or tourist industry? How about the survivors of the disasters, who are by no means homogenous?

The post-disaster aid has led to an exacerbation of the old, and the creation of new, forms of inequities, arising from the prioritization, distribution, and management of aid, both by aid agencies and by national and local governments. Though the aid was fruitful in the short term, in the long term it had unintended consequences on the affected and non-affected communities. This poses certain pertinent questions: What has the aid been expended on? Where has the aid primarily been expended? How has it been expended? Who has it been expended on? What were the unintended consequences of post disaster aid for the communities? These five questions tell us not merely who has benefited from the aid, but the nature and modalities of this benefit, what has been left behind in its wake, and all this within a larger context of society and its institutions.

The following section looks at humanitarianism from a political perspective, and further situates the Asian Tsunami and post-disaster aid and its consequences in the disaster discourse.

1 Contextualizing Post-disaster Aid

Humanitarianism is associated with the humane and the positive, partaking the notion of 'human rights'. Humanitarianism is constructed around helping of 'victims', especially the victims of natural disasters. Humanitarian space is seen as an environment, where humanitarians can work without hindrance and follow the principles of neutrality, impartiality and humanity (Spearin cited in Hilhorst and Jansen 2010). Even though operating in highly politicized circumstances, humanitarian action is often rendered as being non-political (Minear and Weiss 1995, 17). However, humanitarianism is deeply shaped by the political sphere and involves mundane political activities, which have political underpinnings, like getting visa permissions and duty free entry of supplies (Warner 1999). Scholars have cautioned that humanitarian space needs to be understood in informal daily interactions, in socially negotiated arena. In practice, human resources and implementation of relief projects takes place through subtle power processes, defying the very principles of impartiality, neutrality, and independence, coordinated by institutional interests, the local socio-cultural fabric, and the prevailing power differentials among the aid recipients (Hilhorst and Jansen 2010).

Neo-liberal policies in the globalized capitalist world are animated by the underlying principles of high consumption, counter to sustainable development. Development Aid for development programs presently adds up to 100 billion USD annually (Riddell 2007). Globally, this relentless flow of humanitarian aid along with the top down development program initiatives would continue until and unless a conscious effort is made by governments to channelize such aid into more focused intervention rather than succumbing to temptations of an un-mindful development.

Policy changes leading to privatization, the rollback of social-safety nets and government responding to radical social engineering, all led to the generation of increasing inequality. It was found in the case of the post-Katrina reconstruction (Schullar 2008, 19), and was also witnessed during the reconstruction in case of the Asian Tsunami as evident from previous studies as well as from the contributions in the current volume.

International aid also serves as a geopolitical tool that donor states use to influence the beneficiary government and their people, and the strong attachment to international assistance shapes its outcome and efficacy (Hyndman 2011, 8). Further, Klein (2007) uses the concept of 'disaster capitalism' to demonstrate how international aid might function to promote and empower a range of private neoliberal capitalist interests, showcasing the pro-capitalist development agendas, which are pushed in the state of shock. In the neo-liberal era, public authorities tend to overlook the needs of the common people by increasingly compromising the standards set by global corporations and international financial institutions. There are instances where International policies privilege NGOs as the favoured recipients of international support. However, few NGOs have fostered an environment of minimal accountability and the politicians to their local constituencies, such as in Haiti in the aftermath of the disastrous earthquake in 2010, which led to erosion of the state's capacity to act in the time of crises and created unintended consequences (Zanotti 2010).

Scholars have questioned the 'Naturalness' of natural disasters. The fact is that many disasters are composite, which means that naturally induced effects are difficult to separate from anthropogenic ones (Alexander 1999, 9). It is the social interface with the natural events which lead to disasters. Inherently, disaster becomes an integral social phenomenon. It is important to understand that disasters have significant influence, which affect the socio-economic, organizational, political, and environmental domains in complex ways (UNISDR 2007; Wisner et al. 2004) and thus, planning and implementation of reconstruction and rehabilitation (RR) measures also need to incorporate such factors into account.

Disaster assistance and recovery need to be articulated at the ground level, by critically interrogating the complex interconnections of power, rights, and justice to ensure human security beyond mere survival (Gunewardena and Schuller 2008, 8). Thus, the need to critically study aid, which tends to undermine resilience and result in dependency of those affected by disasters.

Coyne (2013) in his provocative book *Doing Bad by Doing Good: Why Humanitarian Action Fails?* uses economics and cites the examples of the disaster in Haiti and the Hurricane Katrina aftermath proffering why often humanitarian efforts that intend to do good, end up doing either nothing or, on the contrary, causing harm. He provides important insights into the limits of humanitarian action, where he does not blame one particular actor, but instead points to a host of actors who could be held accountable: the government, military officers, and NGO officials. Reliance on the government funding often partakes of a tendency towards linear technological thinking, as opposed to organic thinking, which fails to recognize the inescapable complexity of socio-economic conditions (Coyne 2013, 57).

Coyne proposes that one must ensure a general environment of economic freedom of the local people and not a continual attempt to micro-manage the world according to the desire of experts (Coyne 2013, 180).

Speaking about the efficiency of the local governments, Athukorala (2012) opines that effective absorption of aid depends on the ability of authorities to engage local organizations, local communities, and international donors. Instead of dominant state led humanitarian action, he proposes an alternative; of establishing an environment of economic freedom. He proposes to steer clear of the government aid programs and instead encourages direct assistance or assistance delivered via private organizations. However, this proposition seems to be unrealistic given that major funding for reconstruction comes from the government; humanitarian aid by the non-governmental and philanthropic organizations is temporary, lasting for two to three years, whereas the reconstruction and rehabilitation takes much longer, sometimes a decade.

The Sendai framework emphasizes an action plan, which is effective in response and reconstruction, and promotes a sustainable model of economic and social development, in consultation with the people to "Build Back Better".[1] Sustainable recovery, as prepared by 2005–15 Hyogo Framework for Action, emphasizes on 'consultation' with the people concerned. However, there is no mention of 'community participation', which often is only focused to be just token or consultative. However, effective and true community participation could result out of collaboration from planning to implementation and give primacy to bottom up approach. Nakagawa and Shaw (2004) recognize the role of social capital trust, social norms, participation, and network in the successful and speedy recovery and enhancement of collective action post-disaster.

Given this background, the following section looks at the generous aid which flowed post Asian Tsunami, and how the principles of impartiality, neutrality and independence have been compromised.

2 Asian Tsunami Aid: Colossal and Generous Funding

Studies have shown that the responses of the international community to humanitarian crises have been increasingly and highly selective (Binder 2009). According to Hyndman (2011, 5), what shapes international response to crises, whether human-made (war/conflict) or tectonically derived (Earthquake/Tsunami), are post-colonial ties, cold war allegiance and diaspora influence.

What triggers aid flow or international relief is the severity of the disaster, the number of people killed or affected, the average income and related economic

[1]Sendai Framework For Disaster Risk Reduction, Adopted by UNISDR, on 18 March 2015 at the Third UN World Conference, UN Doc. 43291, https://www.unisdr.org/files/43291_sendaiframeworkfordrren.pdf.

indices of the affected country, and the amount of media coverage that the disaster generates. Geographical proximity to countries providing aid, the existence of common languages and colonial ties are other factors (Strömberg 2007). Hyndman (2011) opines that the tsunami of 2004 drew the maximum aid, when compared to the earthquake in Kashmir and Hurricane Katrina, due to the "optics" of the disaster: the videos of foreign tourists dying in the tsunami.

Public response to the 2004 Asian tsunami was enormous and it received the most generous funding ever recorded in the history of post-disaster aid. However, there are no accurate figures on the amount of aid money which poured in from international agencies, individual donors, and respective governments. Different sources gave different figures as shown in Box 1.1.

Box 1.1—Aid Collected and Distributed

- The UN central relief fund raised 6.25 billion USD to assist 14 countries. The UN appeal raised 1.25 billion USD in the first six months of the beginning of 2015, also partly funded by UNICEF child protection work in the region. UN data shows that Japan was the single biggest donor committing to USD 500 million and deploying 100 emergency workers.
- According to 'The Tsunami Report Card', prepared by Inderfurth et al. (2005), the whole world raised 13.4 billion USD and 92 countries provided assistance; US government pledged a total of 857 million USD and their individual and corporate donations totalled 48 billion USD.
- In UK, the Disaster Emergency Committee (DEC) appeal received 392 million pounds in public donations (Kweifio-Okai 2014).
- Another source says 14 billion USD were pledged for the relief and recovery of tsunami-affected countries, with over 5.5 billion USD of this coming from private sources, such as individuals and organizations.
- European commission's humanitarian aid contributed 154 million USD in response.
- Country wise distribution of aid in million USD, shows that the maximum amount of aid was received by Indonesia (1000), followed by Sri Lanka (651.6), India (150.6), Maldives (91.4), Thailand (29.8), Somalia (14.6), Burma (4.8), and Seychelles (4.2) (Financial Tracking Service, UN Office for the Humanitarian Affairs) (www.unocha.org/2015 appeal).
- However, 229 million USD went in food provision. The World Food Program delivered 110,000 tonnes of food, which fed about 2 million people in the first 5 months of the tsunami.
- 187 million USD went to shelter and emergency supplies. The biggest portion of the UK DEC funding went into the construction of long-term shelters for survivors; 13,700 houses were built in Aceh alone (Kweifio-Okai 2014) and thousands of shelters were constructed in India and Srilanka.

Transparency International Sri Lanka Chapter (2009) stated that nearly 472 million USD in tsunami aid has been unaccounted for and more than 603 million USD has been spent on projects unrelated to the disaster.[2]

There is no accurate disaggregate data on how much of the above funds have been expended and on what and for whom, though there are broad indications. There are also gaps in information about how much funds were demanded, how much were actually released and how much were utilized. For example, in India, the four affected states and Andaman and Nicobar Islands had demanded, from the Central Government, INR 117.96 billion for relief and rehabilitation works. However, the Centre sanctioned INR 36.44 billion, of which the Planning Commission earmarked INR 16.07 billion for long-term reconstruction. In the end INR 17.59 billion was actually released by the Centre, and only INR 10.75 billion were utilized. The states could only use less than 10% of the sum they had originally demanded. This misutilization got the attention of the media and even the Controller and Auditor General (CAG) who carried out a story on 'India Today', saying, INR 2.29 billion was misutilized.[3] The report by CAG released to the Parliament in January 2007, revealed that the relief and reconstruction operations were marked by a combination of political indifference and bureaucratic inefficiency.

Humanitarian aid, when approached critically, often reveals that, in a globalized, technology-driven, post-modern era, transnational agencies have limited or no control over the aid once dispensed. The state has been found to have limited or no accountability to their people and civil society organizations, especially under these circumstances. However, not all the changes that follow a disaster are negative, some of the positive developments following the Asian Tsunami also need recognition.

3 Positive Changes After Disasters

The establishment of buffer zones, resettlements of colonies, development of early warning systems, and the creation of new organizations and institutions were witnessed in different countries. However, literature reveals that these buffer zones were not strictly followed and were un-regulated in Sri Lanka and India.[4] The resettlement was not free of controversies either. However, the development of early warning systems was a positive change. Sri Lanka established a new Disaster Management Act on 13 May 2005 and the Disaster Management Centre (DMC) became

---

[2]AFP, 'Tsunami aid goes missing,' *The Sydney Morning Herald*, 28 December 2009. http://www.smh.com.au/world/tsunami-aid-goes-missing-20091227-lgb4.html.

[3]'Govt. pours in relief, monetary compensation for 2004 Tsunami victims', *India Today*, 01 January 2007. http://indiatoday.intoday.in/story/tsunami-hit-the-indian-costline-in-tamil-nadu-2007/1/156482.html.

[4]Buffer zones were created for safety, and communities were settled 500 m away from the high tide line, but the same was not regulated, as many tourist spots were built closer to the sea shore, breaching the coastal regulation zone limits.

operational in September 2005 with the duty to oversee disaster management through the application of a multi-hazard approach (Birkmann et al. 2010).

A similar development in India was the enactment of the Disaster Management Act, 2005.[5] It emphasizes a paradigm shift from a relief-centric to a holistic and integrated cyclic approach, covering prevention, mitigation, and preparedness. The National Disaster Management Authority (NDMA) was set up under the Chairpersonship of the Prime Minister of India. Followed by the State Disaster Management Authority (SDMA) under the chairpersonship of the Chief Minister and at the district level, the District Disaster Management Authority (DDMA) is looked after by the District Commissioner, as per the guidelines laid down by NDMA and SDMA. After the super cyclone in Orissa and Bhuj Earthquake in Gujarat, both states have well-planned and functional state disaster management authorities. In recent years, many other states have established disaster management plans at district levels. Moreover, the National Disaster Response Force (NDRF) has been created, which is actively involved in the rescue and evacuation mission post-disaster, all over the country. The latest development is the National Disaster Management Policy and Plan (2016).[6]

In India, after the Gujarat earthquake 2001, many initiatives taken by the government became global best practices. Institutional system, legal framework, public policy, and community-led long-term recovery and rehabilitation, with the feature of risk-reduction were some of these practices. The herculean task of rehabilitation and reconstruction was completed in record time, both in rural and urban areas. This became a global best practice for 'building back better' (Shaw et al. 2003). However, despite these best practices in Gujarat, they were not replicated in post-tsunami India. In Indonesia, an agency called Reconstruction and Rehabilitation Agency of Aceh and Nias (Badan Rehabilitasi dan Rekonstruksi; BRR) was established and has been operating since 2009. In 2006, a national plan was developed and issued by National Development planning agency (BAPPENAS) and National Coordinating Management Agency (BAKORNAS). A new Disaster Management Law (UU24/2007) was issued in the year 2007. The National Disaster Management agency was founded in October 2007 (Birkmann et al. 2010).

Though positive changes could be seen in organizational structures in all the countries affected by tsunami, in reality, strategies for vulnerability reduction were ambivalent. These strategies tended to undermine livelihoods, disrupted social networks leading to social tension between relocated and host population in several places. Most of the funds appeared to be disbursed on the administration costs, expat salaries, and overheads of the INGOs. There was practically no transparency and reporting of how funds are spent by the agencies (Birkmann et al. 2010, p. 8).

[5]Government of India, National Disaster Management Authority, The Disaster Management Act 2005, 53. http://www.ndma.gov.in/images/ndma-pdf/DM_act2005.pdf.

[6]Government of India, National Disaster Management Authority, National Disaster Management Policy and Plan, 2016. http://www.ndma.gov.in/en/policy.html.

4 Disaster Aid in Post Asian Tsunami: Issues and Concerns

Commendable humanitarian responses were witnessed and without such assistance, it would have been impossible for the countries to cope. This section flags some of the major issues and challenges in Rehabilitation and Reconstruction: too many actors, different salary components of staff, inappropriate use of aid, unaccountability, creation of new inequality and imbalances as a result of post disaster aid, and the like.

While a lot of attention has been paid to relief and recovery operations, little research has been undertaken about the characteristics and processes of change, especially its direct and indirect impacts (Birkmann et al. 2010) on the communities during the course of reconstruction and rehabilitation. Thus, it becomes pertinent to understand that such well-intentioned humanitarian responses to major disasters have led to various unintended consequences. This section will look into some of the major issues and concerns raised by various scholars on the consequences of disaster aid post the Asian Tsunami.

This is not to negate the benefits of the post-tsunami rehabilitation effort. In conflict-ridden places like Aceh, peace deals were made. The government and those involved in the free Aceh movement were brought to the negotiating table, thus commencing a peace process; 'a peace deal' (Kweifio-Okai 2014). Around forty key aid agencies joined hands to form Tsunami Evaluation Coalition (TEC), a collaborative process to evaluate key elements of relief and recovery.

Telford and Cosgrave (2007) had put forth the joint evaluation of international response to the Indian Ocean Tsunami in a Synthesis Report. The report states that the allocation of funds was not based initially on the assessment of the need. Though they were generally satisfied with the relief phase, the affected communities felt that their local leaders had failed them in ensuring equity and inclusion. Press coverage of the tsunami in Sri Lanka (Telford and Cosgrave 2007, 72) gave 86 instances of inappropriate aid, 57 instances of corruption, 18 instances of the waste of resources, and 17 cases stating lack of consultation with the communities. The Tsunami Evaluation Coalition (TEC) Report 2006 (cited in Telford and Cosgrave 2007) further highlighted a lot of critical gaps and limitations in the processes of rehabilitation across the countries.

International agencies and donors had shown a poor understanding of the complexity of recovery and development after the Asian Tsunami, failing to apprehend the disaster through its situated-ness in the spheres of economics, politics, social structure, and culture. The TEC Report (2006, 74; cited in Telford and Cosgrave 2007) stated that their 'arrogance and ignorance' resulted in false starts, waste, and inappropriate aid. Further, context insensitivity, a proclivity for quick fixes animated by the philosophy that "one-size fits all", complicated the recovery process (TEC Report 2006; cited in Telford and Cosgrave 2007). The report stated that, overall, it could be viewed as a missed opportunity; given the amount of aid, there could have been better reconstruction and rehabilitation.

Though, on one hand, post-tsunami aid led to relief and recovery on an unprecedented scale, on the other hand, it also led to the exacerbation of the old and the creation of new forms of inequities. These forms of inequity arose because of the way in which the aid was prioritized, distributed, and managed, both by the aid agencies and by the national and local governments. It was observed that the aid had been fruitful only in the short-term, but re-examining it in the long term revealed the inadvertent consequences it had on the affected and the non-affected communities and their mutual relations.

After the tsunami, there were multiple agencies working in all the affected countries in reconstruction and rehabilitation, with very little coordination, which led to chaos. A directory prepared by UN Humanitarian Information Centre for Sri Lanka (HIC) in January 2005 listed a total of 890 agencies, which were involved in post-tsunami operations in Sri Lanka alone. Similarly, in other countries there were multiple agencies too. Lack of coordination and the resultant chaos were important issues of concern. A large number of international aid organizations with massive funds posed a huge coordination challenge for the national governments at all levels. Though foreign aid helped in the reconstruction processes, excessive competition between the aid organizations hindered effective utilization. The limited aid absorption capacity of the respective governments (Athukorala 2012) was another challenge, which shows that not all the governments have the wherewithal to effectively utilize funds after disasters; this has serious implications in designing policies and programs for dealing with disasters as part of national development strategies.

The scholarship on humanitarian aid also shows the large variation in the salary brackets of the staff that had been hired in Sri Lanka. There were reports of strong resentment, occurring within the organizations. Some International NGOs (INGOs), recruited top-level expatriate staff from European countries, where the agency headquarters were located, mid-level staff from India and low level local staff from Sri Lanka, thereby creating a hierarchy, which was evident in many aid organizations in Sri Lanka and across the world (Lund and Blaikie 2013).

Another important issue is with reference to housing, especially in assessing housing damages and defining reconstruction policy. The social aspects of housing provisions are important in post-disaster reconstruction. Oliver-Smith (1990) contends that post-disaster, housing construction need to avoid rebuilding erstwhile structures, which would perpetuate patterns of inequality and exploitation.

The above mentioned issues and concerns can be seen against the backdrop of what Oliver-Smith had stated, speaking of the 1970 Peruvian earthquake: “First came the earthquake, then came the disaster”. He elucidates the effects of disaster with regard to how inequality is expressed in reconstruction efforts (Oliver-Smith 2002, 9–10). The other cognate expressions used by scholars include “second tsunami” (Silva 2009), “the third wave”, “complex disasters” and “second mowing” (Singh 2009). Each one has tried to emphasize that it is not the disaster event, per se, which is solely destructive, but often the post-disaster interventions that prove to be bigger disasters.

5 The Scope and Structure of the Volume

Research papers in this volume contextualize humanitarian aid and action within the broader context of culture, economics, politics, and social relationships grounded in the historical experiences. The authors in this volume, mostly from social science backgrounds, conducted empirical studies, providing an in-depth understanding of the social processes involved in the attempts to rehabilitate disaster-affected people across nations. This volume discusses the current paradigms of reconstruction and rehabilitation that guide the global humanitarian response, and re-examines the aid and its unintended consequences in response to the Asian tsunami of 2004.

The scope of the volume is limited to the exploration of the consequences of humanitarian aid after the Asian Tsunami, and especially attempts to analyze the effects of long-term reconstruction and rehabilitation. It examines the processes involved in reconstruction and rehabilitation, its conduct, effectiveness, delivery, accountability, and the incumbent ethics. The papers in this volume are divided into three sections, though some of the papers cut across themes, while meeting their objectives.

5.1 Section 1: Aid, Conflict and Differential Distribution

The first six chapters are organized under the theme of 'aid, conflict, and differential distribution' and they seek to address the issues of existing ethnic and political conflict, and inequality, with regard to how these get further aggravated through aid. The exception to this theme is the case of Aceh, where the peace negotiations were in progress when the tsunami hit, and culminated in the signing of a peace accord after the disaster. The chapters by Randall Kuhn and V. T. Thamilmaran address the conflicts between the Sri Lankan Government and the Liberation Tigers of Tamil Eelam (LTTE), and how humanitarian aid has led to further divisions. Chapters by Sunil Santha and Luke Juran, based on the impact of the tsunami in South India, highlight how the caste and occupation based conflicts, which existed before the tsunami, have been further aggravated post-tsunami, as an effect of the aid and its interventions. However, Patrick Daly has critically engaged with the local Acehnese social, cultural, and political processes in recovery.

Randall Kuhn's chapter, 'An Ethno-Political Accounting of the Sri Lankan Tsunami: Impact and Recovery' analyses the role that foreign donors played in the domestic ethno-political landscapes in Sri Lanka. The hopes that the tsunami would facilitate rapprochement between the Sinhalese-dominated government and the LTTE were soon overshadowed by the reality of renewed conflicts. The widening gaps in the levels of foreign assistance were to the detriment of the more severely affected Tamil and Muslim minority communities in the eastern and the northern parts of the island. This chapter combines data from the UNDP development assistance database to provide an in-depth account of the magnitude of post-Tsunami foreign assistance, the role of foreign donors in distributional inequities, and the tragic consequences of the disaster–conflict nexus. While the power bases of both the ruling United National Party (UNP) and the LTTE received foreign assistance nearly twice than that of national average, the war and tsunami-ravaged regions of the Eastern Province received as little as half the national average.

The next chapter Thamilmaran's 'Learning from Nature: Responding to Natural Disaster and the Need for a Multi-Faceted Approach in Law' addresses the legal dimension and politics of aid in Sri Lanka where even humanitarian issues were not exempted from the plague of secrecy and unaccountability. His chapter explicates the de-linking and disjunction between spontaneous humanitarianism, reflected in the care and affection that individuals from all communities showed to each other in the initial phases, and the inadequacy of the country's majoritarian polity to transform such goodwill into a concrete platform for an equitable recovery process, through an effective sharing of resources. The post-tsunami reconstruction process created a negative impact on identity relations in the country, replacing the original spirit of humanitarianism that was demonstrated by the average citizen. The experiences of other countries, in their relief efforts, highlighted a culture of political patronage, where most of the aid and its resources were accessible to those who have political linkages.

Disaster victims are generally viewed as being completely disenfranchised, dependent people, lacking agency, and in dire need of relief and rehabilitation. After the tsunami, undermining traditional wisdom, very little to no attempt was made to identify and use the existing pool of local capabilities and resources for facilitating recovery. Sunil Santha's study in Tamil Nadu, in the chapter 'Social Interfaces in Disaster Situations: Analysing Rehabilitation and Recovery Processes among the Fisherfolk of Tamil Nadu after the Tsunami in India' shows that traditional fisherfolks' understanding of the significance of maintaining mangroves and manmade sand dunes with vegetal barriers in preventing the intrusion of seawater into the villages was undermined in planning the recovery. External agencies expected the 'tsunami victims' to conform to the formal plans of rehabilitation. However, the very same formal organizations failed to recognize local capabilities and coping strategies of the affected communities. Thus, a striking feature of the social interface in this particular context is that, though the goal of all the actors (namely, traditional fisherfolk, state agencies, and NGOs) was disaster recovery, it was often characterized by the prevalence of incompatible beliefs, values, norms, knowledge, and practices.

Luke Juran, in the next chapter titled 'Disaster, Development, and Water: The Reconstruction and (Re)Fabrication of Hazardous Waterscapes in Post-Tsunami India', examines the physical and social products of water-related components of post-tsunami reconstruction in South India. The status of water quality and quantity, along with 'colonization' of scarce water resources, has been examined. He argues that the tsunami-affected people possess little control over the waterscapes which were provided to them post disaster. People had very few opportunities to challenge the local processes of rehabilitation, resulting in constrained agency leading them to accustom and acculturate with the given situation, and succumb to the water sources made available to them in their residential areas.

Governments and the international humanitarian sector are increasingly promoting the need to create post-disaster reconstruction efforts at local levels to increase inclusiveness and accountability. However, in spite of extensive rhetoric, evaluations of post-disaster situations suggest that this ideal often remains unfulfilled. Patrick Daly, in the chapter 'An Applied Anthropological Perspective on Localizing Post-Disaster Aid: Lessons from Post-tsunami Aceh, Indonesia', based on his decade-long research among the communities of Aceh, puts forth a long-term perspective on the recovery and elaborates a cohesive reflection on the relationships between the processes and structures of aid in a localized social and cultural context. He illustrates this through a discussion of cultural spaces, leadership structures, legal systems, and mobilization of resources. He suggests that localizing aid requires detailed understanding of and sincere engagement with a wide range of local political and social infrastructure. He argues that this is a critical step to better synchronize reconstruction efforts with the community in recovery, which in turn leads to increased efficiency.

5.2 Section 2: Build Back Better? Relief Materials, Livelihoods, Shelters, and Discontentment

In disaster discourse, the most used phrase is 'build back better'. Often, with the best of intentions to build back better, the policies and technical solutions tend to fail to achieve the intended outcomes. The failure is attributed to the institutions; while many types of institutional arrangements and trajectories can enable development, as examples around the world demonstrate, many other 'best practices' in one context fail in the other. In some cases, rapid progress comes about suddenly, seemingly unexpectedly. Because of this diversity of paths and perils, it becomes essential to uncover the underlying drivers of policy effectiveness and the interventions carried thereupon.

This section contains five papers, three of which are situated in Andaman and Nicobar Islands in India and one in Sri Lanka and the other in Banda Aceh. The papers question the rhetoric of 'Build Back Better'. They narrate the experiences of people affected due to the inappropriate and insensitive top down interventions in the process of rehabilitation except for the experience of Aceh with a positive recovery and development. The paper by Jane Dunlop points to a different take on reconstruction and rehabilitation in Aceh; it highlights the peace negotiations post

Asian tsunami and how the humanitarian aid helped in reconciliation after a long drawn conflict.

Papers on the Andaman and Nicobar Islands discuss how the inappropriate utilization of humanitarian aid led to dependency and increased vulnerability. Simronjit Singh's paper discusses a more devastating "complex disaster", a "second mowing" after the actual "simple disaster" All three papers by Singh, Abid, and Reddy show that top-down policy interventions dominated by local bureaucracy and humanitarian aid agencies, reshaped and worsened Nicobarese and 'settler' life, making them consumerist, dependent, and vulnerable. The cash flow led to loss of self-organizing capacities of Nicobarese, increased dependency, changing the socio-cultural and socio-ecological interactions.

Based on the research from pre-and-post tsunami Nicobars, in the chapter 'The Sustainability of Humanitarian Aid: The Nicobar Islands as a Case of "Complex Disaster",' Simron Jit Singh, Marina Fischer-Kowalski and Willi Haas describes the process by which the disaster response contributed to the systematic elimination of the resilient attributes inherent in the Nicobarese society. The aid intervention seen as "second mowing", impacted their socio-cultural variables, subsequently affecting their entire ecological system. It had serious consequences for the future sustainability of the archipelago.

The study by Mohammed Abid, in the chapter titled 'The Tsunami and its Aftermath: Resilience and Rupture of the Social Fabric among the Nicobarese', of the Nicobarese tribes of Katchal, seeks to explore issues related to the nature of loss and displacement (territorial/habitat, psychosocial and cultural) and the strategies adopted by them in coping with this unprecedented situation. It also seeks to understand the multi-dimensional impact (ideological, economic, social) of the multiple interventions of the state and the non-state players in providing 'aid' and 'assistance' to the affected Nicobarese and its implications for the future of islands. It shows how a certain amount of effort to understand the social organization of the communities would have helped improve efforts towards rehabilitation, which now have become irreversible.

In the same manner, Sunita Reddy too, in the next chapter 'New Andamans: The Aftermath of the Tsunami in the Andaman and Nicobar Islands', historically traces the settlement patterns in the Union Territory of Andaman and Nicobar Island, making it a 'Mini India', by the Indian Administration due to its strategic location. After the tsunami, major issues of land ownership and illegal settlements came to the fore. The reconstruction and rehabilitation packages were imposed, with only token consultation with Nicobarese and no dialogue or discussion with the 'settlers'. The flaw in the rehabilitation package was the linking of food security to shelters. The major thrust was on making the islands a world-class tourist destination and thus the local elite were benefitted. While the contractors reaped the benefits of reconstruction and rehabilitation, the survivors were left with stigmatized, yet envied, 'tsunami houses' without food and work for the affected families.

Nisha Arunatileke's chapter, titled 'Post-Disaster Housing: Lessons Learnt from the 2004 Tsunami of Sri Lanka', specifically focuses on the issue of post-tsunami housing construction in Sri Lanka. Lack of capacity of the government to efficiently distribute aid to the people and the places in dire need, led to discontentment. Further, the lack of coordination between the donors, and lack of moderation and leadership by the state, increased wastage and reduced the effectiveness of aid. However, the government's approach was to involve the beneficiaries in the housing program from the outset, and provide assistance for rebuilding with minimum disruptions to communities in 'owner-driven' rebuilding. In case of the 'donor-driven' housing strategy, communities which were closer to the seashore, were shifted to other locations and the minimal involvement of the beneficiaries in the program led to much dissatisfaction.

It is very important to recognize the social dimension of housing and develop a post-disaster resettlement plan, which is ecologically and culturally sensitive. Revival of livelihoods should be given more emphasis in the rehabilitation process. For early recovery, it is important to assess damages timely and accurately. It is also important to do assessment from the peoples perspective, on their physical, social, cultural needs by maintaining communication and interaction with the community. For important decisions like relocation, housing design, construction technology, and cultural heritage conservation is important in reconstruction. More importance should be given to the revival of livelihoods, where community organization and participation is pertinent.

Contrary to the above-mentioned papers, Jane Dunlop, in the chapter titled 'Land, Natural Resources, and Sustainable Development in Aceh, Indonesia: Confronting Inequalities through Post-tsunami and Post-Conflict Recovery' addresses a positive development in the case of Aceh, which had endured nearly 30 years of armed conflict when the tsunami devastated its coastline. The recovery efforts after the tsunami led to cooperation amongst diverse stakeholders in the peace and recovery efforts and prompted a renewal of peace talks. A historic Memorandum of Understanding between the Government of Indonesia (GoI) and Gerakan Aceh Merdeka (GAM; the Free Aceh Movement) was signed to resolve the conflict, by accommodating Aceh's desire for greater autonomy. The Indonesian government allocated increased central government funding to support Aceh's development, and allowed the region a greater share of profits derived from the region's extensive natural resources. Further, the international aid agencies advocated that the local governments be included in housing policy, leading to one of the most equitable solutions in the division of post-Tsunami resources in Aceh. True to the words, 'build back better", the Government of Aceh went for structural reforms, and articulated its vision for Aceh's recovery and development.

5.3 Section 3: Coping Strategies, Systems, and Aid Effectiveness

Out of the five papers, the first two papers in this section address the coping strategies initiated by the communities by the construction of memorials of dead and via religious coping mechanisms, though these are complicated in the politics of aid. The papers, one by Monika Lindberg Falk and the other by Edward Simpson and Malathi de Alwis address how religion and construction of memorials played an important role in coping with the disasters respectively. The remaining three papers in this section address the institutional systems, recovery, impact assessment, monitoring systems, and understanding of aid effectiveness.

Monica Lindberg Falk, in the chapter 'Aid, Religion and Recovery in Post-tsunami Thailand' pays special attention to religion and relief work in Thailand, and evokes questions about why and how religion became significant, speaking of faith-based aid and ethical concerns in post-tsunami Thailand. Falk's narratives of the role of Buddhist monks, in helping people cope with the losses of the tsunami, shows the psycho-social care they provided to the community in Phang Nga and helped them to recover from trauma. Monks provided the survivors with a sense of coherence, rendering their traumatic experiences more comprehensible, meaningful, and manageable as a shared challenge. Religion played an important role in community resilience. The teachings of Buddhism, among them the idea that helping others is the process of helping one self, were the best ways to cope with the losses, which helped in the healing process.

Edward Simpson and Malathi de Alwis's chapter, 'Memory Interventions: Gujarat and Sri Lanka', addresses the politics of memorialization, comparing the Bhuj earthquake in India to the tsunami in Sri Lanka, and how these are inseparable from reconstruction initiatives and the broader political currents. Challenging the notions of erecting 'memorials' in the name of those who are dead, it is the state which controls and benefits and not really the survivors. Memorialization, as part of reconstruction projects, primarily reflects the agenda and interventions of an outside agency in the reconstruction process, especially when local people are unenthusiastic about public memorials. Since the memorials are also linked to international funding, the impetus to build or even complete were stalled once the funding reduced or stopped. Building of memorials has been controversial, as memorials built from 'aid' money are acts of control by the state, whereas memorials built by the people can be acts of resistance. The paper therefore questions, how relevant is it to convert personal losses into public memorials?

In the next chapter, titled 'Conflicting Paradigms and the Danger Discourse: Re-thinking Indian Disaster Management Framework in the Post-tsunami Era,' Pravin Kushwaha critically looks at the conflicting paradigms and the complex discourse in India's disaster management approach in the Post-Tsunami era. Adoption of the 'Paradigm Shift' approach in the enactment of the Disaster Management Act (DMA), and the formulation of the National Policy on Disaster Management, are the major initiatives that unfolded a new era of disaster management in India. The new initiatives are aimed at developing a comprehensive mechanism of disaster governance in India, yet the focus is merely on post-event responses. In particular, the notion of 'technology-driven strategy' that often dwells upon hazard-centric practices continues to dominate the new approach. The discourse of production of disaster vulnerabilities as an outcome of various socio-technical as well as socio-political processes is yet to take centre-stage in policy formulations. In fact, many interventions in the post-tsunami phase were lopsided in favour of deploying the 'exotic methodology' of damage assessment. Consequently, responses that followed the tsunami were framed as solution-defined problems to be resolved by experts and aid-giving agencies, rather than by using the local and specific social contexts.

Lakshman Dissanayake came up with a methodological approach developed for Sri Lanka, Tsunami Recovery Impact Assessment and Monitoring System (TRIAMS), which he details in the chapter 'Tsunami Recovery Impact Assessment and Monitoring System (TRIAMS): Methodological Approach Developed for Sri Lanka'. TRIAMS was developed to identify any pockets not covered by existing recovery programmes, addressing unmet needs, and preventing inequities from being created in the allocation of tsunami-related resources. It was expected that TRIAMS implementation would keep governments, donors, aid agencies, civil society, and other stakeholders aware of the progress and recovery efforts in tsunami-affected areas. This measure aimed to enable them to provide assistance to recovery programmes based on the analysis of collected data, through a qualitative approach, with an active participation of affected communities and those involved in the recovery process. Regular consultations with the affected communities would provide a unique opportunity for those who make interventions to assess their contribution with regard to relief, reconstruction, and rehabilitation activities. It would also refine or modify their strategies for future activities accordingly. It is important therefore, to develop such tools in general for assessment and monitoring of aid effectiveness for any disaster in future.

In the final chapter, 'Tsunami Aid and its Effectiveness', P. G. Dhar Chakrabarti addresses aid and its effectiveness. The decade-long experience with tsunami aid and the gaps from pledges to releases, from releases to expenses, from expenses to recoveries, and from recoveries to resilience, elucidate the key lessons learnt on aid and its effectiveness. It highlights the importance of a comprehensive framework, which is useful for the management of humanitarian aid from the inception to the execution: from emergency relief and rehabilitation to long-term recovery and reconstruction; from reconstructing infrastructure and livelihoods to building resilience of communities and countries. This framework is useful for policy makers and practitioners at all levels, from global to national and from national to local. The paper talks about the relevance of the ongoing initiatives of different stakeholders—the aid givers, the aid managers, and the aid receivers in making the international humanitarian assistance more useful and effective.

Effective use of aid depends on effective governance. It is important in the long run to have effective democratic government systems along with inbuilt evaluating and monitoring systems in place for increasing accountability, transparency, and effective reconstruction and rehabilitation. More recently in 2015 and 2016, there has been some positive global development in Humanitarianism, which are discussed in the following section.

6 Comparative Experiences and Lessons Learnt

Reflecting on all the fifteen papers, this section compares the experiences of different countries post Asian Tsunami through the lenses of social scientists. With a critical perspective on the aid and its consequences on the lives of the affected communities, the book re-examines the current paradigms of reconstruction and rehabilitation, guiding the global humanitarian response. Multiple organizations in operation, without proper coordination and without proper recovery and rehabilitation plans, across nations, led to chaotic conditions in the beginning. The collated experiences across countries demonstrated un-coordination, duplication, disproportionate distribution, and corruption, leading to unintended consequences of inequity, dissatisfaction, and increase in vulnerability.

Despite overwhelming international aid, there were no sustained efforts evident from the experiences across countries; most external organizations left 3–5 years after the tsunami, with very few remaining to support and oversee long-term rehabilitation, thereby prolonging the time to rehabilitate after such a disaster (Reddy 2012). The voluntarism and humanitarian response to the tsunami, motivated by sympathy towards the affected populations, quickly disappeared and gave way to high technological interventions with limited sensitivity to local cultures and environment. Most projects were 'supply-driven' rather than need-driven, animated by 'one-size-fits-all' approach, where new concepts, ideas, schemes and projects from the outside were introduced, undermining the coping capacities of the Nicobarese in the Andaman and Nicobar Islands and Sri Lankan tsunami survivors. These projects did not take into account the socio-cultural milieu of the people.

As was seen in the immediate scenario after the tsunami, communities across nations extended help, indicating that the community acted as the first responder, till other help arrived. With the coming of the humanitarian response, the local community initiatives gradually fizzled out, as they were not recognized and were, in fact, undermined. This was witnessed in Sri Lanka, the Andamans, and Aceh.

The experiences across countries in this volume show group solidarity among the local communities in the initial stages, which slowly declined, and competition, jealousy, and heartburn, which gradually turned into resentment towards beneficiaries of aid. The papers in this volume bring out inter and intra- community conflicts arising out of aid distribution, some of which were inherent in the pre-disaster scenario, and some which manifested post disaster. Differential distribution of relief materials and cash compensation led to discontent, conflict, and despair among the people. A common feature across nations was also the growing resentment and apathy exhibited by those non-affected towards the tsunami survivors.

The experiences of rebuilding and reconstruction in Aceh reveal certain similarities with the case of Nicobarese and 'settlers' in Andaman Islands, wherein the physical and social spaces for women were lost with the establishment of new settlements which led to the women feeling isolated and insecure. Housing was the most important issue and there was dissatisfaction with the quality and location of the houses constructed. It also led to jealousy among the people of the community who were not granted what they had expected. The tsunami exposed the hidden issues of land disputes and resource management related to water, land, and forest, be it in Aceh, the Andamans, Chennai, or Sri Lanka.

The papers also show that, across nations, aid was mostly expended on reconstruction works, building infrastructure and housing, whereas the revival of livelihoods and rehabilitation of people were not given appropriate attention. Survivors struggled to get jobs, livelihood opportunities and to improve their living conditions. The papers in this volume also point out how humanitarian aid did more harm than good, undermined coping capacities and community resilience, and ushered in conflicts, jealousy, and dissatisfaction. The unintended consequences of unprecedented humanitarian aid flow were the inequities, social disruption, and loss of cultural heritage, which brought about irreparable and irreversible changes in the social fabric. Reconstruction of permanent shelters took a very long time. Possibly cash grants and cash transfers could have helped people to use as per local needs, rather than supplies, which were either not needed or arrived too late. An important lesson learnt was that people needed training and building material, or cash, to rebuild their own houses according to their social and physical needs. However, governments across nations were apprehensive about the capacities and the agency of the tsunami-affected to build their own lives.

Most of the papers in the volume prove that a wide gap existed between the "bureaucratic norm" and "emergent norm", concepts adopted from Schneider (1992). According to the 'bureaucratic norm', the states and the humanitarian agencies see the aid and the rehabilitation packages as successfully completed and delivered with few limitations. However, from the emic perspective—the lived experiences of the disaster survivors—the 'emergent norm' reveals that the rehabilitation was grossly insufficient and did not match with the local needs and expectations. 'One size fits all' and a top-down modernization approach led to largely insensitive interventions to local realities; socio-cultural, geographical, and political. Disasters were also sites for corruption and black-marketing as demonstrated in a few papers.

The papers situated in the Indian context raised issues of structural inequalities on the lines of caste, class, ethnicity, and religion. There are papers which speak about political conflict (internal politics) against the backdrop of the Sri Lankan tsunami rehabilitation. These findings bear resemblance to previous studies by scholars (Telford et al. 2006) with respect to how aid had aggravated social tensions due to lack of community engagement, along with the lack of transparency.

Thus, all the papers in this volume have similar conclusions: that reconstruction and rehabilitation must take into account the long-term needs of the affected, as well as the complex social, cultural, ecological, economic, and political contexts in which the disaster took place. A failure to do so often results in more harm than caused by the disaster itself.

7 Positive Global Development in Humanitarianism

While this book documents the unintended consequences of post-disaster aid on affected communities, there has been a realization across the globe, after the tsunami, and other disasters in the past decade, that the international funding mechanism should be overhauled to improve accountability, predictability, impartiality, and transparency.

Howden (2009) has developed a humanitarian logistics information systems, which can improve humanitarian supply chain of physical aid to the beneficiaries, help reduce duplication, promote the sharing of information, and reduce corruption and market distortions.

In 2015, all countries with humanitarian response planned to prepare 'humanitarian response monitoring frameworks', using online financial tracking service platforms. To ensure more effective response, the findings, and recommendations presented in the periodic monitoring reports need to be adjusted to close as many gaps as possible. Although millions of people have been beneficiaries of aid, problems of access still hinder the delivery of humanitarian assistance. Shortage of funds to meet global needs remains a persistent problem. The Global Humanitarian Response Plan Status Report shows that only 26% of aid requirement is fulfilled, with a shortfall of USD14 billion. Thus, it is imperative to use the aid fund judiciously and reach the neediest of the world (Global Humanitarian Overview Status Report 2015, 23).

For a more effective humanitarian response, the World Humanitarian Summit (WHS) has committed to increase pooled funds under the Central Emergency Response Funds (CERF) and Country Based Pooled Funds (CBPFs). It also envisages ensuring transparency and efficiency, complying with donor conditions and prioritizing humanitarian needs without undue burden on humanitarian actors (Thomas 2017).

The Sustainable Development Goals of UN focus on "ensuring that no one is left behind" and calls for disaggregated data that pinpoint where specific population groups are lagging behind and reach out to them (UN Economic and Social Council Secretary General, SDG Report 2016).

The annual meeting in 2017 of the humanitarian program highlights the increase of 'digital humanitarianism' in recent years, using a range of emerging technologies. These include satellite imaging, participatory mapping, read-write web, big-data analysis, artificial intelligence, crowd source translation, social media, and mobile technology, all of which can help anticipate and respond to crises. Data can be used in humanitarian crises to effectively identify the needs of individuals and improve decision-making (World Economic Forum 2017, 7).

The key concepts behind the reconstruction policy should be to complete collaborative partnership with the community, sustainability, and risk reduction. It needs to be beyond the rhetoric of "community participation" with tokenism (Jha and Duyne 2010). A handbook by Podger (2013) provides a very relevant aspect of context specific understanding of reconstruction tasks and how to undertake them. These should be developed for each country under the reconstruction and rehabilitation policy. International, national, and local partnerships in reconstruction, monitoring and information management, evaluation, and mitigation of risk of corruption have been provided in this handbook (Podger 2013).

8 Suggestions and Recommendations

After due deliberations, discussions, and debates on the effectiveness of aid after the Asian Tsunami in scholarly writings, there are suggestions and strategies, which have evolved and some are yet to evolve, in order to effectively handle the reconstruction and rehabilitation in case of ensuing major disasters.

Based on the assessment of the Tsunami Evaluation Coalition (TEC) and other evaluations, Telford and Cosgrave analyzed funding capacity and quality, recovery, and ownership. He proposed a fundamental reorientation of international disaster response rooted in sustainable Disaster Risk Reduction (DRR) and recovery based on national ownership. He assessed that a few international organizations had managed well but many had not. The urgency to spend money visibly worked against making the best use of local and natural capacities (Telford and Cosgrave 2007, 21).

Telford and Cosgrave (2007, 23) recommended projects that would support the affected communities' own relief and recovery projects. He further proposed that aid agencies needed to be transparent and accessible to the affected populations, and that they should support the local and national authorities to manage and coordinate. Community participation, in the truest sense of collaboration, and joint implementation of programs should be the norm, which would lead to less social disruptions. The rehabilitation packages should be socially, culturally, and ecologically sensitive. Others have suggested that the tagline of many groups including even the latest Sendai framework 'Build Back Better' could be substituted by a slogan such as 'Build Back Safer' wherein 'safer' provides a clearer goal than 'better' (Kennedy et al. 2008).

Bosher and Dainty (2011) argue for embedding resilience at both the project and the institutional levels. Making sustainability of the projects part of the development agenda should be emphasized, along with increasing the capacities of the local governments to plan and execute disaster rehabilitation process. The exit policy of the project team should be well thought of, considering budget and time flexibility for reconstruction and rehabilitation (Shaw et al. 2003).

As most governments have emergency response systems in place, there is a need to have long-term policy for recovery, reconstruction, and rehabilitation, which government officials should undertake as part of a regular capacity-building program of the department of personnel and training. In times of big disasters, local governments become dysfunctional; there is a need of a permanent project management team as a recovery unit, for carrying out reconstruction and rehabilitation at the right moment.

The politics of aid and aid distribution reflect the agendas of geopolitical powers and the power of the local political elite in most affected countries. Addressing issues of power become very important in the disaster discourse. The Asian Tsunami, although seen as a lost opportunity, can instead be more fruitfully viewed as a lesson learnt, which should be remembered while handling disasters in the future. The effort of this volume is to trace what went wrong, in the process of evaluating a decade-long rehabilitation, and how it is important to realize and recognize the big mistakes, misses, and gaps. It is also important to involve the right actors—such as social scientists, especially anthropologists and sociologists—who understand social structures, community resilience, local contexts, and their needs, aiding in better mitigation of disasters.

It is important to put appropriate policies, plans, systems, and actors in place. The colossal humanitarian response in the wake of the tsunami showed the humane side of people and organizations, which should not be undermined by repeating the same mistakes of mismanagement and inappropriate response; we may lose out on such a humane response in future, to the peril of humanity.

References

Alexander, D.E. 1999. *Natural Disasters*. London: Kluwer Academic Publishers.

Athukorala, P.C. 2012. Indian Ocean Tsunami: Disaster, Generosity and Recovery. *Asian Economic Journal* 26 (3): 211–231.

Binder, M. 2009. Humanitarian Crises and the International Politics of Selectivity. *Human Rights Review* 10 (3): 327–348.

Birkmann, J., P. Buckle, J. Jaeger, et al. 2010. Extreme Events and Disasters: A Window of Opportunity For Change? Analysis of Organizational, Institutional and Political Changes, Formal and Informal Responses After Mega-Disasters. *Natural Hazards* 55 (3): 637–655.

Bosher, L., and A. Dainty. 2011. Disaster Risk Reduction and 'Built-In' Resilience: Towards Overarching Principles for Construction Practice. *Disasters* 35 (1): 1–18.

Coyne, C. 2013. *Doing Bad by Doing Good: Why Humanitarian Action Fails*. United States: Stanford University Press.

Global Humanitarian Overview. 2015. *Status report*. United Nations Office for the Coordination of Humanitarian Affairs. UNOCHA, 23. Available: https://www.unocha.org/sites/unocha/files/GHO-2015.pdf. Retrieved on 12 May 2016.

Gunewardena, N., and Mark Schuller. 2008. *Capitalizing on Catastrophe: Neoliberal Strategies in Disaster Reconstruction*. New York: Altamira Press.

Hilhorst, D., and B.J. Jansen. 2010. Humanitarian Space as Arena: Perspective on the Everyday Politics of Aid. *Journal Development and Change* 41 (6): 1117–1139.

Howden, M. 2009. How Humanitarian Logistics Information System can improve Humanitarian Supply Chains: A View from the Field. In *Proceeding of the 6th International ISCRAM Conference—Gothenburg, Sweden*, May 2009. Available: http://www.iscramlive.org/ISCRAM2009/papers/Contributions/241_How%20Information%20Systems%20Can%20Improve_Howden2009.pdf. Accessed on 12 May 2016.

Hyndman, J. 2011. *Dual Disasters: Humanitarian Aid After the 2004 Tsunami*. Sterling, VA: Kumarian Press.

Inderfurth, K.F., D. Fabrycky, S.P. Cohen. 2005. The Tsunami Report Card. *Foreign Policy*: 57–68. Available: http://foreignpolicy.com/2005/12/06/the-tsunami-report-card/. Accessed on 12 May 2016.

Jha, A.K., Duyne, J.E. 2010. Safer Homes, Stronger Communities: A Handbook for Reconstructing After Natural Disasters. World Bank Publications, Available: https://openknowledge.worldbank.org/handle/10986/2409. Accessed on 12 May 2016.

Kennedy, J., J. Ashmore, E. Babister, and I. Kelman. 2008. The meaning of 'build back better': evidence from post-tsunami Aceh and Sri Lanka. *Journal of contingencies and crisis management,* 16(1), 24–36.

Klein, N. 2007. *The Shock Doctrine: The Rise of Disaster Capitalism*. New York: Macmillan.

Kweifio-Okai, C. 2014. Where Did the Indian Ocean Tsunami Aid Money Go? *The Guardian*. Available: http://www.theguardian.com/global-development/2014/dec/25/where-did-Indian-ocean-tsunami-aid-money-go. Accessed on 12 May 2016.

Lund, R., and P. Blaikie. 2013. *The Tsunami of 2004 in Sri Lanka: Impacts and Policy in the Shadow of Civil War*. UK: Routledge.

Minear, L., T.G. Weiss. 1995. *Humanitarian Politics, Headline Series No. 304*. Foreign Policy Association, 17, Available: http://files.eric.ed.gov/fulltext/ED392696.pdf. Accessed on 12 May 2016.

Nakagawa, Y., R. Shaw. 2004. Social Capital: A Missing Link to Disaster Recovery. *International Journal of Mass Emergencies and Disasters* 22 (1): 5–34.

Oliver-Smith, A. 1990. Post-Disaster Housing Reconstruction and Social Inequality: A Challenge to Policy and Practice. *Disasters* 14 (1): 7–19.

Oliver-Smith, A. 2002. Theorizing Disasters: Nature, Power, and Culture. In *Catastrophe & Culture: The Anthropology of Disaster*, ed. S.M. Hoffman and A. Oliver-Smith, 23–47. Santa Fe: School of American Research Press.

Podger, O. 2013. Adapting Professional Practices for Post-Disaster Reconstruction. *Project Management World Journal* 2 (4): 1.

Reddy, S. 2012. *Clash of Waves: Post Tsunami Relief and Rehabilitation in Andaman and Nicobar Islands*. Delhi: Indos Publishers.

Riddell, R. 2007. *Does Foreign Aid Really Work?*. UK: Oxford University Press.

Schneider, S.K. 1992. Governmental Response to Disasters: The Conflict Between Bureaucratic Procedures and Emergent Norms. *Public Administration Review* 52 (2): 135–145.

Schullar, M. 2008. Deconstructing the Disaster after the Disaster: Constructing Disaster Capitalism. In *Capitalizing on Catastrophe: Neoliberal Strategies in Disaster Reconstruction*, ed. N. Gunewardena and M. Schuller, 17–28. New York: Altamira Press.

Shaw, R., M. Gupta, and A. Sarma. 2003. Community Recovery and Its Sustainability: Lessons from Gujarat Earthquake of India. *The Australian Journal of Emergency Management* 18 (2): 28.

Silva, K.T. 2009. 'Tsunami Third Wave' and the Politics of Disaster Management in Sri Lanka. *Norsk Geografisk Tidsskrift-Norwegian Journal of Geography* 63 (1): 61–72.

Singh, S.J. 2009. Complex Disasters: The Nicobar Islands in the Grip of Humanitarian Aid. *Geographische Rundschau International Edition* 5 (3): 48–56.

Strömberg, D. 2007. Natural Disasters, Economic Development, and Humanitarian Aid. *The Journal of Economic Perspectives* 21 (3): 199–222.

Telford, J., and J. Cosgrave. 2007. The International Humanitarian System and the 2004 Indian Ocean Earthquake and Tsunamis. *Disasters* 31 (1): 1–28.

Telford, J., J. Cosgrave, R. Houghton. 2006. Joint Evaluation of the International Response to the Indian Ocean Tsunami: Synthesis Report. *Tsunami Evaluation Coalition*. Available: http://www.sida.se/contentassets/1756188e06354b6286c76aeb0afdaf2e/joint-evaluation-of-the-international-response-to-the-indian-ocean-tsunami_3141.pdf. Accessed on 16 Mar 2017.

Thomas, M. 2017. *Understanding Humanitarian Funds: Growing Beyond Country Based Pooled Funds*. Norwegian Refugee Council, Norwegian, Ministry of Foreign Affairs. Available:

https://www.nrc.no/globalassets/pdf/reports/pooled_funds_2017_16mar2017_web_v2.pdf. Accessed on 16 Mar 2017.

UN/ISDR. 2007. Disaster Risk Reduction: 2007 Global Review. *Consultation Edition, United Nations International Strategy for Disaster Reduction*. Available: http://www.preventionweb.net/globalplatform/2007/first-session/docs/session_docs/ISDR_GP_2007_3.pdf. Accessed on 16 Mar 2017.

Warner, Daniel. 1999. The politics of the political/humanitarian divide. *International Review of the Red Cross* 81 (833): 109–118. International Committee of the Red Cross.

Wisner, B., P. Blaikie, T. Cannon, et al. 2004. *At Risk: Natural Hazards, People's Vulnerability and Disasters*. UK: Routledge.

World Economic Forum. 2017. The Future of Humanitarian Response. *World Economic Forum Annual Meeting*. Available: http://www3.weforum.org/docs/WEF_AM17_Future_Humanitarian_Response.pdf. Accessed 16 Mar 2017.

Zanotti, L. 2010. Cacophonies of aid, failed state building and NGOs in Haiti: setting the stage for disaster, envisioning the future. *Third World Quarterly* 31 (5): 755–771.

Part I
Aid, Conflict and Differential Distribution

An Ethno-political Accounting of the Sri Lankan Tsunami Impact and Recovery

Randall Kuhn

Abbreviations

LTTE	Liberation Tigers of Tamil Eelam
CFA	Cease Fire Agreement
DAD	Development Assistance Database
DAP	Dollars per Affected Population
GoSL	Government of Sri Lanka

1 Introduction

It is, by now, conventional wisdom that the global response to the Indian Ocean Tsunami of 26 December 2004 was unprecedented in its scale and complexity. Foreign governments, international organizations, multilateral institutions, and private citizens allocated and delivered more financial and logistical assistance than had been delivered following any natural disaster, before or since (Flint and Goyder 2006; Houghton 2007). Foreign donors also became more deeply involved in direct service-delivery and reconstruction in unprecedented ways, most notably in the area of housing reconstruction. While the scale of these efforts has led, overall, to an equally unprecedented level of achievement in measurable reconstruction outcomes, it has also led to questions about favouritism and discrimination in global humanitarian response and delivery (Olsen et al. 2003; Drury et al. 2005; Telford and Cosgrave 2007).

Many have sought to explain why the outpouring of support was so much greater after this disaster than others, such as the earthquake that affected Kashmir in 2005 (Flint and Goyder 2006). One anecdote suggests that more of the money initially

R. Kuhn (✉)
UCLA Fielding School of Public Health and California Center for Population Research, Los Angeles, USA
e-mail: kuhn@ucla.edu

S. Reddy (ed.), *The Asian Tsunami and Post-Disaster Aid*,
https://doi.org/10.1007/978-981-13-0182-7_2

allocated for relief of Hurricane Mitch, which affected several Central American countries in 1998, was actually reallocated, and spent in South and South East Asia after the tsunami than in Mitch-affected areas themselves. A less widely discussed bias in the delivery of post-tsunami assistance lies in the regional distribution of tsunami relief aid within affected countries and the role of political and ethnic disputes in altering the flow of foreign donor assistance in this and other crises (Uvin 1998; Le Billon 2000; Grewal 2006). Although the magnitude of assistance was so great that even the most forsaken areas received considerable transfers, the Tsunami Evaluation Coalition nonetheless raised concerns about bias within the affected countries:

> Overall, with so many funds available, most people seem to have received some assistance, and many have received more than they ever expected and have been able to sell the surplus.
>
> However, this does not necessarily mean that assistance has been impartial, particularly in the reconstruction phase (Flint and Goyder 2006: 34).

This paper will address these concerns in the context of Sri Lanka, where the tsunami occurred at a delicate moment in its internal politics (ICG 2008; Le Billon and Waizenegger 2007; Ofstad 2002). Since 1983, Sri Lanka's Sinhalese-dominated government and the LTTE had been engaged in a civil conflict in the predominately Tamil- and Muslim-populated north and east of the island nation. The conflict had led to the death of 65,000 people, and the displacement of one million, or 5% of the population, before a Cease Fire Agreement (CFA) was signed in 2003 (Wilkinson 2003). The greatest impact of the disaster was felt by many of the same Tamil and Muslim communities on the east coast of the country (ADB 2005). While there were hopes that the tsunami would unite these forces and reinforce the CFA, the opposite came to pass (Le Billon and Waizenegger 2007). The CFA collapsed, in practice in mid-2006, and formally in January 2008 (ICG 2008). Whether the disputes over tsunami-relief allocations caused the ceasefire to collapse, or whether this was likely to occur anyway, regional and ethnic variations in the flow of Tsunami relief and the progress of reconstruction on the ground remain a highly contested issue in Sri Lanka (Hyndman 2007; IANS 2007). Anecdotal stories abound and a number of studies report systematic inequities in finance or outcomes (Grewal 2006; Flint and Goyder 2006). Yet, no study has taken a detailed look at specific dimensions of the funding response in Sri Lanka, and few have done so in any context (Fengler et al. 2008).

The paper will analyze the distribution of foreign assistance using the Development Assistance Database (DAD) made available by the United Nations Development Programme (UNDP) to analyze the provincial distribution of post-Tsunami assistance in the Sri Lankan context. As discussed below, while Sri Lanka's provinces do not cleanly reflect the battle lines of the ongoing conflict, provincial boundaries do reflect some combination of conflict exposure, insurgent control, and ethnic composition. Provincial-level data offer a level of methodological simplicity and clarity not afforded by an ethnic analysis, which would require a level of quantitative complexity not well suited for this volume. The conceptual power of the provincial-level analysis is enhanced by the categorization of the contested Northern Province into areas controlled by the government and the

LTTE. After addressing the global discourse on aid delivery in conflict-affected societies, a stylized understanding of the meaning of the provincial distinctions in use and their relationship to the ongoing ethnic and political conflict is offered in this paper. The subsequent section characterizes the level of tsunami damage in each province according to a variety of measures of impact. The next section then reports on an analysis of provincial donor assistance allocations and expenditures, with specific attention to variations by sector, by funding agency, and over time. To characterize the relationship between tsunami impact and donor assistance, a new indicator referred to as donations (or dollars) per affected population, or DAP is developed. This measure enables standardized between-group comparisons while maintaining the intuitive appeal of capturing the value, to foreign donors, of human lives.

2 Humanitarian Assistance and Conflict

In the aftermath of the aid-fuelled conflicts of the 1990s, in places such as Rwanda and Sudan, considerable concern was voiced about the role of foreign donors in supporting discrimination against minority groups, in fuelling insurgency and insecurity, or in supporting military build-up (Gasper 1999; Le Billon 2000; Cliffe and Luckham 2000; Fox 2001; Barnett 2005). Of particular concern has been the increasing tendency for modern humanitarian operations to span the entire continuum from emergency relief to long-term political and economic development (Fox 2001; Goyder et al. 2006). This early recovery approach was formalizaed in the Sphere Project's Humanitarian Charter and Minimum Standards in Disaster Response (Sphere 2004; Tong 2004). Many contest that whereas emergency relief can be carried out in a largely apolitical fashion, developent activities can be more readily co-opted by geopolitical powers and the local political elite (Buchanan-Smith and Maxwell 1994; Weiss 1999; Easterly 2002; Hilhorst 2002; Freeman 2004). As will be further demonstrated in this paper, a great majority of the post-tsunami assistance provided to Sri Lanka, even in the earliest stages of relief, was conducted outside the emergency relief sector (Houghton 2007; Telford and Cosgrave 2007).

The Sphere Project has also placed an unprecedented emphasis on accountability and measurement in humanitarian operations that has facilitated creative and innovative analyses of the conduct, effectiveness, and ethics of aid delivery (Benini et al. 2008; Fengler et al. 2008). It is hoped that this analysis will continue this process by reconciling realistic estimates of the regional balance of tsunami devastation against the financial accounts of foreign donor agencies. It will be possible to quantify regional differentials in the extent of foreign assistance, and to explore a number of potential pathways explaining these differences. First, an ongoing conflict may hamper the delivery of post-disaster assistance to affected areas (Slim 1997; White and Cliffe 2000). Second, donors may avoid providing assistance to areas under the immediate control of insurgent groups such as the LTTE (Smillie and Minear 2003;

Uvin 1998). Third, ruling governments may divert assistance away from areas affected by conflict or areas populated by minority groups (Macrae et al. 1997; Smillie 2001). Politically motivated aid distributions may benefit insurgent-controlled areas as well as government-controlled ones. To the extent that insurgent groups control territory or wield considerable political influence, these groups, and the areas they control, may themselves impact the flow of assistance. Aid flows to insurgent-controlled areas, potentially at the expense of ethnic minority areas that remain under government control.

Humanitarian assistance may also accumulate to privileged or dominant population groups for logistical reasons that, while still problematic for operational efficacy, do not imply politically motivated bias or manipulation. Areas populated by dominant ethnic groups may be characterized by higher levels of development and higher wages and prices, thus, perhaps, justifying greater investment (Morris and Wodon 2003). Politically dominant areas may also happen to be more accessible to points of entry for finance, materials, and managers; what is often referred to as 'tarmac bias' (Waters 2001; Benini et al. 2008). These possibilities are addressed in the context of provincial variation in donations per affected populations and in variations by sector, funding agency, and time.

3 Population and Civil Conflict in Coastal Sri Lanka

In 2002, Sri Lanka's population composition by ethnicity/nationality was 74% Sinhalese (primarily Buddhist), 18% Tamil (primarily Hindu), and 7% Muslim (Sri Lanka Census 2008). Widespread civil conflict began in 1983, with the LTTE seeking an independent homeland in the Northern Province, with a largely Tamil population, and the Eastern Province, with a mixed population of Tamils and Muslims, who are mostly Tamil-speaking (Wilkinson 2003). The 1980s and 1990s saw significant territorial gains by the LTTE as well as significant relocation of Sinhalese and Muslim populations from the Northern Province. At the time of the tsunami, Government of Sri Lanka (GoSL) forces had retaken Jaffna District in the Northern Province, which could be reached from the Sri Lankan mainland either by sea or by travelling through Mullaitivu and Kilinochchi Districts, which remained under LTTE control. Mullaitivu and Kilinochchi maintain parallel administrative and ministerial structures, with most public services remaining under GoSL control, while a separate police, military, and courts are administrated by the LTTE. The analysis will proceed at the level of the province, taking further account of the de facto separation of the Northern Province into government-controlled and LTTE-controlled areas. Figure 1 shows the ethnic population composition of the affected divisions of Sri Lanka, broken down by district. The map suggests considerable, though not complete, provincial segregation of Sri Lanka's three main ethnicities: Sinhalese, Tamil, and Muslim.

The Southern Province is overwhelmingly Sinhalese, with a sizeable minority of Muslims, principally those of Malay origin (Sri Lanka DCS 2001). The Western

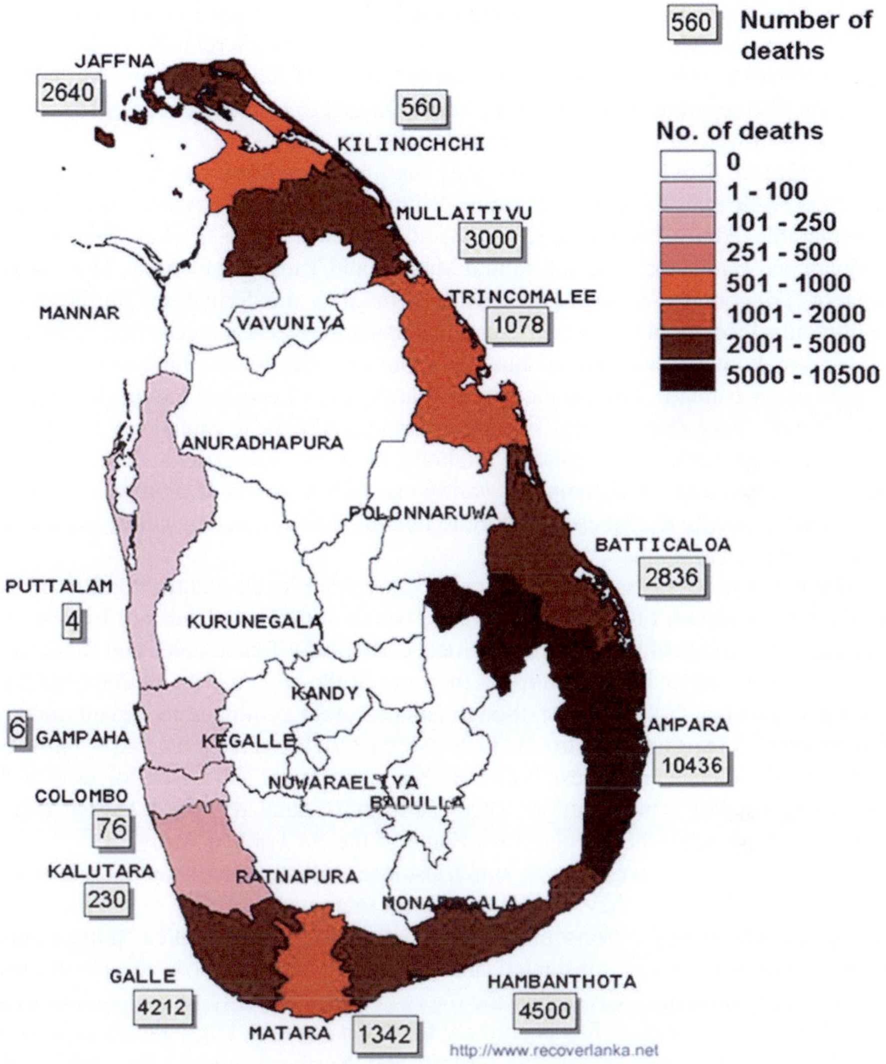

Fig. 1 Tsunami deaths by district, National Disaster Management Centre. *Source* Sri Lanka NDMC (2005)

Province, particularly the capital district of Colombo, has a considerably more mixed population, though, as we will see below, it accounted for relatively few tsunami victims. With the exception of an attack in 2006 on a naval installation in the southern city of Galle, LTTE attacks on the Sinhalese heartland have been limited to the capital city of Colombo.

The Northern Province has an overwhelming majority of Tamils following the displacement of Muslim populations during the civil war (Hasbullah 2001; Brun

2003). This is true in government-controlled Jaffna District and in LTTE-controlled Kilinochchi and Mullaitivu. Most of the Northern Province remains under the de facto governance control of the LTTE, including all of Kilinochchi and Mullaitivu Districts. We separated these out from Jaffna District to measure the extent to which assistance flowed to insurgent-controlled areas.

The Eastern Province is, thus, the only highly affected province with significant numbers from each major ethnic group. Within the Eastern Province, there is also considerable ethnic heterogeneity at the district level. The coastal divisions of Ampara and Batticaloa have substantial Muslim and Tamil populations. The coastal area of Trincomalee is majority Muslim, but 30% are Sinhalese. Furthermore, community-level population and tsunami-impact statistics suggest that even this ethnic breakdown does not adequately represent the ethnic dimensions of the impact of the tsunami. For instance, more than 31.6% of tsunami victims in Ampara were Tamil. Very few tsunami victims in Trincomalee were Sinhalese. A detailed accounting of nationwide ethnic variations in donor allocations depends on a detailed accounting of community variations within Ampara, alongside tests to distinguish specific hypotheses relating to ethnic, political, ecological, or economic forces driving variations.

The paper proceeds with analysis at the provincial level, identifying five of the distinct areas shown on the map in Fig. 1, which shows the death toll by district. Substantial portions of the Eastern Province, particularly Trincomalee and Batticaloa Districts, were under LTTE control at the time of the tsunami. As of 2008, all had returned to formal GoSL control, though isolated areas continued to remain under a mix of dual LTTE–GoSL control (ICG 2008). Eastern Province remained a setting of widespread violence and unrest, both between government and LTTE forces as well as among smaller armed factions (ICG 2008). The area remained largely under military rule through the Special Task Force of the Sri Lankan Army.[1]

Provincial breakdowns of tsunami-assistance will not facilitate finely-grained tests of ethnic bias, but they will offer answers to a number of key questions. First, donor allocations to the Western Province will address the so-called 'tarmac bias' towards providing more substantial aid allocations to areas close to airports and other aid infrastructure. Given that the overwhelming majority of the victims were outside the capital region, this is unlikely to play a big role. High aid allocations to the Southern Province, relative to Eastern and Northern, could offer indications either of political bias, though its relative proximity to the capital may also indicate tarmac bias. Finally, the most important goal of this study is to assess the specific conflict-related dimensions of assistance. Of particular interest is whether aid flows were more affected by the fighting itself, or political bias against working with the insurgent group. In this case, the data offer a fairly clear distinction between LTTE-controlled areas of Northern Province and the highly affected government-controlled areas of Eastern Province, where fighting has continued.

[1]By the middle of 2009, however, the civil war in Sri Lanka was over with the defeat and surrender of the LTTE.

4 The Tsunami and Its Victims

The tsunami made significant impact in each of Sri Lanka's 14 coastal districts. Figure 2 depicts the regional pattern of damage in terms of the official death toll, based on police reports (Sri Lanka NDMC 2005). The impact was felt first, and, by most measures, most severely, in areas to the east, and, therefore, directly facing the source of the wave in Sumatra. The detailed data on tsunami death toll, homes destroyed, and affected population shown in Table 1 offer a fairly consistent sense of provincial variation in impact (Sri Lanka DCS 2005). On all dimensions, the Eastern Province accounts for the greatest amount of damage, the Southern Province is next, with somewhat less damage in the remaining three provinces. It is important to note that these figures reflect raw numbers of victims, not rates. Given large coastal populations in Southern and Western Provinces, rates of impact were actually much higher in Eastern and Northern Provinces. For instance, tsunami-related death rates were of the order of 20 per 1,000 population in affected divisions of the Eastern Province; 30 per 1,000 in the Northern Province, and only 10 per 1,000 in the Southern Province. Death rates in the Western Province were negligible. Official death toll estimates produced by the National Disaster Management Centre shortly after the tsunami indicate a total death toll of 31,183, with 46% in the Eastern Province and 32% in the Southern Province. Subsequently, the government Department of Census and Statistics conducted village-level surveys, which indicated a lower death toll of 21,660, with proportional reductions in all areas except the LTTE-controlled Northern Province. In this study, provincial

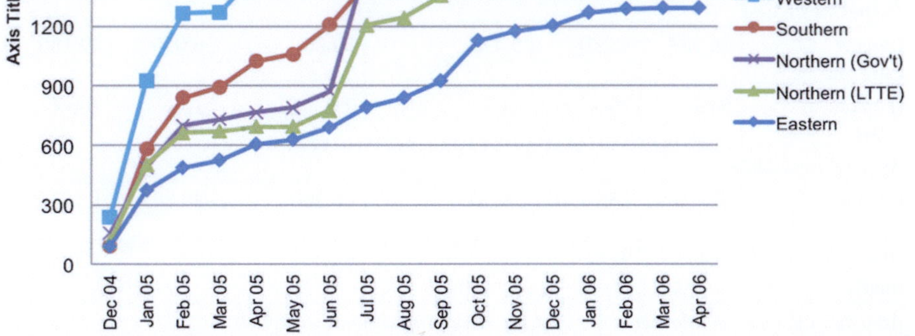

Fig. 2 Average donor commitment per adjusted affected population, by month of project initiation and province. *Source* See Table 3

death tolls are captured by the average of these two figures, which suggests that 46% of deaths were in the Eastern Province, 32% in the Southern, 13% in the LTTE-controlled Northern Province, and 7% in the government-controlled Northern Province.

A second key estimator of tsunami impact comes from the number of homes destroyed, also collected as part of the census survey. This estimator is particularly important in the Sri Lankan context, since most donor expenditure went towards the reconstruction of homes. Housing damage estimates suggest an even greater concentration of damage in Eastern Province, with 50% of all destroyed homes located there, compared to 19% in Southern Province, and 11% in LTTE areas. Finally, the census survey estimated the total population affected by the tsunami, a category that includes individuals who suffered in terms of their own health, home, livelihood, or lost relatives. This estimator further increases the concentration of impact in Eastern Province, which accounted for 58% of the affected population, compared to 22% in Southern Province.

The last set of figures in Table 1 offers a weighted average of tsunami impact based on all three indicators. A linear regression model, using data for all divisions

Table 1 Tsunami impacts according to death toll, homes destroyed, and affected population, by district

	Eastern	Southern	Northern (LTTE)	Northern (Gov)	Western	National
Death toll						
Official	14,489	10,130	3,560	2,640	364	31,183
(Percent of total)	46%	32%	11%	8%	1%	100%
Census	9,720	6,874	3,423	949	694	21,660
(Percent of total)	45%	32%	16%	4%	3%	100%
Average	12,105	8,502	3,492	1,795	529	26,422
(Percent of total)	46%	32%	13%	7%	2%	100%
Homes destroyed						
Census estimate	24,783	9,230	5,425	3,686	6,056	49,180
(Percent of total)	50%	19%	11%	7%	12%	100%
Affected population						
Census estimate	671,185	253,860	79,705	68,260	80,845	1,153,855
(Percent of total)	58%	22%	7%	6%	7%	100%
Adjusted affected population (weighted for death toll, homes, affected)						
Weighted average	617,044	266,746	107,847	75,480	86,739	1,153,855
(Percent of total)	53%	23%	9%	7%	8%	100%

Source Sri Lanka NDMC (2005), Sri Lanka DCS (2005)

nationwide, estimated the association between total aid commitment and each of these three measures of tsunami impact. The results indicated that the coefficient for each additional death was 13 times as large as that for each affected individual, while the association for an additional home destroyed was 17 times as large. Accordingly, the weighted average counts each death 13 times, each destroyed home 17 times, and each affected person once. The weighted figures are normalized to reflect the total estimated affected population of 1,153,855. The resulting estimates suggest about 53% of the affected population is in the Eastern Province, 23% in the Southern Province, and 7–9% in other areas.

The immediate aftermath of the disaster saw an outpouring of global publicity and assistance as well as a brief period of cooperation between the GoSL and the LTTE. As part of the ongoing ceasefire, the GoSL and the LTTE signed a joint Post-Tsunami Operational Management Structure (P-TOMS), ostensibly establishing a joint management and oversight structure for tsunami relief. The Department of Census and Statistics, for instance, conducted a census of affected populations and structures in all affected areas, including those controlled by the LTTE.

While tales of inter-communal cooperation were widespread, concerns persisted about the re-emergence of conflict, about the ability of foreign donors to work with the GoSL and the LTTE together, and about pre-existing sources of social and political disadvantage in communities in the east. On 15 July 2005, the Sri Lanka Supreme Court ruled key provisions of P-TOMS to be unconstitutional, setting the stage for a new round of conflict over tsunami assistance. The election of Mahendra Rajapaksa as President in November 2005 was followed by further escalation in tension. Military activity increased considerably, particularly in Trincomalee and Batticaloa Districts in the Eastern Province, each of which has been the site of GoSL recovery of territory coupled with sustained humanitarian crises. It is estimated that 6,000 have died in the conflict since November 2005 (ICG 2008). The next section will address the relationship between these damage estimates and provincial aid contributions.

5 Down the Rabbit Hole: Development Assistance Database in Practice

In the follow-up to the UN's Millennium Declaration of 2000, the Paris Declaration on Aid Effectiveness established protocols for ownership, harmonization, alignment, results, and mutual accountability in development assistance. In accordance with these goals and those of the Sphere Project, aimed at achieving minimum standards and accountability in humanitarian operations, one of the first acts of the Office of the UN Special Envoy for Tsunami Recovery was to implement a regional Development Assistance Database (DAD) tracking project funds for Indonesia, Sri Lanka, Thailand, and Maldives (DAD 2008). More detailed national databases are

Table 2 Donor assistance database description and exclusions from analysis

	Commitment	Disbursement	Projects	Multi-province projects	Province allocations
All projects	USD 3,361,276,130	USD 1,844,606,011	1,021	191	1,612
Tsunami-related only	3,355,970,114	1,842,724,510	1,009	188	1,592
Specific province allocation	2,277,245,699	1,058,296,623	907	152	1,422
Excluding northwest province	1,963,721,820	1,054,735,864	902	152	1,399

Source DAD (2008)

maintained by the lead tsunami-relief agency of each participant country. The regional database includes a rich, comprehensive record of all major donor commitments to Sri Lanka. This analysis covered funding data from December 2004, the month of the tsunami, to the end of April 2008.

Although DAD includes built-in aggregation tools, aggregated data do not accurately represent individual projects that cover multiple districts. This analysis is based on listings of individual projects and coding of drill-down data linked to each project in the DAD. Table 2 indicates that the total database includes 1,021 projects. These projects were budgeted for a total cost of USD 3.36 billion, but only USD 1.84 billion was actually disbursed. Top-level project records could indicate one or more districts or sub-districts receiving support. Projects could also indicate multiple funding and implementing agencies. The drill-down pages provide detailed percentage allocations of the costs for each project, but not for disbursements or other accounting indicators. For example, project LKA/000025, Food Assistance to Tsunami Victims had a total cost of USD 68,054,484, of which USD 48,860,073 was eventually disbursed. According to the top-level record, the project supported all 12 highly-affected districts, including the LTTE-controlled areas, and the project was funded by 17 donors, including most of the major bilateral organizations, one major international organization (American Red Cross), and one multilateral aid agency (ECHO, the European Commission's Humanitarian Aid Office). The drill-down provided the exact breakdown of funds budgeted to each of the 12 districts and the exact contribution of each of the 17 agencies, but it did not indicate whether certain aid agencies were contributing to particular districts.

In cases with multiple districts and funders or implementers, the analytic assumption is, therefore, that district allocations for a particular funder or implementer are proportional to the overall district allocation. Almost no project included multiple funders and multiple implementers. A bigger challenge is that the drill-down indicates district, funder, and implementer allocation of costs, but not of disbursements. It is, therefore, assumed that for each project funds were always disbursed according to the overall rate of disbursement for the project. In the WFP example, since 71.8% of total funds were disbursed, it is assumed that each of the funding agencies disbursed 71.8% of their original commitment, and that each district received 71.8% of their original allocation. As one of the key factors driving low disbursement rates is the conflict itself, it is likely that the estimates overstate disbursements to conflict-affected areas and understate disbursements to other areas, suggesting that results on the effect of the conflict are a lower bound. The focus of the analysis on the province rather than the district level further minimizes this effect, since this assumption will not have to be employed for projects covering multiple districts in the same province. Overall, 19% of the projects in the DAD include allocations to multiple provinces. The 1,021 projects include a total of 1,612 individual province allocations.

The analysis addresses a subset of the projects in the DAD based on the exclusions indicated in Table 2. First, projects that are coded as 'not tsunami-related', totalling 1% of projects and 0.2% of total costs, are excluded. Second, the analysis excludes all projects for which the district allocation is coded either as 'All Districts', 'District to be Determined', or 'District Unspecified'. This exclusion removes only 10% of projects, but 32% of total project cost (USD 1.078 billion) as many of the 'All District' projects were high-value (principally debt relief). In total, 41% of the costs in this category were 'All District' allocations (USD 440 million), while 42% were 'To be Determined' (USD 454 million), and 17% were 'Unallocated' (USD 187 million). By excluding the latter two categories, the assumption underlying the results is that these funds would eventually be allocated and disbursed in proportion to funds already allocated. A more likely result is that these funds were not disbursed at all. Finally, the analysis excludes all allocations made to Puttalam District, the only affected area of the sparsely-populated North-West Province, where tsunami-impact statistics indicate very low rates of damage. Five projects were allocated strictly to Puttalam, and a further 18 multi-district projects included a Puttalam allocation. This includes the single-largest project in the database, a USD 300 million power plant funded by a Chinese loan that was coded as 'Tsunami-related' and happened to be planned for sparsely-populated Puttalam. No funds have yet been disbursed for the project.

The resulting analytic database includes 902 projects with a total cost of USD 1,963 billion and a total disbursement of USD 1.05 billion. These projects covered 1,399 specific provincial allocations. The provincial breakdown of projects, allocations, and disbursements is shown in Table 3. Although DAD also included data on project expenditures, these data sets are not considered reliable and require further verification, and are therefore reserved for future publications. Preliminary analysis indicates that provincial variations in the percentage of disbursed funds that were expended would not alter the general findings of this paper (Kuhn 2008).

Table 3 describes the total donor commitments to tsunami-related projects in Sri Lanka by province. The first two lines indicate the total number of provincial allocations covered in the database. Compared to any estimate of the distribution of tsunami impact, projects appear to emphasize the Southern Province (28%) and the Western Province (13%) at the expense of the Eastern Province (42%), and, to a lesser extent the Northern Province. Subsequent lines suggest that total project cost

Table 3 Donor project allocations and aid commitment by province

Measure	Eastern	Southern	Northern (LTTE)	Northern (Gov)	Western	National
Project allocations	584	392	112	130	181	1,399
(Percent of total)	42%	28%	8%	9%	13%	
Total commitment (in millions of USD)	911	537	168	134	214	1,964
(Percent of total)	46%	27%	9%	7%	11%	
Commitment per affected population	1,477	2,015	1,555	1,770	2,463	1,702
(Relative to national)	0.87	1.18	0.91	1.04	1.45	

Source Authors' dataset compiled from DAD (2008), Sri Lanka DCS (2005), Sri Lanka DCS (2001, 2005)

allocations are not as unfavourable to the Eastern and Northern Provinces, but are still indicative of disproportionate funding for the Southern and Western Provinces. Row 3 indicates the total aid allocation by province, while row 4 indicates the percentage of the total going to each district. Row 5 provides an estimate of the dollars per affected population (DAP) in project allocations to each province, derived by dividing the total province allocation by the adjusted tsunami-affected population shown at the bottom of Table 1. DAP offers an intuitive measure of the amount of aid provided to each tsunami victim in each province that accounts for the mix within each province of those who were merely affected, those who lost homes, and those who died. Nationally, the USD 1.96 billion in aid was allocated to 1.15 million affected people, suggesting a commitment of USD 1,702 per affected person. Compared to this national average, the Eastern Province received only 87% as much (USD 1,477 DAP); LTTE-controlled areas in the Northern Province received 91% as much (USD 1,555 DAP); and government-controlled areas in the Northern Province received 104% as much, or 4% more (USD 1,770 DAP). By comparison, the Southern Province received USD 2,015 DAP, or 18% above the national average, or 36% more per affected person than the Eastern Province. The Western Province received USD 2,463 DAP, or 45% above the national average.

DAP provides one stylized estimate of relative aid response that is only as precise as is the estimate of impact. Yet, almost any plausible weighting of impact would imply that the Southern and Western Provinces received disproportionately more aid, relative to impact, than other areas. As discussed above, housing damage offers perhaps the most reliable source of data (since homes are easy to count) and relates directly to the single biggest donor expenditure category on reconstruction of housing and infrastructure. While 19% of the destroyed homes were in the Southern Province, 27% of aid was allocated there. In the absence of any gold

Table 4 Donor aid disbursement by province

Measure	Eastern	Southern	Northern (LTTE)	Northern (Gov)	Western	National
Total disbursement (in USD millions)	472	301	85	67	130	1,055
(Percent of total)	45%	28%	8%	6%	12%	
(Percent of commitment)	52%	56%	51%	50%	61%	54%
Disbursement per affected population	765	1,127	792	882	1,503	914
(Relative to national)	0.84	1.23	0.87	0.97	1.64	

Source See Table 3

standard measure of damage based on deaths, affected survivors, or lost property, these estimates at least point to a higher flow of aid relative to impact in the Southern and Western provinces. Furthermore, these estimates do not imply that the largely Tamil and Muslim areas of the Eastern Province, almost entirely under government control, received more favourable allocations than either the LTTE- or government-controlled areas of Northern Province. If anything, it would appear that the Eastern Province, which was the most heavily affected by fighting and the tsunami, received the lowest allocations per affected person. It is unclear, based on these results, whether differential allocations reflect bias in aid flows. They may also reflect variations in the cost of aid delivery across region. They might also reflect the greater economies of scale of delivering aid in areas of the Eastern Province where a much more concentrated proportion of the population was impacted.

Table 4 summarizes the disbursement of pledged donor funds to each province in comparison to commitments and to the affected population. Nationally, only about 54% of committed dollars were actually disbursed. Provincial variations in disbursement rate were small, but reflected a further negative impact of the conflict on the Eastern and Northern provinces, where between 50 and 52% of pledged dollars were disbursed, compared to 61% in the Western Province and 56% in the Southern Province. In other words, above and beyond any variations in donor commitments resulting from the conflict, affected areas lost a further 10% of their post-tsunami allocation compared unaffected areas due to the failure to deliver aid. Looking at the two most highly affected provinces, the last line of Table 4 indicates that the Southern Province actually received USD 1,127 disbursed DAP, compared to USD 765 in the Eastern Province, 47% more.

The remaining analysis will seek to explain variations over time, and by funding source, in DAP allocations. DAP results are presented for illustrative purposes, though the principal interpretive goal in presenting these results will be to compare

Table 5 Donor aid disbursement per affected population, by sector and province

Sector	Eastern	Southern	Northern (LTTE)	Northern (Gov't)	Western	Total
Emergency relief and response						
Total (in DAP)	107	117	89	97	156	111
Relative to national average	0.97	1.05	0.80	0.88	1.41	
Social services and infrastructure						
Total (in DAP)	316	567	390	436	740	421
Relative to national average	0.75	1.35	0.93	1.04	1.76	
Health, population, education and environment						
Total (in DAP)	216	191	130	122	228	197
Relative to national average	1.10	0.97	0.66	0.62	1.16	
Finance, trade, and industry						
Total (in DAP)	765	1,127	792	882	1,503	914
Relative to national average	0.84	1.23	0.87	0.97	1.64	

Source See Table 3

relative allocations across time and funding source, so the choice of denominator for the affected population will not enter into the equation.

5.1 Aid by Sector

Table 5 describes aid disbursements by sector, as coded in the DAD. The possibility remains that differential allocations to conflict-affected areas could reflect many things: imprecise estimations of the scope of damage, differences in the cost of aid delivery, or, in the case of the Eastern Province, economies of scale in aid delivery to heavily affected areas. Were this the case, we might anticipate that these provincial differentials would be reflected consistently across sectors. Yet, the results indicate considerable sectoral variation in aid differentials. If differentials were explained by erroneous assumptions about the scope of damage, we might expect to see greater bias towards the Southern and Western Provinces during relief operations. But quite the contrary to that expectation, we see that allocations in the Emergency Relief and Response sector closely mirror the distribution of damage. The sector as a whole accounted for USD 128 million (USD 111 DAP), or 12% of the total disbursement. Disbursements were highest in the Western Province (USD 156 DAP), which may reflect some tarmac bias, as well as the presence of coordination operations in a province that includes the capital. But disbursements per affected person were not notably higher in Southern Province (USD 117 DAP) than in Eastern Province (USD 107 DAP). Relief allocations were slightly lower in the Northern Province (USD 89 in LTTE areas, USD 97 in government-controlled). This deficit is readily explained by the fact that rates of disbursement of committed aid were slightly lower in Northern Province (about 61%) compared to the rest of the country (about 72% in all other areas). After taking this into account, the level of donor commitment per adjusted affected person of the population was almost identical across the most highly affected areas of the Eastern, Southern, and Northern Provinces.

Yet, striking variations persist across other sectors, most notably the largest sector, Social Services and Infrastructure, which includes the massive housing reconstruction programme and accounted for about USD 485 million dollars in disbursements (USD 421 DAP), or 46% of the total. Total allocations were highest in the Southern Province (USD 740 DAP). They were also much higher in the Southern Province (USD 567 DAP) than in any conflict-affected area. SS&I disbursements in the Northern Province, compared to the Southern, were 32% lower in LTTE-controlled areas (USD 390 DAP), and 23% lower in government-controlled areas (USD 436 DAP). More strikingly, SS&I disbursements in the Eastern Province (USD 316 DAP) were barely half of those in the Southern Province, and were even 20% lower than those in LTTE-controlled areas. Only a small percentage of these gaps can be explained by the failure to convert commitments into disbursements, as the disbursement rate in the Eastern Province was 44% compared to 50% in Southern. If these disbursements are compared to the distribution of destroyed homes instead of to an adjusted average, the differential becomes even more extreme. While only 19% of destroyed homes were estimated to be in Southern Province, 31% of the SS&I funds were disbursed there. By contract, in the Eastern Province, 40% of the funds were expended in an area containing fully 50% of destroyed homes. A different pattern emerges in the combined Health, Population, Education, and Environment sector, which includes the DAD categories for Education, Health and Population, Water and Sanitation, and General Environment, and accounted for USD 227 million (USD 197 DAP), or 22% of all disbursements. Average allocations per affected person were actually about 13% higher in the Eastern Province (USD 216 DAP) than in the Southern Province (USD 191 DAP). The Western Province was barely higher than the Eastern (USD 228 DAP). Allocations in both areas of the Northern Province were considerably lower, with LTTE-controlled areas (USD 130 DAP) receiving 34% less than the national average, and government-controlled areas (USD 122 DAP) receiving 38% less. Little of this gap could be explained by the disbursement rate of committed dollars.

The last major sectoral grouping, Finance, Trade, Livelihoods, and Industry, offers the most striking evidence of differential funding towards the Southern and Western Provinces. This sector, which combines Financial Services and Business (USD 47.0 million), Fishing (USD 33.9 million), Agriculture and Forestry (USD 23.5 million), Industry (USD 17.4 million), Tourism (USD 3.1 million), and Trade (USD 1.5 million), accounted for USD 127 million (USD 110 DAP), or 12% of the total disbursement. The sector overwhelmingly favoured the Western (USD 197

DAP) and the Southern Provinces (USD 184 DAP), which together accounted for USD 66 million, or 52% of the sectoral allocation. The FT&I allocation per affected person for the Southern Province was 2.5 times higher than in the Eastern Province (USD 72 DAP), 2.7 times higher than in LTTE-controlled the Northern Province (USD 62 DAP), and 1.6 times higher than in the government-controlled Northern Province (USD 118 DAP). While it is certainly more likely that enterprise and industry sustained more damage in the more developed areas of the Southern and Western Provinces, it is also true that many larger enterprises also received private insurance payments. More importantly, it seems that these economic development initiatives in non-conflict areas may have been financed at the expense of additional housing reconstruction resources in conflict areas. In any case, it should be clear that differentials in aid disbursement were driven more by donor priorities than by differences in the cost of aid delivery.

5.2 *Aid by Funding Agency*

Table 6 offers some insight into how differential aid allocations were shaped by the key categories of donors involved in the tsunami relief and reconstruction effort. Looking first at bilateral aid agencies, which accounted for 39% of the total disbursement (USD 410 DAP), we see a considerable disadvantage for conflict-affected areas, and particularly for areas controlled by the LTTE. LTTE-controlled areas

Table 6 Donor aid disbursement per affected population, by funding agency and province

	Eastern	Southern	Northern (LTTE)	Northern (Gov't)	Western	Total
Bilateral aid agencies						
Total (in DAP)	353	505	294	378	692	410
Relative to national average	0.86	1.23	0.72	0.92	1.69	
International NGOs						
Total (in DAP)	232	385	233	234	499	288
Relative to national average	0.81	1.34	0.81	0.81	1.73	
Multilateral/UN agencies						
Total (in DAP)	173	202	258	262	300	203
Relative to national average	0.85	0.99	1.27	1.29	1.48	
Private donors						
Total (in DAP)	6	35	7	9	12	13
Relative to national average	0.43	2.62	0.52	0.67	0.93	

Source See Table 3

received only USD 294 DAP, or 72% of the national average and less than half of the aid delivered to non-conflict areas. Although government-controlled areas in the Northern (USD 378 DAP) and the Eastern Provinces (USD 353 DAP) fared better, we nonetheless see a fairly consistent bias towards the Southern (USD 505 DAP) and the Western Provinces (USD 692 DAP). The Southern Province received about 40% more aid per affected person than government-controlled areas in the conflict zone, and about 70% more than LTTE-controlled areas. These differentials would look even clearer if the US government and USAID (which avoid supporting the LTTE) were separated out from other bilateral donors, many of whom provided considerable support to LTTE areas. While bilateral aid flows display a bias against insurgent-controlled areas, they also display a greater-than-average bias against all conflict-affected areas.

The second-largest donor category was international organizations including CARE, MercyCorps, and the International Federation of the Red Cross and Red Crescent and its national affiliates, which disbursed USD 330 million (USD 288 DAP), or 27% of the total. These agencies exercised no bias against LTTE-controlled areas, but displayed a substantial differential allocation between conflict and non-conflict areas. The Southern Province (USD 385 DAP) received about 65% more aid per affected person than the conflict-affected areas (jointly about USD 233 DAP), while the Western Province (USD 499 DAP) received more than twice as much per person.

Multilateral agencies, including the European Commission and Asian Development Bank, which accounted for 19% of total disbursements (USD 234 million, or USD 203 DAP), displayed considerably less bias towards non-conflict areas. Multilateral contributions to the Southern Province (USD 202 DAP) were only 17% higher than in the Eastern Province (USD 173 DAP), while contributions to the Western Province (USD 300 DAP) were 73% higher. More strikingly, multilateral contributions, particularly those of the European Union, substantially favoured the Northern Province. The Northern Province (USD 258 DAP for LTTE, USD 262 for government) received about 50% more per affected person than the Eastern Province, and about 28% more than Southern Province.

5.3 Aid Allocations and Disbursements Over Time

Finally, the DAD offers the opportunity to look at the emergence of aid differentials over time. While DAD does not isolate expenditures into specific months, project expenditures can be characterized in terms of the project start date. Table 7 divides the projects into the key phases in the aid process. First comes the relief and rehabilitation phase, running from the date of the tsunami through the end of March 2005. This phase includes far more than the funds categorized under the Emergency Relief and Response sector, with about USD 534 million disbursed (USD 463 DAP), or 51% of the total donor disbursement. Although Emergency Relief and Response allocations largely mirrored the distribution of damage, total allocations

Table 7 Donor aid disbursement per affected population, by funding agency and time

	Eastern	Southern	Northern (LTTE)	Northern (Gov't)	Western	Total
Relief (December 2004–March 2005)						
Total (in DAP)	373	561	409	445	883	463
Relative to national average	0.81	1.21	0.88	0.96	1.91	
Reconstruction pre PTOMS (April 2005–June 2005)						
Total (in DAP)	109	215	109	125	206	142
Relative to national average	0.77	1.51	0.77	0.88	1.45	
PTOMS and housing phase 1 (July 2005–December 2005)						
Total (in DAP)	178	282	215	269	231	215
Relative to national average	0.83	1.31	1.00	1.25	1.07	
Phase 2 and beyond (January 2006–April 2008)						
Total (in DAP)	104	68	59	44	183	94
Relative to national average	1.11	0.73	0.63	0.47	1.95	

Source See Table 3

during this time period did not. Disbursements per affected person in the relief phase were 49% higher in the Southern Province (USD 561 DAP) than in the Eastern Province (USD 373 DAP). Disbursements in Western Province (USD 883 DAP) were 2.4 times higher than in Eastern Province, and about twice as high as in Northern Province. Again, the differences between allocations in the ER&R sector and during this phase as a whole suggest that, in spite of the gradual emergence of more accurate data on the distribution of damage, aid was beginning to be allocated in favour of Southern and Western Provinces.

Phase 2 consisted of the early housing reconstruction phase preceding the establishment of the PTOMS agreement. About USD 164 million (USD 142 DAP), or 16% of all aid was disbursed in this phase. By this time, the Sri Lanka Department of Census and Statistics would have completed their assessments of housing damage and displacement, as well as the more finely-grained death statistics. Yet, the relative aid advantage experienced by the Southern and Western Provinces increased. Aid disbursement per affected person in the Southern Province (USD 215 DAP) and the Western Province (USD 206) was about double that received by Eastern Province (USD 109 DAP). Contrary to the expectation that the absence of a formal agreement on distribution of aid to LTTE-controlled areas would have had a deleterious impact on aid delivery, the LTTE-controlled areas of the Northern Province received about the same disbursement as the Eastern Province (USD 109 DAP). Although allocations to the LTTE-controlled areas were substantially lower, 100% of these dollars were actually disbursed (compared to 65% in the rest of the country), negating any disadvantage relative to other minority areas. The Government-controlled Northern Province received only slightly more (USD 125 DAP). Phase 2 was clearly distinguished by a 2:1 ratio of aid disbursement favouring Southern and Western Provinces over all conflict-affected areas.

Phase 3 marked the signing, and immediate rejection by the Sri Lanka Supreme Court, of the PTOMS agreement, as well as the continuation of Phase 1 of the housing programme. In total, about USD 248 million were disbursed during this period (USD 215 DAP), or 20% of the total. Substantially more was allocated, but only 38% of allocations were actually disbursed. Moreover, there was substantial regional variation in the rate of disbursement (not shown in table), with 46% of allocations in the Southern and Western Provinces being disbursed compared to only 33% in the Northern and Eastern Provinces. In spite of the higher disbursement rate, this phase saw a considerable narrowing of the gap between conflict and non-conflict areas. Contrary to the expectation that the failure of the PTOMS agreement might have led to a serious aid deficit in LTTE-controlled areas, the average disbursement in the LTTE-controlled Northern Province (USD 215 DAP), in fact, mirrored the national average, and was only 25% lower than the Southern Province allocation (USD 282 DAP), and rising above those received by the Western Province (USD 231 DAP). Government-controlled areas of the Northern Province fared even better under the abortive arrangement (USD 269 DAP), receiving well above average and almost as much as the Southern Province. The area that fared worst during this phase was, once again, the Eastern Province (USD 178 DAP). Though it fared relatively better than it had in previous phases, it nonetheless received only 83% of the national average and only 63% as much per affected person as the Southern Province.

The final phase of the establishment of new aid projects ran from January 2006 to the present, with the final DAD data for this analysis coming in April 2008, but very few new projects were initiated after mid-2006. This phase coincided with the escalation of hostilities between government and LTTE forces, as well as the scale-up of the Phase II housing reconstruction programme, focusing on donor-driven efforts to rehouse or relocate individuals whose homes were located in the original coastal exclusion zone. The total disbursement during this phase was USD 108 million (USD 94 DAP), or 10% of the total. As in the previous phase, rates of disbursement were much lower than in earlier phases; 39% nationally. This final phase saw some reversal of the previous pattern of aid imbalance between conflict and non-conflict areas, or at least a drastic reduction in aid to the Southern Province (USD 68 DAP, or only 72% of the average). The total allocation to the Western Province remained high (USD 183 DAP), or about double the national average. This phase was also advantageous to the Eastern Province, the focal point of Phase 2 housing reconstruction, which received USD 104 DAP. Although this was only 11% above the national average, it was 52% more than was received by the Southern Province. The Northern Province received considerably less during the final phase. It is unclear whether this was a result of the conflict (recall that rates of disbursement were highest here), or merely the fact that relatively fewer homes in these areas lay in the coastal buffer zone.

Figures 2 and 3 summarize the pattern of aid commitment and disbursement over time. Figure 3 indicates that adjusted commitments were consistently highest in the Western Province and lowest in the Eastern Province. The final phase of projects did little to narrow this gap. We also see the Southern Province running a clear second place. A dramatic one-time commitment to government- and LTTE-controlled areas of the Northern Province in July 2005 led them to approach parity with Southern Province in terms of accumulated commitments (with government-controlled areas faring better), but the Southern Province once again pulled ahead after the collapse of PTOMS. Through all phases, even prior to the establishment of PTOMS, levels of donor commitment to the Northern Province remained higher than to the Eastern Province.

Figure 3 summarizes donor disbursements over time, and reveals the further effects of the conflict on aid delivery to the Northern and Eastern Provinces. While considerable commitments were made to the Northern Province as a part of the PTOMS process, its collapse and the onset of fighting led to a significant failure to deliver aid in this context. As a result, we see a clear ordering, over time, in terms of the actual disbursement of aid by province, with the Western Province leading the way and the Southern Province in the middle. Government-controlled areas of the Northern Province maintained a consistent advantage over LTTE-controlled areas, which maintained a consistent advantage over the Eastern Province. There is some evidence in the final months of the graph that the Eastern Province gained, particularly in comparison to LTTE-controlled areas.

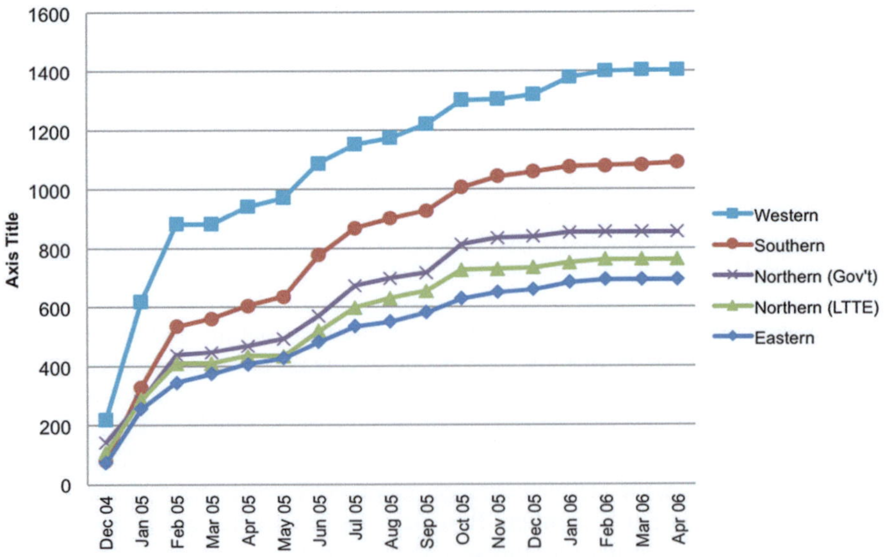

Fig. 3 Average donor disbursements per adjusted population, by month of project initiation and province. *Source* See Table 3

6 Conclusions

The first days following the Indian Ocean tsunami were met with naïve hopes of goodwill emerging between the warring parties in Sri Lanka. As in the Indonesian province of Aceh, the tsunami hit hardest in areas of ongoing civil insurgency. As in Aceh, the tsunami placed considerable pressure on insurgent groups and created considerable incentives for cooperating with the international system. Yet, the similarities ended there. The tsunami devastated the vital population centres of the Free Aceh Movement (GAM), which was already facing strategic disadvantages. In contrast, the stronger overall position of the LTTE within Sri Lanka at the onset of the tsunami was one of many sources of leverage for the LTTE. While the tsunami inflicted considerable damage on the LTTE's coastal bases, the vast majority of the affected Tamil population resided in those areas that had rarely been under the complete control of either LTTE or GoSL.

The analysis identifies a clear pattern emerging in which the combination of, first, differential allocation, and, second, inability to disburse, left conflict areas at a considerable resource disadvantage. There is little evidence that the Northern Province fared worse than the Eastern, and there is evidence that the intention, based on allocations at least, was to provide more aid to the Northern Province. In other words, to the extent that the conflict impacted the delivery of donor assistance, the effects were strongest not in the areas under the control of insurgent groups, but in areas seemingly under the control of no group, where the impact of the conflict and the tsunami had been heaviest. Aid to the Eastern Province was not only less substantial, but much of it came too late, months after the tsunami, and only after the resumption of hostilities. Furthermore, this pattern was not the result of the hostilities themselves, but rather the result of diminished aid allocations. These gaps in assistance were much more pronounced in the allocation of funds for housing and livelihood reconstruction than they were in the allocation of emergency relief. They were also more pronounced in the allocations of bilateral donors and international NGOs than in those of multilateral organizations. While this chapter leaves analysis of donor expenditures and achievements to later publications, it offers considerable insight into donor intentions and offers clues as to their likely achievements.

The dominant storylines of the tsunami recovery will likely include overwhelming generosity and measured success, if not efficiency. Yet, the crisis also raises serious questions about the practice of humanitarian assistance in this century. The dominant concern emerging from the tsunami relates to why other crises haven't garnered as much support as this one. Yet an equally important concern relates to the interplay between civil conflict, national politics, and humanitarian assistance. Why is it that the areas and people with the greatest needs, both prior to and after the tsunami, were also the ones that received the least assistance? Future studies from this project will test possible explanations for this pattern relating to donor bias in favour of nation state actors, so-called 'tarmac bias' towards areas that are more centrally located, or simply careless assessment or allocation. They must also link aid allocation bias to tangible post-tsunami outcomes in the housing and welfare sectors. Finally, they must address the potential role of biased aid allocation and delivery in the recent resumption and decisive government victories in the conflict. The modern humanitarian discourse has moved beyond the naïve notion that aid delivery can be apolitical, but the Sri Lankan experience illustrates the urgent need to adopt targets and measurement tools that could at least monitor the level of aid neutrality in the face of internal and external political demands.

References

Anderson, M.B. 1999. *Do No Harm: How Aid Can Support Peace—Or War*. Boulder, CO: Lynne Rienner.

Asian Development Bank. 2005. *Sri Lanka 2005 Post-Tsunami Recovery Program, Preliminary Damage, and Needs Assessment*. Asian Development Bank, Japan Bank for International Cooperation, and World Bank.

Barnett, Michael. 2005. Humanitarianism Transformed. *Perspectives on Politics* 3 (4): 723–740.

Benini, Aldo, Charles Conley, Brody Dittemore, and Zachary Waksman. 2008. *Survivor Needs or Logistical Convenience? Factors Shaping Decisions to Deliver Relief to Earthquake-Affected Communities, Pakistan 2005–06.*

Brun, Cathrine. 2003. Local Citizens or Internally Displaced Persons? Dilemmas of Long Term Displacement in Sri Lanka. *Journal of Refugee Studies* 16 (4): 376–397.

Buchanan-Smith, M., and S. Maxwell. 1994. Linking Relief and Development: An Introduction and Overview. *IDS Bulletin* 25 (4): 2–16.

Cliffe, Lionel, and Robin Luckham. 2000. What Happens to the State in Conflict? Political Analysis as a Tool for Planning Humanitarian Assistance. *Disasters* 24 (4): 291–313.

Drury, A. Cooper, Richard Stuart Olson, & Douglas A. Van Belle. 2005. The Politics of Humanitarian Aid: U.S. Foreign Disaster Assistance, 1964–1995. *Journal of Politics* 67 (2): 454–73.

Easterly, William. 2002. The Cartel of Good Intentions. *Foreign Policy* 131: 40–49.

Fengler, Wolfgang, Ahya Ihasn, and Kai Kaiser. 2008. Managing Post-Disaster Reconstruction Finance: International Experience in Public Financial Management. *World Bank Policy Research Working Paper No. 4475.*

Ferguson, James. 1994. *The Anti-Politics Machine: 'Development', Depoliticization, and Bureaucratic Power in Lesotho*. Minneapolis: University of Minnesota Press.

Flint, M., and H. Goyder. 2006. *Funding the Tsunami Response*. London: Tsunami Evaluation Coalition.

Fox, Fiona. 2001. New Humanitarianism: Does it Provide a Moral Banner for the 21st Century? *Disasters* 25 (4): 275–289.

Freeman, Paul. 2004. Allocation of Post-Disaster Reconstruction Financing to Housing. *Building Research and Information* 32 (5): 427–437.

Gasper, Desmond R. 1999. Drawing a Line: Ethical and Political Strategies in Complex Emergency Assistance. *European Journal of Development Research* 11 (2): 87–114.

Goyder, Hugh, Cowan Coventry, Jerry Adams, Tania Kaiser, Suzanne Williams, and Ian Smillie. 2006. *Linking Relief, Recovery, and Development (LRRD)—Policy Study*. London: Tsunami Evaluation Coalition.

Grewal, Mandeep Kaur. 2006. *Approaches to Equity in Post-Tsunami Assistance, Sri Lanka: A Case Study*. Report Commissioned by the Office of the UN Special Envoy for Tsunami Recovery (OSE).

Hasbullah, S.H. 2001. *Muslim Refugees: The Forgotten People in Sri Lanka's Ethnic Conflict*. Nuraicholai: Research and Action Forum for Social Development.

Hilhorst, Dorothea. 2002. Being Good at Doing Good? Quality and Accountability of Humanitarian NGOs. *Disasters* 26 (3): 193–212.

Houghton, Rachel. 2007. *The Tsunami Evaluation Coalition: Implications for Practice*. London: Tsunami Evaluation Coalition.

Hyndman, Jennifer. 2007. The Securitization of Fear in Post-Tsunami Sri Lanka. *Annals of the Association of American Geographers* 97 (2): 361–372.

Indo-Asian News Service (IANS). 2007. Tamils Left Out in Lanka's Tsunami Rehab Plans. *Indo-Asian News Service*, 26 Dec 2007.

International Crisis Group. 2008. *Sri Lanka's Return to War: Limiting the Damage*. Asia Report No. 146. Brussels/Colombo: International Crisis Group.

Kuhn, Randall. 2008. Tsunami and Conflict in Sri Lanka (Unpublished manuscript, 2008).

Le Billon, Philippe. 2000. *The Political Economy of War: What Relief Agencies Need to Know*. Overseas Development Institute: London.

Le Billon, Philippe, and Arno Waizenegger. 2007. Peace in the Wake of Disaster? Secessionist Conflicts and the 2004 Indian Ocean Tsunami. *Transactions of the Institute of British Geographers* 32 (3): 411–27.

Macrae, Joanna, Mark Bradbury, Susanne Jaspars, Douglas Johnson, and Mark Duffield. 1997. Conflict, the Continuum and Chronic Emergencies: A Critical Analysis of the Scope for Linking Relief, Rehabilitation and Development Planning in Sudan. *Disasters* 21 (3): 223–243.

Morris, S.S., and Q. Wodon. 2003. The Allocation of Natural Disaster Relief Funds: Hurricane Mitch in Honduras. *World Development* 31 (7): 1279–1289.

Ofstad, Arve. 2002. Countries in Violent Conflict and Aid Strategies: The Case of Sri Lanka. *World Development* 30 (2): 165–180.

Olsen, Gorm Rye, Nils Carstensen, and Kristian Høyen. 2003. Humanitarian Crises: What Determines the Level of Emergency Assistance? Media Coverage, Donor Interests and the Aid Business. *Disasters* 27 (2): 109–126.

Slim, Hugo. 1997. Doing the Right Thing: Relief Agencies, Moral Dilemmas and Moral Responsibility in Political Emergencies and War. *Disasters* 21 (3): 244–257.

Smillie, I. 2001. From Patrons to Partners? In *Patronage or Partnership: Local Capacity Building in Humanitarian Crises*, ed. I. Smillie, 175–193. Bloomfield: Kumarian Press.

Smillie, I., and L. Minear. 2003. *The Quality of Money: Donor Behavior in Humanitarian Financing: An Independent Study*. Somerville, MA: Tufts University.

SPHERE Project. 2004. *Humanitarian Charter and Minimum Standards in Disaster Response*, 2004 edition. http://www.sphereproject.org/handbook/pages/navbook.htm?param1=0.

Sri Lanka Department of Census and Statistics (DCS). 2001. *Census of Population and Housing*. Colombo, Sri Lanka: DCS.

Sri Lanka Department of Census and Statistics (DCS). 2005. *Statistics of the Census of Buildings and Persons Affected by the Tsunami—2004*. Colombo, Sri Lanka: DCS.

Sri Lanka National Disaster Management Centre (NDMC). 2005. *Tsunami Impact Data*. Colombo, Sri Lanka: NDMC.

Telford, John, and John Cosgrave. 2007. The International Humanitarian System and the 2004 Indian Ocean Earthquake and Tsunamis. *Disasters* 31 (1): 1–28.

Tong, Jacqui. 2004. Questionable Accountability: MSF and Sphere in 2003. *Disasters* 28 (2): 176–89.

United Nations Development Program (UNDP). 2008. *Development Assistance Database*. Bangkok: United Nations Development Program.

Uvin, Peter. 1998. *Aiding Violence: The Development Enterprise in Rwanda*. West Hartford, Connecticut: Kumarian Press.

Waters, T. 2001. *Bureaucratizing the Good Samaritan: The Limitations to Humanitarian Relief Operations*. Boulder, CO: Westview Press.

Weiss, Thomas G. 1999. Principles, Politics and Humanitarian Action. *Ethics and International Affairs* 13: 1–32.
White, Philip, and Lionel Cliffe. 2000. *Matching Response to Context in Complex Political Emergencies: "Relief", "Development", "Peace-Building" or Something In-Between?*
Wilkinson, Ray. 2003. The Road to Recovery: Sri Lanka Emerging from the Ruins. *Refugees* 1 (130): 4–13.

Learning from Nature: Responding to Natural Disaster and the Need for a Multi-faceted Approach in Law

V. T. Thamilmaran

1 Introduction

The Boxing Day tsunami of 26th December, 2004 struck Sri Lanka with all its fury. The casualties of the tsunami of the Indian Ocean were the highest in Sri Lanka, in proportion to its population. The tsunami devastated almost 70% of the country's coastline and more than 30,000 people were killed, while half a million were displaced (http://factsanddetails.com/asian/cat63/sub411/item2545.html). 4.4% of Sri Lanka's Gross Domestic Product was lost in less than an hour on that fateful morning of the Boxing Day, 2004. For an economy that was already burdened with the cost of the civil war, the damage was devastating. The recovery from this natural disaster, as expected, has been slow. At the time of collecting the data for this work, according to the Government of Sri Lanka (GoSL), three years after the tsunami, 85% of the reconstruction work had been completed.

This chapter is a brief attempt at explaining the link (or de-linking) between the individual's response to the calamity caused by the tsunami and the response/or commitment of the State to an equitable, long term, recovery process. The chapter will also briefly comment on the demonstration of generosity by individuals, which made an impact on the political leadership of that time, and outline the impact of the tsunami on the different communities in the Northern and Eastern Provinces of Sri Lanka. Finally, this chapter will assess the overall nature of the State's response to the reconstruction process. In evaluating the State's response, particular attention is focused on the unavoidable question of the State's representative character in reflecting the will of the society. In this regard, it will be argued that an unbridgeable gap exists between civil society and the State, and that the existence of this gap became evident in the immediate aftermath of the disaster. This gap highlights the extent to which Sri Lanka has negated the will of the people.

V. T. Thamilmaran (✉)
Faculty of Law, University of Colombo, Colombo, Sri Lanka
e-mail: tmaran25@gmail.com

S. Reddy (ed.), *The Asian Tsunami and Post-Disaster Aid*,
https://doi.org/10.1007/978-981-13-0182-7_3

2 The Immediate Response to the Tsunami: A Short-Lived Bubble of Spirited Humanitarianism

The spontaneous outpouring of concern and the empathy from all cross-sections of Sri Lankan society, following the tsunami, was a rare manifestation of the spirit of humanitarianism in the country. Irrespective of ethnicity, religion, caste or creed, people were reported to be helping those affected by the tsunami. There were even reports of combatants from either side participating together in immediate relief work.

A columnist for the Daily Mirror, writing in the aftermath of the disaster, commented:

> The mighty wall of the Tsunami wave has killed Sinhalese and Tamils, Buddhists, Hindus, Muslims and Christians, Buddhist monks and Catholic nuns, and supporters of the SLFP, UNP, JVP, and TNA. Joint burial services by Buddhist monks, Catholic priests, and Muslim Moulavis were held in many places. The shared grief by communities should help us come together and rebuild Sri Lanka.[1]

The leaders from both sides of the Sinhala–Tamil divide were also quick to comment on this wave of humanitarianism. The then President Mrs. Chandrika Kumaratunga, in her address to the country following the tsunami, noted:

> [I]t is not possible to deal with a massive natural calamity of this magnitude separately as Sinhalese, Tamils, or Muslims. We must all stand together. It is not possible for any one community alone to rebuild this country…Therefore, it is my earnest appeal that in the sacred task of rebuilding this country from this massive destruction, we all stand together as one country and one people, irrespective of all differences, transcending the boundaries that divide us. In a country where every aspect of life has been politicized, the building of such a feeling of unity may not be an easy task. However, this is a good opportunity to achieve such an end, even though difficult. It is my belief that however great the tragedy we face today, we Sri Lankans possess the courage and strength to overcome it.[2]

The LTTE[3] also, in turn, expressed similar sentiments. Mr. Soosai, Head of the Sea Tigers, who spearheaded the LTTE's tsunami immediate relief programme said:

> The Sri Lankan government should not look at this as a Sinhala or Tamil issue. It should see it as a human tragedy and help. We should now ensure that the people rescued from the Tsunami devastation are protected…than let the enormity of the tragedy make us inactive. The government of Sri Lanka should consider our people also as human beings.[4]

---

[1]Keith (2005).

[2]'Unite to Assist the Affected–CBK', *The Island*, 30 December 2004. http://www.island.lk/2004/12/30/news11.html.

[3]The Liberation Tigers of Tamil Eelam.

[4]'Tamils are also human beings—LTTE Commander', *Tamilnet*, 31 December 2004, http://www.tamilnet.com/art.html?catid=13&artid=13789.

In retrospect, it is evident that these hopes and aspirations did not survive for long. The great divide between the State and the society arguably did not allow for the transformation of the small wave of humanitarianism into a forceful tide of confidence and trust building. In other words, the structure and the nature of the State, which was embedded as sacrosanct within the Southern Sri Lankan polity, did not allow for this transformation to take place. It is viewed as the inherent weakness of the structure of the State that operates as a barrier against bringing the people of the country together even for humanitarian purposes.

3 The Tsunami and the Minorities Affected by the Civil War

> Displaced three times by the conflict, Thamilagan, 30, is from Maruthankerny in Jaffna. Situated just 50 metres from the coast, his village lost 189 people to the tsunami. Thamilagan describes the sound of "fire crackers and a heavy popping noise" outdoors on the day of the tsunami. The cracks around his front door and windows started to let in water. "Before I knew what was happening, the walls caved in and a rush of water washed away everything in sight," he says. "My father-in-law grabbed the twins but could not hold onto them as the force of the water was too strong."
>
> Floating nearly 1 km from his home, Thamilagan eventually tried to make his way back but his foot had been crushed by falling bricks and he could not walk. Found later in the afternoon by the police, Thamilagan was taken to Jaffna Teaching Hospital where he remained for three days without word as to the whereabouts of his family, before a relative delivered the terrible news. Now severely scarred, crippled with grief and suffering from depression and survivor's guilt, Thamilagan asks: "Why am I the only one alive?" Having earned his living as a fisherman, Thamilagan has a usable boat and work is currently underway to clear the mass of debris blocking the harbour. But the painful memories linger and returning to the sea "will only remind me of those I lost". (Jeffels 2005, 37)

In less than an hour, the tsunami took away a number of lives of people that is equivalent to half the number of lives that the civil war took twenty years to take. The natural disaster is over, the misery caused is being overcome, but the human disaster continues unabated. Many of those affected in the North and East, like Thamilagan, had been, and are being, subject to multiple-displacements. These displacements include not only those that happened before and during the tsunami but also those which happened after the tsunami. Military operations in 2006 in the East re-displaced many tsunami affected people. Daya Somasunderam, a Professor of Psychiatry, formerly attached to the University of Jaffna, in a study scientifically establishes a well-known characteristic of the people in war torn areas in the Northern area: a sort of expertise in surviving under extraordinarily stressful conditions (Somasunderam 2007). Somasunderam establishes that these persons have become capable enough to deal with complex emergencies from their previous experiences. While the changes after the tsunami were not so prominent, Somasunderam notes a worrisome impact on the society which has become more dependent, passive, silent, without leadership, mistrustful, and suspicious.

Additional adverse effects, Somasunderam adds, include the breakdown in traditional structures in family, institutions, and familiar ways of life, and deterioration in social norms and ethics.

Furthermore, according to Frerks (2006), the combined effect of the civil conflict and the tsunami in Sri Lanka seems to be the result of a "multiple disaster and complex political emergency".

> Whereas the impact of the tsunami per se is huge, it is further compounded by a number of other factors. These include the conflict; the prevailing levels of vulnerability; poverty and underdevelopment in the regions affected; the political culture and the system of governance prevailing in the Island-state. Those factors are obviously intertwined and Sri Lanka's condition can be properly denoted as a 'multiple disaster' or a 'complex political emergency', where various factors converge towards catastrophe. The experiences in the immediate post-tsunami period strikingly underline this analysis. (Frerks 2006, 75)

Recovery and reconstruction work has been hampered in the North and East of Sri Lanka owing to the heightened military contestations between the government and the LTTE during the period of 2006–07. According to the government, out of the yet to be constructed 15,000 houses, most of the houses were intended for the North and East by the end of 2007. In Mullaitivu (a district in the North, under LTTE control) according to the Government Agent, only 50% of the required housing reconstruction had taken place by 2007.[5] In contrast, it was reported that in some districts in the South, notably in Hambantota, the native district of President Mahinda Rajapakse, houses had been constructed in excess, providing for even those who are being called 'indirectly affected'.[6] In the North and East, on the other hand, agencies involved in recovery work have reported on the difficulty in finding or transporting construction materials. Due to the renewed war in the later part of 2007, some of the tsunami reconstruction work already completed were targeted by the warring factions and subjected to severe damage.

One of the specific problems arising from the response to the tsunami in the North and East has been with regard to keeping the equity between the assistance provided to conflict-affected internally displaced peoples (conflict IDPs) and the tsunami-affected internally displaced peoples (tsunami IDPs). In a report commissioned by the UN Special Envoy on Tsunami Recovery it was noted that:

> Support for conflict IDPs has focused mostly on helping them cope with constraints rather than on re-establishing the potential for independent and assertive living. Tsunami recovery has been shaped by a commitment to restore full independence as quickly as possible to survivors, as well as building back better prospects and opportunities where possible. The reasons this policy imperative has been so much stronger in the post-tsunami rather than post-conflict context are at least partially related to the unfinished, man-made nature of the

[5]Puthinam, 'Mullaiyil 50 vizhukadu veeduhale meelamaikappattulana: Imalda Sukumar" (Only 50% of the houses have been reconstructed in Mullaitivu: Imalda Sukumar), 24 December 2006 accessed at http://www.puthinam.com/full.php?22GmYcc3pW24d71k202JRZ4d37fN0bDbH2e2KRK3b3aJoe.

[6]'Tsunami Billions at Sea', *Sunday Times*, 23 December 2007. http://www.sundaytimes.lk/071223/News/news0001.html.

Table 1 Summary of assessed reconstruction needs and funding commitments

Regions	Proposed Share of $2.2 billion total funding required (%)	Share of $2.1 billion commitments allocated as at March 2006 (%)
North	19	7
East	45	30
South	26	15
West	10	6
Commitments that are non-region specific	N/A	41

Source Grewal (2006, 18)

> conflict as opposed to the exogenous origins, immediacy and breadth of impact of the tsunami. These differences have impacted on political and bureaucratic incentives to deliver. The media coverage and the vast funds available with fewer conventional policy conditions have additionally strengthened both the political impetus and feasibility of enhanced delivery in the post-tsunami context. (Grewal 2006)

Also of concern was the political ethnicization of the humanitarian assistance that was made available for tsunami IDPs, and the resulting inequity that was identifiable between the different districts affected by the tsunami. The report to the special envoy noted that there was a disparity in the allocation to the different regions. The following table extracted from the report speaks for itself. (The total amount of funding required for the national recovery programme was assessed to be USD 2.2 billion. The money pledged stood at USD 2.1 billion) (Table 1).

The above Table clearly shows the disparities in allocation to the North in comparison to the other regions.

The following figure further illustrates the inequity prevailing in the delivery, by a comparison of the process of housing schemes (Fig. 1).

The report, having analyzed various indicators, concluded that, within the tsunami IDP group, the affected communities in the East and, particularly, the North have experienced a slower pace of progress than those in the South and West of the country. It identified institutional constraints, in specific the unitary characteristic of government, as one of the reasons for the inequity in the division of the funds received.

4 The Politics of the Response of the State

There are two probable reasons as to why it was important to have made the LTTE a party to the recovery and reconstruction process. One is the logical reasoning based on the principle of subsidiarity, which recognizes that people at the local level know what is best for them. It has been accepted, through best practices in disaster management, that decentralized structures are important, and even essential, for effective disaster management and reconstruction (see, for example, Ravi 2005). India's post-tsunami response, in terms of swiftness and effectiveness, was arguably

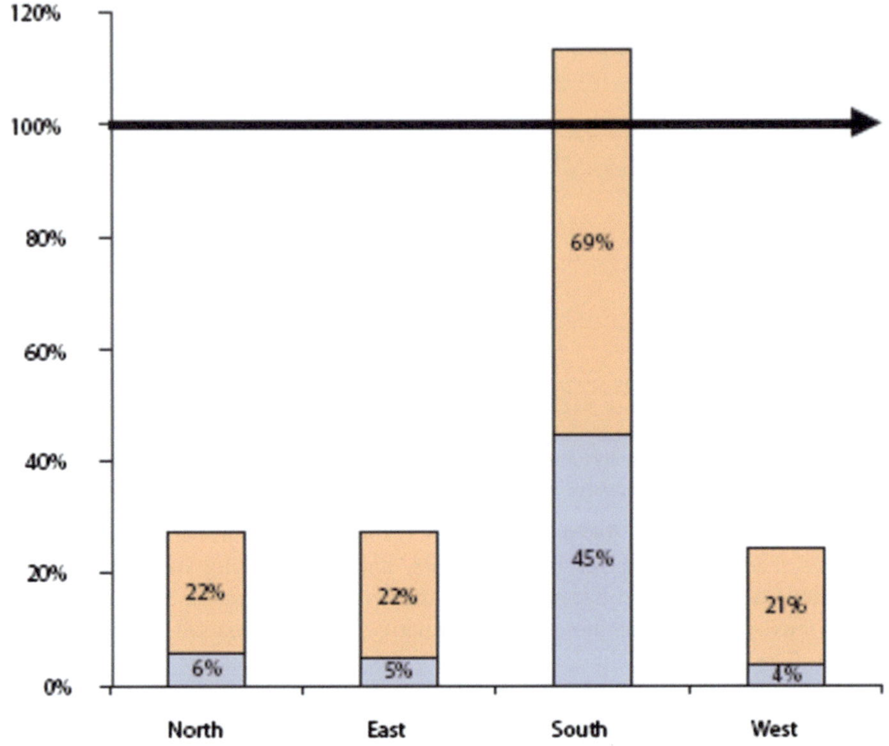

Fig. 1 Regional Breakdown of progress in housing construction made through the donor driven scheme as on April 2006 (percentage of needs being met or in the process of being met). *Source* Grewal (2006, 21)

better in comparison with that of Sri Lanka, for the reason that it has a federal form of government. It is also politically obvious that a community's needs are best taken care of by its own political structures, since they have a direct link, in terms of interest and affection, in the welfare of the community. Frerks, too, points out that the principle of subsidiarity has not been taken seriously in the reconstruction efforts in Sri Lanka.

> …there has been a nearly total lack of communication, consultation and information vis-à-vis the local population, leading to frustrations, misapprehensions, and accusations. Especially, there was a lot of insecurity regarding the reconstruction plans in the future as the government has been slow and indecisive on many important issues and decisions to be tackled in this connection. This issue also relates to the (new) donor principle of 'subsidiarity' that amounts to working at the lowest possible (administrative) level. (Frerks 2006, 89)

The other reason as to why the inclusion of the LTTE would have been important and relevant is that it would have been a recognition of a fact in political reality. The non-functioning provincial and local government structures in the LTTE-controlled areas, the de facto government of the LTTE in these areas, and the wide influence the LTTE had in areas in the North and East not under its control (despite the dislike that may be entertained of the organization) made it an important player in the recovery process. Immediately following the tsunami, the LTTE had established district-level task forces through its Planning and Development Secretariat involving government representatives, NGOs, and INGOs which were in charge of the relief and recovery programmes. It was a matter of fact that they had control over the recovery programme[7] and any agreement was probably about formalizing and widening representation of these structures and securing funds for its programme.

Immediately after the tsunami, in an unprecedented move, the President was reported to have invited the late S. P. Tamilchelvan, the then Political Head of the LTTE, to sit with the Government in a Special Task Force to oversee the tsunami recovery programme. The invitation set in motion a six-month-long process of negotiations, facilitated by the Norwegian government, which, in June 2005, resulted in the signing of the Post-Tsunami Operational Management Structure (P-TOMS). The P-TOMS went through a protracted negotiation process for many reasons, including the commitment of the President and the LTTE leadership to finalize an agreement.[8] This was possibly a rare extension of good will humanitarianism that was generated following the first few days of the tsunami. Additional factors, such as pressure from the donors, were crucial in the parties being thrust towards arriving at an agreement. From the very beginning, the Government and the LTTE took pains to explain that these negotiations were independent of the peace process. Mr. Balasingham was reported to have said, in January 2005, that the negotiations were only between the Peace Secretariats of the Government and the LTTE, and was confined to matters of rehabilitation and reconstruction.[9] Harim Peiris, the Secretary to the President termed the negotiation process as one for 'humanitarian peace' and not for 'political peace'.[10] There could have been two reasons for this: firstly, the parties were determined to demonstrate that whatever was 'conceded' by agreeing to P-TOMS should not be seen as a permanent compromise and that they were free to take different positions regarding the resolution of the national question; and secondly, it could have been that the parties wished to prevent extremist elements from interpreting the P-TOMS agreement as impacting

---

[7]These structures for coordination were set up within two days after the tsunami, whereas the Government took seven days to come up with a coordinating body. See Glasius (2005).

[8]The agreement set up three committees (A High Level Committee, A Regional Committee and District Committees with representation from the Government, LTTE and the Muslim parties) to manage the recovery process in six districts affected by the tsunami in the Northern and Eastern Provinces. A regional fund was set up to which all local and foreign aid received and allocated for the six districts, based on need-assessments conducted by the Government, were to be pooled in.

[9]Balasingham (2005).

[10]'All foreign assistance should be directed through state, says Peiris (2005).

some of the burning issues relating to the negotiation process of the ethnic conflict, such as the Interim Self Governing Authority proposals of the LTTE, in the event of which there could have been possibly a deadlock in arriving at an agreement.[11]

However, this 'de-linking' did not achieve its purpose. The Janatha Vimukthi Peramuna (JVP), the ultra-nationalist and Marxist party, then a constituent of the ruling coalition, and the Jathika Hela Urumaya (JHU), the Buddhist monks' party, opposed the proposal for a joint mechanism from the outset, even threatening violent revolt. Steeped in majoritarian unitarism, and guided by an ideology of Sinhala Buddhist chauvinism, the parties saw the agreement, which intended to equitably share resources amongst the tsunami affected populace, as an agreement that would hurt the sovereignty and territorial integrity of the country. If one were to be asked to identify two words that have been consistently part of the Sinhala nationalist rhetoric to block any meaningful process at arriving at a solution, they are undoubtedly, 'sovereignty' and 'territorial integrity'. It seems as if the imagination of the Sri Lankan polity is stuck with the orthodox Austinian-Diceyan conception of sovereignty, which is clearly outdated even in their country of origin (Elliott 2001). As Dayan Jayatilleke, Sri Lanka's Ambassador to the UN, phrased it, the P-TOMS agreement would have been a "surrender of legitimate monopoly over part of our maritime areas…an unconscionable erosion of even relative sovereignty. It is a process of dissolution of the borders of the Sri Lankan state and therefore of the state itself".[12]

The donor and NGO community support for the agreement was construed as the West's contribution to the erosion of the Sri Lankan State's sovereignty. The minority-phobia and the 'Western-phobia' makes for a curious mix in Sri Lanka as a result of the nationalistic, pseudo-Marxist, and ethnic-chauvinistic mix found within the Southern Sri Lankan polity.[13]

The P-TOMS agreement was challenged by the JVP in the Supreme Court of Sri Lanka, a court which, throughout Sri Lanka's post-independence history, was not well known for safeguarding the civil rights of the minorities. The court interestingly dismissed the petitioner's contention that the government cannot enter into an agreement with a terrorist organization by reference to the ceasefire agreement or the MoU of 2002 between the GoSL and the LTTE. It even set aside the argument that the P-TOMS agreement does not have its base in a statute and hence was *ultra vires*. Citing Hart, the Chief Justice refused to impose mechanical jurisprudence

---

[11]Uyangoda (2006) argues the importance of maintaining a link between the post-tsunami recover process and the peace process. He notes: "Sri Lanka's post-tsunami recovery process necessitates new ways for constructive handling of the ethnic conflict as well as the peace process in what (may be termed) 'a conflict and peace sensitive' recovery framework". Further, he argues, "a sustainable recovery process can productively be linked to a program of state reform conceived in a framework of 'deep federalization'" (Uyangoda 2006, 341).

[12]Jayatilleka (2005).

[13]It must be mentioned that there was significant opposition from the Muslim political leadership for not having included them in the negotiation process. Exclusion of the Muslims continues to be a recurring fault of Sri Lanka's peace processes.

saying that the legislators could not have envisaged all future happenings. When one reads through the judgment it is possible that he or she could have believed that the sequence of the reasoning would be leading to the upholding of the constitutionality of the agreement by the court. However, it did not turn out to be so. The court found the setting up of a regional fund to be managed by a multi-lateral agency to which all local and foreign aid was to be pooled in, as disturbing the parliament's control of finance. What appeared to be a departure from legal formalism and acceptance of political reality, when the court refused to apply mechanical jurisprudence on the question of P-TOMS not having a statutory base, took a 180° turn on the question of the handling of funds. This is quite typical of Sri Lankan courts, known to be professors of, to borrow Imtiaz Omar's pharaseology, an uncoordinated, discordant jurisprudence (Omar 1996, 154). Following the judgment of the Supreme Court, the P-TOMS died a quick death. It amounted to yet another agreement, in Sri Lanka's pathetic history of negotiations, which was torn apart.

This leads to a difficult question that the minorities have been unable to ignore in Sri Lanka: How is Sri Lanka going to achieve a peaceful transition towards a non-majoritarian form of democracy, by employing the tools of the system that one wants to change, such as a majoritarian parliament and an executive and a court system entrenched in the majoritarian democratic value system?

Ana Pararajasingham (2006) identifies three consequences that have arisen due to the failure of the GoSL to implement the P-TOMS agreement: Firstly, this failed agreement is evidence that the GoSL, acting on its own, is not capable of being party to a negotiated settlement to the conflict; secondly, he argues that the disregard by the GoSL of the sufferings of the tsunami victims of the North and East, demonstrates that the government has lost "the moral if not the legal right to rule them" (Pararajasingham 2006, 98); thirdly, the failure to implement the P-TOMS, in turn, adds justification for the call for self-determination by the Tamils. In the words of Verena Graf, Secretary General of the International League for the Rights and Liberation of Peoples, "(t)he developments during the last three years compounded by the post-tsunami experiences raise the spectre that time is running out; that there is no hope for the Tamils within a united Sri Lanka, that their only chance lies in fighting for external self-determination" (quoted in Pararajasingham 2006, 98: note 24).

Regardless of the failure of the P-TOMS agreement, the donors and international agencies found ways to work in the North and East through local NGOs, INGOs, and government departments that had a presence in the area.[14] It was no secret that the activities of the above three were under the overarching control of the LTTE, and its development arm, the Tamil Rehabilitation Organization (TRO). Hence, whether the Southern extremist forces liked it or not, the LTTE exerted strong influence on how the tsunami reconstruction process was carried out in the North

[14]There has always been a 'consultative relationship' between International Financial Institutions like the World Bank and the LTTE. See Burke and Mulakalan (2005).

and East. However, there is no denying that more work with less circumvention could have been done if the P-TOMS had been implemented.

5 Issues of Good Governance

Apart from the socio-political issues which emerged after the tsunami, serious questions are being raised in relation to the fundamentals of good governance in Sri Lanka. The manner in which the organs of the government conducted themselves, with regard to the basic principles of transparency and accountability in disbursing the funds received for disaster recovery projects, not only invited harsh criticism from many people, cutting across socio-political divides, but also served as an eye-opener, prompting the need for a constitutional structure with proper checks and balances.

There was also a complaint by an opposition Member of Parliament that the then Prime Minister, Mahinda Rajapakse had transferred money to what was named as the ‘Helping Hambantota Fund’ out of funds received by the government of Sri Lanka from various international aid agencies for tsunami reconstruction and rehabilitation and that the transfer was tantamount to misappropriation.[15] Sri Lankan media alleged that the money sent by international donors to the Prime Minister’s national relief fund was credited to the privately run ‘Helping Hambantota’ fund, and that the fund was set up to help the constituency from which Prime Minister Mahinda Rajapaksa hails.[16]

Although legal proceedings were initiated to look into the matter, it came to an abrupt end with the funds in question being reverted back to the Treasury. Moreover, the police officers who conducted the investigations were charged with violation of fundamental rights of the Prime Minister, by the highest court of the land.[17] The judgment of the case has been severely criticised by many scholars for lacking in impartiality and for abuse of judicial power.[18]

Even after three years, allegations were still being levelled against State institutions, regarding their involvement in the mismanagement of the tsunami funds running into billions of rupees. One of these was the damning report issued by the international watchdog, Transparency International Sri Lankan Chapter in December 2007. According to this report, a huge amount of the funds so far

[15]Mr. Kabir Hashim, Member of Parliament from the United National Party, filed a case against the then Prime Minister over the alleged irregularities in handling the relief funds.

[16]“More Questions than answers for PM’s ‘Helping Hambantota’ account”, *Sunday Times*, 31 July 2005, accessed on 02 February 2012.

[17]‘Helping Hambantota- no further probe necessary,’ *Lanka Truth*, 19 January 2006. http://www.Lankatruth.com/full_story/2006/jan/20060119.html.

[18]See critique by Pinto Jayawardena (2012).

Table 2 Total amount of aid received in Sri Lanka

Receiver	Donors (amount in SLR)		Total amount in SLR
	Local*	Foreign	
Government of Sri Lanka	2,728,212,540	2,014,634,908	4,742,847,448
NGO	10,912,467,277	20,643,109,202	31,555,576,478
Others	4,123,889,239	7,836,164,933	11,960,054,171
Total	17,764,569,056	30,493,909,043	48,258,478,097

Source Central Bank Annual Report-2007 (Bank Supervision)

received have either gone missing or have been misused for 'other purposes' than the recovery projects.[19] It was disclosed that nearly 53 million Sri Lankan Rupees (SLR) had gone missing and was unaccounted for. The report further added that of the SLR 241.5 billion promised by the donors only SLR 122.2 billion was given to the implementing agencies. According to this report, some of the donors withdrew their commitments as the progress was quite unsatisfactory. What is so strange, and, in fact, supports the contention of lack of transparency, is that, while the first-year report on the progress made after the tsunami had been issued jointly by the GoSL and development partners, in the second year the report was issued in the name of the Sri Lankan government alone (Table 2).

The issue of transparency became further compounded with the revelation by the former President Chandrika Bandaranaike Kumaratunga. She called on the government to explain the fate that befell a sum of over SLR 2,500 million received by the country as tsunami aid which was lying in a Presidential account at the Central Bank when she left the office. She left office after 11 months of the devastation caused by the tsunami. Her disappointment over the disbursement of funds came following the statement issued by the World Bank that 15,000 families of tsunami victims were still without permanent houses.[20]

The prevailing political culture therefore has defined the outcome of the reconstruction process in Sri Lanka. This has been noted by Frerks, who identifies the political culture in Sri Lanka as a contributing factor for the gaps in the tsunami reconstruction process. One aspect of this political culture is "patronage politics" where State resources and funds are distributed through a patronage system. Frerks points out that this patronage system includes an ethnic component "as most political parties are run and controlled on the basis of ethnicity" (2006, 83). Subsequently, he argues that the tsunami reconstruction process had a negative impact on the identity relations in the country, which replaced the original spirit of humanitarianism that was demonstrated by the average citizen (2006, 83).

[19]'Rs. 53 billion of disbursed tsunami funds missing-TISL', *The Island*, 6 December 2007. http://www.island.lk/2007/12/30/news11.html.

[20]'CBK queries fate of Rs. 2,500 mn tsunami funds left in Central Bank,' *The Sunday Leader*, 30 December 2007. http://www.thesundayleader.lk/2007/12/30/NEWS.HTM.

Frerks also points out that Sri Lanka is heir to a "history of mismanaged relief" (2006, 85). Even though the immediate response to the tragedy was commendable, the GoSL failed to sustain the process. Due to the failures in the administrative mechanism and the culture of patronage politics, Frerks argues that aid not only became politicized but became "a much wanted political commodity itself" (2006, 85), and that the politicians saw aid as a resource to be exploited for political popularity.

In comparing the initial outpouring of voluntary assistance by people and the long-term response of the GoSL to the tragedy caused by the tsunami, Frerks identifies several lessons that can be learned (2006, 87). Two of these lessons are relevant to the current discussion. One is that the mismanagement of aid, and political disagreements regarding the management of the aid, effectively diffused the spirit of solidarity and comity which was evident in the concerted efforts made by the public, cutting across all mundane and parochial differences. The other is that the availability of aid and assistance at the regional level in the aftermath of a disaster can result in social dynamics that would be defined in terms of ethnic identity.

One can safely assume that the above matters confirm the belief that the inevitable cleavage caused between the State and society encompasses all areas of social life of the people of the country, and that even humanitarian issues are not exempt from this plague of secrecy and unaccountability. Perhaps the jurisprudence developed through public interest litigation in India, whatever the arguments as to its purported limits, could serve as a vital tool in arresting the situation and bridging the gap between the people and the state.

6 Conclusion

This chapter has sought to explain the de-linking between the spontaneous humanitarianism reflected through the care and affection that individuals from all communities showed to each other, and the inadequacy of the imagination of the country's majoritarian polity to transform such a show of goodwill into a concrete platform for an equitable recovery process through a reasonable sharing of resources. The explanation lies in the socially protracted nature of the conflict, stuck in ideological semantic debates over sovereignty and territorial integrity. The entrenched effect of majoritarian unitarism continues to blur all opportunities that come forth for genuine long term partnership between the constituent peoples of Sri Lanka.

References

Balasingham, Anton. 2005. Pirapaharan's top Priority is Reconstruction, not Politics. *Tamilnet*, 28 Jan 2005. http://www.tamilnet.com/art.html?catid=13&artid=14100. Accessed 01 Feb 2012.

Burke, Adam, Anthea, Mulakala. 2005. *Dnors and peace building*, vol. 2, p. 27.

Elliott, Mark. 2001. *The constiutional foundations of judicial review*. Hart Publishing, p. 44

Glasius, Marlies. 2005. *The European union and Sri Lanka: Tsunami response and long term policy*. Centre for Global Government , London School of Economics and Political Science. Accessed 01 Feb 2012.

Frerks, Georg. 2006. *The Politics of Post-Tsunami Reconstruction in War-torn Sri Lanka*. Sri Lanka: Centre for Just Peace and Democracy.

Jayatilleka, Dayan. 2005. Every P-TOM, Dick and Harry. *The Sunday Island*, 10 July 2005. http://www.island.lk/2005/07/10/features1.html.

Jeffels, Lyndon. 2005. 'Six months on: Facing Fears.' *Forced Migration Review* 36–37 (Special Issue. Tsunami: Learning from Humanitarian Responses). http://www.fmreview.org/sites/fmr/files/FMRdownloads/en/tsunami.pdf. Accessed 01 Feb 2012.

Keith, Noyhr. 2005. Tsunami's Wall of Destruction Should Heal Lanka's Divisions. *Daily Mirror*, 01 Jan 2005. http://www.dailymirror.lk/2005/01/01/opinion/1.asp.

Omar, Imtiaz. 1996. *Rights, emergency and judicial review*. Martinus Nijhoff, p. 154.

Pararajasingham, Ana. 2006. *The failure of joint mechanism - causes-consequences and conclusions*. Centre for Just Peace and Democracy, Sri Lanka.

Peiris, Harim. 2005. *Tamilnet*. 26 Jan 2005, http://www.tamilnet.com/art.html?catid=13&artid=14084. Accessed 01 Feb 2012.

Pinto Jayawardena, Kishali. 2012. Pistol Whipping Justice in Sri Lanka and the Truth about Helping Hambantota. 21 Oct 2012. www.colombotelegrph.com/2012. Accessed 10 Dec 2012.

Ravi, Faiia. 2005. From disaster to opportunity: Tsunami relief as a study in the devolution of power. In *Challenges of linking post-tsunami reconstruction with peace building*. Centre for Policy Alternatives.

Somasunderam, Daya. 2007. Collective Trauma in Northern Sri Lanka: A Qualitative Psychosocial-Ecological Study. *International Journal of Mental Health Systems* 20017 (1): 5. http://www.ijmhs.com/content/1/1/5. Accessed 01 Feb 2012.

Uyangoda, Jayadeva. 2006. *Ethnic conflict, the state and tsunami disaster in Sri Lanka in 06 Inter-Asia cultural studies*, p. 341.

Social Interfaces in Disaster Situations: Analyzing Rehabilitation and Recovery Processes Among the Fisherfolk of Tamil Nadu After the Tsunami in India

Sunil D. Santha

1 Introduction

This paper stresses the need to incorporate rehabilitation and recovery processes within the framework of vulnerability reduction. By going against the dominant top-down approaches of looking at disasters as natural events, and, proposing technical solutions for mitigation and recovery, this paper highlights the need to revisit vulnerability reduction from a people-centred perspective. Analyzing the recovery processes that followed the Indian Ocean tsunami 2004, this paper examines three important aspects amongst the traditional fisherfolk in Tamil Nadu.

The pre-impact factors that have significantly contributed to the vulnerability of the coastal fishing communities affected by the tsunami are examined first. The nature of social interfaces that has emerged among the affected population in the context of relief, rehabilitation, and reconstruction are then analyzed. Subsequently, this paper examines whether the rehabilitation and recovery interventions have contributed to vulnerability reduction or towards reinforcing the hazard risk of the affected population. This paper is part of a larger, long-term study that was conducted to examine the processes contributing to coastal vulnerability in south India.

This paper draws considerable insights from the political economy model to analyze community vulnerability. The progression of vulnerability during the pre- and post-impact phases of the tsunami is analyzed by applying the Pressure and Release (PAR) Model, and the Access Model advocated by Wisner et al. (2004). Accordingly, recovery should not only be seen as rebuilding infrastructure and housing, but should also incorporate the economic, political, and social needs, paving the way for more resilient livelihoods. In this regard, each and every household affected by the disaster should not only be able re-establish its

S. D. Santha (✉)
Centre for Livelihoods and Social Innovation, School of Social Work,
Tata Institute of Social Sciences, Mumbai, India
e-mail: sunilds@tiss.edu

S. Reddy (ed.), *The Asian Tsunami and Post-Disaster Aid*,
https://doi.org/10.1007/978-981-13-0182-7_4

livelihood, physical assets, and patterns of access, but also should be in a better position to resist or withstand the shocks of possible extreme future events. However, this requires reversing the dynamic pressures and root causes that contributed to the disaster in the first place (Wisner et al. 2004). Reversing the root causes could include creating constructive changes in structures of domination and social relations, and generating sufficient access to land, credit, employment, and information.

Contrary to popular assumption, social recovery is never a linear process and cannot be assumed to pass through certain pre-assumed fixed stages (Rubin et al. 1985; Berke 1993). The process of recovery is actually complex due to added uncertainty, unforeseen events, and even the reproduction of vulnerability in itself. In many cases, the most vulnerable households and individuals do not recover after an impact (Wisner et al. 2004). The present paper is based on the assumption that rich insights on social recovery and vulnerability reduction can be gathered by examining the social interfaces that emerge among different actors in the rehabilitation and recovery phase. Social interfaces are defined as "critical points of intersection between different social fields, domains, or life worlds, where social discontinuities based upon differences in values, social interest and power, are found. Interfaces typically occur at points where different, and often conflicting, life worlds or social fields intersect" (Long 2001, 177). Analyzing social interfaces in natural resource management, studies have shown that the experiences and perspectives of multiple actors involved in resource management merge through the processes of cooperation, accommodation, or conflict to generate newly emergent forms of organization, practices, and understanding (Cleaver 2001; Santha 2007a). Different actors and organizations perceive and define disasters in their own way. Issues of power attain significance in this context. Power could be considered as both the ability to control or command, and make changes or influence a particular course of action.

It has been observed that unequal access to resources and the gap between those who have greater access to resources (who are also in a position to exercise power and thereby influence the operations system in charge of reconstruction), and those who do not have any access to resources, increases at a greater proportion after the disaster event (Wisner et al. 2004).Short-sighted reconstruction projects have often resulted in the disruption of ethnic communities, their cultures, their coping strategies, and the environment, further weakening the potential for local livelihood in the face of future disasters (Chairetakis 1992). The root causes and the dynamic pressures leading to unsafe conditions could, thus, persist while the disaster unfolds and also during the recovery phase. The following section will now analyze the factors that have led to progression of vulnerability during the pre-impact phase among the tsunami-affected coastal fishing communities in Tamil Nadu.

2 The Progression of Vulnerability: The Pre-impact Phase

This section is primarily based on the assumption that the analysis of the Indian Ocean tsunami as a disaster requires us to trace the linkages between the impacts of the tsunami on local communities with a series of social factors and processes that generates vulnerability. Explaining the progression of vulnerability, Wisner et al. (2004) have identified three sets of links that connect the disaster to these processes: (1) the root causes, (2) dynamic pressures, and (3) unsafe conditions. This section will now analyze these linkages to the gradual progression of vulnerability among the traditional fisherfolk of Tamil Nadu.

2.1 Root Causes

Wisner et al. (2004) explain root causes as follows:

> …the root causes are an interrelated set of widespread and general processes within a society and the world economy. They are 'distant' in one, two or all of the following senses: spatially distant (arising in distant centre of economic or political power), temporally distant (in past history), and finally, distant in the sense of being so profoundly bound up with cultural assumptions, ideologies, beliefs and social relations in the actual lived existence of the people concerned that they are 'invisible' and 'taken for granted'…. (2004, 52)

The root causes could be economic, demographic, and political processes, which could affect the allocation and distribution of resources among different stakeholders. These processes are also a function of certain ideological orders such as economic, social, and political structures, legal definitions and enforcement of rights, gender relations etc. The root causes often reproduce the exercise and distribution of power in a society, and is very much linked to governance, the functions of the State, the rule of law, and administrative capabilities (Wisner et al. 2004, 52).

Analyzing the existing literature on problems related to coastal fisheries in south India, it has been observed that the root causes of vulnerability among the traditional fisherfolk are not only economic, demographic, and political processes, but also ecological insecurity. Exploring the relation between the policies of the government and degradation of the coastal environment, Salagrama (2001) has observed that the current ecological and social crisis began with the modernization of marine fisheries, which he classifies into four crucial phases. The first phase of modern fisheries management in India began in the late nineteenth century, with reforms introduced by colonial rulers. During this phase, the policy makers believed that the fishing industry was rather primitive, with simple technology, and bound by custom and ignorance. It was taken for granted that the industry could be strengthened only by the initiative of the government officers, so the Department of Fisheries was established during this period. The second phase commenced in the 1950s with the assistance of developed nations like the United States. This is said to be the period when entrepreneurs invested in sophisticated technologies in modernizing the fisheries. Outsiders, by and large, who had the economic capacity to invest in the new technology, entered this sector. Trawling, industrial fishing, and aquaculture began to emerge, encouraged by generous state subsidies and loans. In the third phase, from the late 1970s to the early 1990s, capital financing was done for advanced technology. This period also witnessed a gradual displacement of the traditional industry, and artisanal fishermen, along with the degradation of vulnerable resources like mangroves and coral reefs. Mechanization in fisheries, through the introduction of trawlers and purse seines, to increase the capacity of production, and development of cold storage and canning facilities so as to pack, store, and facilitate export of fish and other marine products, were all significant interventions of the modernization phase. Salagrama (2001) observes that this modernization process has affected the artisanal fishing communities, specifically the east coast, displacing them from their traditional means of survival.

The state governments subsidized the purchase of trawlers heavily and supported endeavours such as the reclamation of wetlands during the modernization drive. Gadgil and Guha (1995) have observed that the mechanization drive in the fisheries sector initially boosted the fish catch as well as the supply of fish to the inland towns. It also improved the position of shrimps in the export market, as a major source of foreign exchange. However, in the long run, this process resulted in overfishing and decline in fish production, placing the life of many traditional fisher-folk in peril. Subsequent to the modernization processes, the supposedly homogeneous traditional fishing communities of the coast have been divided into the traditional and modern sectors, fragmenting the tightly-knit structures of the traditional fishing community (Platteau 1984). Moreover, the mechanization drive and the opening up of the export market in the fisheries sector have also encouraged migrant workers to switch to fishing, pressurizing the traditional fishing economy.

The centres of state power were always located far away from the traditional fisherfolk on the periphery. The former often crafted policies and implemented programmes that pushed the latter away from the benefits of mainstream development. In due course, the duties of the nation state in meeting development needs, executing resource management functions and ensuring social welfare has become, in essence, the practice of defining and regulating the allocation of privileges, whereby state agencies act as paternalistic providers. Local communities, on their part, pursue entitlements (licences, savings-cum-relief schemes, pensions, etc.) as the legal–institutional means through which they can affirm identity, establish politico–legal status, and achieve agency in relation with the nation state and other stakeholders (Jentoft 2000; Sunil 2007).

2.2 Dynamic Pressures

These are the processes and activities that 'translate' the effects of root causes, both temporally and spatially, into unsafe conditions. These are more contemporary or immediate conjectural manifestations of general underlying economic, social, and political patterns (Wisner et al. 2004). Dynamic pressures could include population growth, agrarian failure, food insecurity, epidemics, rapid urbanization, natural resource degradation, climate change, conflict situations, foreign debt, and certain structural adjustment programmes.

With respect to the coastal fishing communities in India, one could identify various dynamic pressure agents such as the transformation of the common property nature of marine resources to open access nature with modernization, use of inappropriate technology (such as purse seine and trawlers), booming demand for certain species such as shrimps and tuna, state subsidies for mechanization, and increase in population pressure (due to the presence of both artisanal and trawler fishermen) in the inshore waters, and so on, which all have led to over-fishing on the coast. This has resulted in declining productivity and income of fishermen, which manifest in the increasing economic disparity between fish workers and owners of fishing craft and gear (Kurien and Achari 1994).

Seasonality and irregularity of income are prime characteristics of the traditional fishery. Amidst gradually increasing resource scarcity, the traditional fisherfolk are also forced to compete with many other stakeholders such as the non-traditional fisherfolk (people who have shifted mainly from the agriculture and allied sectors to fishing), mechanized trawler fishermen, and foreign fishing vessels. The fisherfolk are also in an unfortunate situation, as they do not find enough opportunities to sell their labour to other sectors. The construction industry, for example, considers the fisherfolk as highly unskilled for masonry or carpentry work. Thus, the opportunities for livelihood diversification among the traditional fisherfolk are also much restricted.

The traditional fisherfolk are also tied to a very exploitative fish-trading network of fish merchants, moneylenders, and business capitalists. Studies have shown that, in spite of satisfactory per capita income, fishermen are forced to borrow money at high rates of interest from moneylenders and traders, with the condition that they have to sell the catch at a predetermined price, which is exploitative (Samar and Meher 2003). Though the government established fish workers' cooperatives to eliminate the role of intermediaries in the fish trade, in practice, the traditional fisherfolk are least represented in these institutions. People from non-fishing communities make decisions in these organizations. The state agencies, such as the cooperatives, consider themselves as providers of all benefits (such as loans for fishing gear and craft) to fisherfolk, and the latter as receivers of aid. However, over time, these organizations have become platforms to serve the vested interests of different political groups rather than the true stakeholders, the traditional fisherfolk. Fishermen lack trust in such cooperatives and believe that such arrangements have facilitated the entry of outsiders in the decision-making arrangements for the fishery sector (Sunil 2007). It needs to be understood that such experiences are stored in the actors' memories and will considerably influence any future intervention strategies (Giddens 1984). Structures, according to Giddens, perpetuate relationships in a society through memory traces. He describes memory in the following terms.

> ...Memory (or recall) is to be understood not only in relation to the psychological qualities of individual agents but also as inhering in the recursiveness of institutional reproduction. Storage here already presumes modes of time-space control, as well as a phenomenal experience of 'lived time' and the container that stores the authoritative resource is the community itself...The storage of authoritative and allocative resources may be understood as involving the retention and control of information or knowledge whereby social relations are perpetuated across time-space... (Giddens 1984, 261)

Studies have explored the effectiveness of village-level institutions among the coastal fishing communities. Community-level institutions such as the fisher councils exist across fishing villages, perhaps in different structural forms catering to various functions such as meeting livelihood needs, resource sharing and asset allocation, coping with resource scarcity, conflict resolution, or maintaining social order. These community-level institutions are also culturally knitted to the local knowledge systems, coping strategies, and information-sharing mechanisms of the community. However, in reality, what exists is a kind of *institutional bricolage*: a complex mix of informal and formal, traditional and modern, or community-based and state-managed institutions (Cleaver 2001, 26). In addition, as explained earlier, most of the interventions made by the state government and other development organizations have often neglected the existences of such community-level institutions, creating new levels of power and inequality. In this background, the economically and politically marginal groups, such as the traditional fisherfolk, are more likely to stop trusting their own methods of community organization and self-protection, and to lose confidence in their own local knowledge and coping strategies (Wisner et al. 2004).

Various dynamic pressure agents, such as ecological degradation and scarcity of resources, exploitative market networks, severe indebtedness, limited opportunities for livelihood diversification, failure and lack of trust in development interventions made by the state, and the erosion of traditional community-level institutions, have influenced the progression of vulnerability to unsafe conditions. The following section will examine those unsafe conditions that the traditional fisherfolk were exposed to before the tsunami.

2.3 Unsafe Conditions

These conditions are the specific forms in which the vulnerability of a population is expressed in time and space in conjunction with a hazard (Wisner et al. 2004). Unsafe conditions can be apparent in the physical environment, local economy, social relations, and the public institutions. Due to the very nature of their occupation, the traditional fisherfolk are forced to stay very close to the sea. However, due to various factors cited earlier (the root cause and the dynamic pressures), these communities were unable to afford safe buildings and also lacked effective preparedness or protection mechanisms of the state. The fishermen continued to pursue their livelihoods in the dangerous sea, competing with many other stakeholders for scarce fish resources. On land, the traditional fish vendors, who are mostly women from the fishing households, had to compete with the male-dominated merchandise, and political and market forces in unhealthy and unhygienic conditions to earn a living (Santha 2007b). Thus, the traditional fisherfolks' access to tangible resources, such as cash, safe shelter, food stocks, and fishing equipment, were limited. Their social networks were also often restricted to their own community activities and their access to centres of power and governance were limited. The exposure of the communities to early warning systems were more related to storms during the monsoons, and the awareness level of the community with respect to an earthquake, tsunami, fire, or spread of epidemic diseases such as cholera, dysentery or typhoid, were inadequate. The culture of the state agencies to respond to such events was also in the form of providing some relief measures *after* the impact. The emphasis of state agencies was mostly focused on disaster *response*, rather than *preparedness*.

The disaster that followed the Indian Ocean tsunami was an intersection of two opposing forces: (1) those processes that generate vulnerability on one side, and (2) the natural hazard event on the other side, the tsunami itself. Though the scale and magnitude of the tsunami were very high, the fisherfolk were equally exposed to threat of cyclones and flooding, fire outbreaks in fishing colonies and spread of epidemics. The progression of vulnerability and the summary of the above discussions are depicted in Fig. 1.

Having discussed the progression of vulnerability among the traditional fisherfolk, the next section will analyze the social interfaces that emerged in the context of tsunami rehabilitation and recovery programmes.

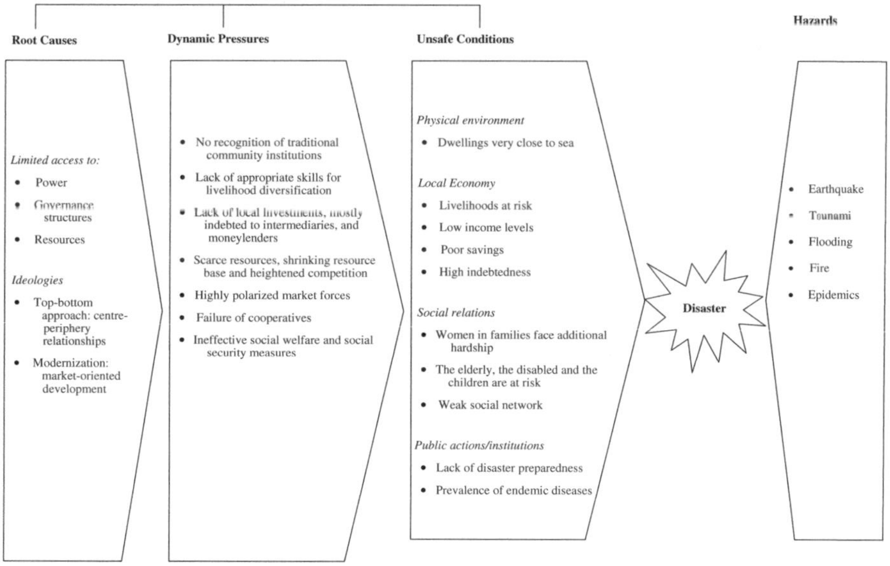

Fig. 1 The PAR model: progression of vulnerability among the traditional fisherfolk in South India

3 Social Interfaces in Tsunami Rehabilitation and Recovery in Tamil Nadu

This section analyses the nature of the relationship that exists between the traditional fisherfolk and other actors involved in the rehabilitation and recovery operations in the aftermath of the tsunami. This relationship can be understood by an analysis of the social processes of conflicts, cooperation, or accommodation, that take place between different actors in recovery. By analyzing social situations and their interfaces, this section attempts to see how the experiences and perspectives of multiple actors, involved in disaster recovery and vulnerability reduction, merge through processes of accommodation and conflict to generate new forms of organization, practices, and understanding. The different situations of interface could be understood by looking at the processes emerging with respect to strengthening access to resources, rebuilding livelihoods, providing relief and designing rehabilitation, and mainstreaming development. These interfaces are discussed in detail in the following sections.

3.1 Strengthening Access to Resources: Customary Property Rights and the Deregulation of the Coastal Environment

Access involves the ability of an individual, family, group, class, or community to use resources which are directly required to secure a livelihood in normal, pre-disaster times and their ability to adapt to new and threatening situations (Wisner et al. 2004, 94). For the traditional fisherfolk, both the sea and the shoreline are two important resource bases that help them to meet their day-to-day livelihood needs and are also significant parts of their cultural and social spaces. Though the state has the legal ownership of these resources, these spaces are a kind of customary property for the communities. Rules governing the access to and use of these common pool resources were highly localized and complex.

It was discussed earlier how the modernization processes in fishery affected the common property nature of the marine resources. However, the resource that was less affected when compared to the marine resources was the shoreline and the adjoining open spaces that the community owned in private or as communal property.[1] The fishing community used these resources on land to rest their canoes and craft and dry their nets. In many places, these resources also served as platforms for fish trade. The very sense of communal ownership of the place gave the fisherfolk the 'power' to bargain for a better price for their catch. These resources were the 'bridge' between their homes and the ocean, which also helped them to take

[1]It has to be noted that land use changes in many coastal regions were visible several years before the tsunami, due to growth of the real estate market and tourism. However, in most fishing villages, the 'common pool' nature of seashore and adjoining open spaces prevailed.

important livelihood decisions based on their local knowledge and experiences. For example, staying close to the sea aided the fisherfolk to take important decisions like whether to sail out for fishing on a particular day/time or not; and in ascertaining weather conditions, the probability of a storm, the strength and directions of the wind, the availability or non-availability of light, or in the selection of a fishing route. These open spaces also acted as one among the important platforms for fisherfolk to share knowledge and information among one another.

These situations have considerably changed after the tsunami. Today, different actors involved in rehabilitation perceive 'safety' differently. For the traditional fisherfolk, staying close to the sea always ensured a secure livelihood, amidst competition and resource scarcity. For them, the first step towards safety has been to resurrect their livelihood. However, for the state officials and other actors in development organizations, 'safety' is being defined in terms of the *distance from the sea*. Existing Coastal Regulation Acts are used to support their arguments, and, if necessary, they are diluted to fit in the 'new' understanding of safety (Biswas 2007). Fishing communities are relocated further away from the sea, thus exposing their culture and sources of livelihood to newer forms of risk.

The community leaders see this strategy of relocation as 'back door' activities by the power centres to allocate resources to outsiders, namely, real estate builders and tourist resort lobbies. These situations exemplify the lack of trust that the affected fishing communities have in the state actors. Such experiences add to their previously existing memories of state interventions (the failure of fish workers' cooperatives, for instance), resulting in aggravated relationships of conflict between different actors. Such relocation strategies also result in cultural chaos and new kinds of conflicts; say between the relocated fishing communities and the host communities.

3.2 Rebuilding Livelihoods: Ecological Insecurity and Resource Use Conflicts

The recovery operations have never tried to address the issues related to resource scarcity, overcrowding, and over-exploitation in the fisheries sector. Instead, the steps to rebuild livelihoods were aimed at providing catamarans or boats, engines and fishing gear, without analyzing the social structure, previously-existing norms of fishing, and the conflicts that existed in the fishery. The non-governmental organizations (NGOs) and state agencies that facilitated the allocation of these livelihood resources imposed their own working arrangements and norms of sharing these resources. These external organizations decided how many fisherfolk should share a catamaran, boat, or fishing net. In some villages, where the fisherfolks' councils were strong, allocation was done based on previously existing sharing norms. However, in most of the affected fishing villages, efforts were also made to allocate one boat to each and every fisherman. There are instances where fishermen who received fishing craft sold them to outsiders, as they were not able to manage the resources within the newly defined patterns of ownership.

Conflicts over resources have also worsened, as many communities, which were not traditionally fishing communities, have also received fishing craft and gear as aid. Earlier, these communities used to work as labourers with the traditional fishing communities. They seldom owned craft or gear. Today, many among the non-traditional fisherfolk have also become owners of craft and gear. This has intensified the competition and subsequent conflicts among resource users in the sea. Fisherfolk observe that they get very little catch these days due to over-crowding and scarcity of resources.[2]

Traditional fisherfolk also complain about the quality of fishing equipment provided to them. Based on the particular season, availability of fish, or fishing grounds, fishermen choose different types of fishing gear to capture fish. However, in many villages, organizations have supplied fishing gear without taking into account the intricacies of such fishers' knowledge and practices. Fisherfolk are also apprehensive about the maintenance of fishing craft that were supplied by some NGOs.

3.3 Providing Relief: Defining the Affected, Identities and Conflicts

A considerable transformation is also occurring in the identity of fishermen following the recovery processes. Fisherfolk usually identify a true fisherman as one who: (1) fishes for his livelihood, (2) maintains the craft and gear very well, (3) is able to enhance his fishing skills from experience and appropriate resources from the environment in a fair manner, (4) is responsible towards his family, and (5) spends his income that God gives in a responsible way (Sunil 2007). However, in the aftermath of the tsunami, the external actors identify fisherfolk as disaster victims, who have lost their assets, and as helpless human beings who require external aid. According to Sivaraman:

> the focus of government and non-governmental organisations in rehabilitation has been on dealing with the problems of 'tsunami survivors', meaning who were touched by salt water on the fateful day. All others living in the same context, however vulnerable, have been deemed *irrelevant*. Therefore, many poor communities with very low development indicators have been bypassed in the distribution of relief and material aid. (2007, 43)

---

[2]It needs to be understood that many of the poor among the non-traditional fishing communities were the worst affected by the tsunami. Many among them have lost their livelihoods, equipment, and other resources, and these communities were not given any priority in rehabilitation. From the perspective of these communities, the fisherfolk unions have snatched away a larger share of their due benefits. The argument that this paper proposes is that recovery interventions have intensified the conflicts between different resource users. In addition, neither have the root causes of vulnerability been addressed in any of the cases.

Resources are, thus, allocated through mechanisms such as 'lottery methods', without analyzing the cultural conditions that aid livelihoods and resource use. These methods might distribute resource equitably for the present, but, in the long run, they could hamper the social fabric and sustainable resource use arrangements. Consequently, there are chances that the traditional fisherfolk adopt the criteria as described by the state or external agencies to define themselves as 'helpless disaster victims' rather than linking their identity with the resource and occupation.

3.4 Designing Rehabilitation and Mainstreaming Development

One of the most obvious shortcomings of disaster response in the aftermath of the tsunami has been the complete lack of understanding about what was happening to the coastal communities before the tsunami. Any intervention in terms of recovery and development has not taken into account the poor socioeconomic conditions of the affected communities, which have contributed to their distress. A thorough analysis of the interventions made by development organizations also lacks an understanding of their history, culture, and local-level politics (Sivaraman 2007).

As mentioned earlier, the 'tsunami victims' are looked at as completely dependent people in need of relief and rehabilitation. There has been little attempt to identify and use existing local capabilities and resources for recovery. For instance, traditional fisherfolk understand the significance of maintaining mangroves and manmade sand dunes with vegetational barriers in preventing the intrusion of seawater into the villages. However, modern science along with top-down bureaucratic and market-oriented decision-making arrangements, meant that community practices were neglected, and large sea walls worth millions of dollars were proposed to be constructed. As a result of such approaches, the traditional fisherfolk were not able to strengthen and develop their own knowledge and skills, which could have provided them with additional income opportunities (Biswas 2007).

Critics observe that there has been a gross violation of human rights with respect to the rehabilitation of the affected population (Martin and Sivaraman 2007). Approximately 60% of the population affected by the tsunami in Tamil Nadu still live in temporary shelters (Chaudhry 2007). Almost three years after the disaster, people were still forced to stay in worn-out tar-sheeted sheds in low-lying areas, which were like hot ovens during the summer, and flooded during the rainy season. These habitats were infested with snakes and rodents, creating new environments of risk for the children, the aged, and women in the households. In addition, these structures were more prone to fire accidents that led to the migration of many victims to other places in the state. Analyzing recovery from a rights-based approach, Chaudhry's (2007, 20) major question was, "why after two years, the majority of tsunami survivors are still living in temporary shelters, without sufficient information on when and where they will be allotted permanent housing?"

The sanitation and hygiene norms have been poorly implemented, creating newer forms of public health risks. The existing toilet facilities do not serve the purpose of the whole community, so communities have given up using them. Such approaches have neither taken into account the local culture nor have incorporated in the planning as part of a 'community-oriented' process. Prior to and after the tsunami the community is seen as individualistic receivers of privileges/patronage offered by the state or development organizations.

Examining the interface that has emerged, one could see that the traditional fisherfolk have become more dependent on the state and external agencies for building their livelihood opportunities (which is contradictory to their occupational characteristic of desire for freedom). However, in a temporal-space dimension, they have gradually lost trust in these interventions. As Chaudhry has rightly pointed out:

> even when it is their legally enforceable human right to demand rehabilitation, security, adequate housing, health care, food, education, livelihood and information, they are still being denied. The affected communities fear to resist as any demand made may not even give what is being given. When people are subjected to such prolonged despair, anything provided in the form of 'rehabilitation' or 'housing' tends to generally be accepted. (2007, 20)

4 Discussion: Disaster Recovery, or Re-creating Vulnerabilities?

Social interfaces in a post-disaster setting are thus characterized by a bricolage of dynamic local culture and apparently static or insensitive formal institutions. The latter, be it the state or the NGOs, do not often recognize the existence of the local, and, in this context, the resultant outcome is usually a continuous improvization or rejection of the externally designed recovery plans. This process has significantly affected the effective rehabilitation processes in the region. It has also weakened the social fabric and resource use arrangements among the fishing communities, adding new dynamic pressures to their vulnerabilities.

The external agencies expect the 'tsunami victims' to conform to the formal plans of rehabilitation. However, the very same formal organizations fail to recognize the local capabilities and coping strategies of the affected communities. Thus, a striking feature of the social interface in this particular context is that though the goals of all the actors (namely, traditional fisherfolk, state agencies, and NGOs) is disaster recovery, it is characterized by the prevalence of often-incompatible beliefs, values, norms, knowledge, and practices between them. In this context, conflicts between fishing communities and other stakeholders emerge when these external interventions threaten/contradict those components of culture that serve the purpose of sustaining livelihood needs. Fishing communities accommodate differences, if their source of livelihood is not threatened. In this process, alteration and rejection of formal processes turns out to be a regular feature, adding to the inefficiencies of disaster recovery.

In the context of the social situations discussed above, it is essential to examine whether the rehabilitation and recovery interventions have contributed to vulnerability reduction, or to reinforcing the hazard risks of the affected population. Rehabilitation has been taken for granted as the mere building of temporary shelters. However, the linkages of vulnerability to other development needs and resources, such as availability of educational or health-care facilities, and public infrastructure like good roads and transport, are not looked into. These needs existed even before the tsunami, and have been neglected even after the impact.

In the name of disaster recovery, the fishing communities have been pushed further down the ladder of power and access to governance structures. Decisions are made on their behalf by culturally insensitive bureaucrats in the State organizations, by project officers of development agencies and a few influential community leaders. The recommended standards of relief and rehabilitation were not taken into account. In addition, the resource-use conflicts have worsened the livelihood crisis, as more users now compete for a scarce fishery resource.

To sum up, the disaster recovery processes that followed the Indian Ocean tsunami 2004, were not incorporated in the framework of vulnerability reduction. Instead, these interventions themselves have acted as dynamic pressures and have worsened the unsafe conditions that existed prior to the tsunami. Today, the communities are exposed to multiple dimensions of hazard, including public health risks. The simultaneous degeneration of the community life and the sudden exposure to different mainstream cultures in itself act as catalysts in transforming the local capacities to vulnerabilities. There is a long way to go before each and every household affected by the tsunami re-establishes its livelihood, physical assets, and patterns of access to resources. It is an even more distant dream that these communities will be in a better position to resist or withstand the shocks of extreme future events.

References

Berke, P., J. Kartez, and D. Wenger. 1993. Recovery after disaster: achieving sustainable development, mitigation and equity. *Disasters* 17 (2): 93–109.

Biswas, N. 2007. Blue battle lines. In *miles 2 go: Tsunami, rehabilitation, future*, ed. M. Martin and S. Sivaraman. New Delhi: Indiadisasters.org.

Chairetakis, A. 1992. Past as present: History and reconstruction after the 1980 earthquake in campania, Southern Italy. Paper presented at the 51st annual meeting of the society for applied anthropology, Memphis, TN.

Chaudhry, S. 2007. Two years, still waiting. In *miles 2 Go*, ed. M. Martin and S. Sivaraman.

Cleaver, F. 2001. Institutional bricolage, conflict and cooperation in usangue, Tanzania. *IDS Bulletin* 32 (4): 26–35.

Gadgil, Madhav and Ramachandra Guha. 1995. *Ecology and equity: The use and abuse of nature in contemporary india*. New Delhi: Penguin Books.

Giddens, Anthony. 1984. *The constitution of society: Outline of the theory of structuration*. Cambridge: Polity Press.

Jentoft, S. 2000. The community: A missing link in fisheries management. *Marine Policy* 24 (1): 53–60.

Kurien, J., and T. Achari. 1994. Overfishing and the coastal commons: Causes and consequences. In *social ecology*, ed. R. Guha. New Delhi: Oxford University Press.

Long, Norman. 2001. *Development sociology: Actor perspectives*. London and New York: Routledge.

Martin, M. and S. Sivaraman (eds.). 2007. *miles 2 go: Tsunami, rehabilitation, future*. New Delhi: Indiadisasters.org.

Platteau, J-P. 1984. The drive towards mechanisation of small-scale fisheries in kerala: A study of the transformation process of traditional village societies. *Development and Change* 15 (1): 65–103.

Rubin, C., M. Saperstein, and D. Barbee. 1985. Community recovery from a major natural disaster. Monograph No. 41, Boulder: University of Colorado.

Salagrama, V. 2001. Coastal area degradation on the east coast of india: Impact on fishworkers. Paper presented at the forging unity: Coastal communities and the indian ocean's future conference, October 9–13, Chennai, India.

Samal, K.C., and S. Meher. 2003. Fishing communities on chilka lake: Comparative socio-economic study. *Economic and Political Weekly* 38 (31): 3319–3325.

Santha, D.S. 2007a. State interventions and natural resource management: A study on social interfaces in a riverine fisheries setting in Kerala, India. *Natural Resources Forum* 31 (1): 61–70.

Santha, D.S. 2007b. Conflicts in fish trade: A study among the riverine fishing communities in Kerala, India. *ISA E-Bulletin* 8: 22–35.

Sivaraman, S. 2007. Apocalypse Here and Now. In *miles 2 go*, ed. M. Martin and S. Sivaraman.

Sunil, D.S. 2007. Culture, formal institutions and natural resource management: A study among the riverine fishing communities in Kerala, South India. Doctoral dissertation, Chennai: Indian Institute of Technology, Madras.

Wisner, Ben, Piers Blaikie, Terry Cannon, and Ian Davis. 2004. *At risk: Natural hazards, people's vulnerability and disasters*. London and New York: Routledge.

Disaster, Development, and Water: The Reconstruction and (Re)Fabrication of Hazardous Waterscapes in Post-tsunami India

Luke Juran

1 Introduction: The Historical and Theoretical Context

It has been seven years since the 2004 tsunami unexpectedly hit the coast of otherwise serene South India. Residents awoke to a cool morning breeze. It was cloudy. It was the monsoon season. It rained that day, a welcome reprieve from the sun that ordinarily scorches the region. The anniversary of the tsunami was commemorated with prayers, flowers, and speeches. Newspaper articles marked the day by recounting stories of survivors and narrating the contemporary lives of the disaster affected. But one article, in particular, stood out.

On that day, 26 December 2011, R. Venkatramani, in an article in *The New Indian Express* entitled 'Better amenities amidst bitter memories', chronicled a tale of successful rehabilitation in Nagapattinam District, Tamil Nadu. According to the article, a rosy post-tsunami existence prevailed for virtually all affectees:

> Today, though the painful memories of the tidal attack still continue to haunt the affected families, there is perceptible improvement in their lifestyle.
>
> Death and destruction apart, the tsunami aftermath has brought a lot of positive changes in the lifestyle of coastal communities.
>
> Tsunami-hit hamlets have undergone total transformation of infrastructure development. Fishing harbour, fish landing centres, fish auction halls, cold storage facilities and fish drying yards have virtually wiped the scars of tsunami destruction.

This research was supported by a Fulbright-Nehru Fellowship.

L. Juran (✉)
Department of Geography and Virginia Water Resources Research Center,
Virginia Tech, Blacksburg, US
e-mail: ljuran@vt.edu

S. Reddy (ed.), *The Asian Tsunami and Post-Disaster Aid*,
https://doi.org/10.1007/978-981-13-0182-7_5

> Many villages are equipped with new school buildings, vocational course institutes, computer training institutes, knowledge centres, water purification plants, multi-purpose halls, community halls, metal roads and new bridges.
>
> The affected families have been provided with new houses and the fishermen have become proud owners of fibreglass boats. Many fishermen have bought big steel boats utilising the subsidised loans provided by banks.
>
> Prior to the tsunami, women of the area did not know anything other than fish vending. Now, through SHGs [self-help groups], they handle hi-tech fish dryers, make footwear and leather articles, pickles … and the list is endless.

Temporarily retreating from its patronizing language, the article briefly mentions that "tones of discontent" exist as a result of houses being situated far from the sea and waterlogging in the rainy season. However, the article swiftly brushes the all-too-real concerns aside in its closing paragraph: "Despite these voices of discontent, seven years after the tsunami, the fishing villages have undergone a positive transformation." Nobody wishes the article to be true more than the affectees themselves. However, evidence on the ground does not necessarily match what was reported.

While a 'transformation' did take place, its composition more nearly mirrored a 'disaster after the disaster.' Scholars have applied this concept to denote post-disaster processes of anthropogenic origin that exacerbate, and sometimes overshadow, the impacts of the initial disaster. In other words, government and non-profit mismanagement, poor resettlement and reconstruction, and failures in neo-liberal approaches coalesce to produce a second disaster that sometimes eclipses the preceding event. Disasters after the disaster have been detailed after the Gujarat earthquake by Virmani (2008), Hurricane Katrina by Button and Oliver-Smith (2008), Adams et al. (2009), theoretically by Schuller (2008) and Juran (2015). Many critiques stem from 'disaster capitalism' and the dominant humanitarian agenda of 'build back better', and much of the literature has focused on the 2004 Indian Ocean tsunami (see Kennedy et al. 2008; Lyons 2009; Juran 2015). Adding to the discourse, this paper critically examines the post-tsunami scenario in India through the lens of 'water'. A principal theme is that the hazardous pre-disaster context was merely refabricated in reconstruction processes to create a hazardous post-tsunami waterscape of manmade, engineered, proportions, where issues of quality, quantity, and access abound.

2 Methods

This paper employs a socio-geographical approach to investigate the long-term products of water-sector reconstruction in Nagapattinam District, Tamil Nadu, and the neighbouring territory of Karaikal District, Puducherry. Fourteen newly built post-tsunami villages (selected randomly from 35) were investigated across six field visits from 2008 to 2016 (Fig. 1). The villages, seven in each district, average about 200 households and have been occupied since approximately 2007. The villages are both urban and rural (as delineated by political jurisdiction and governance) and

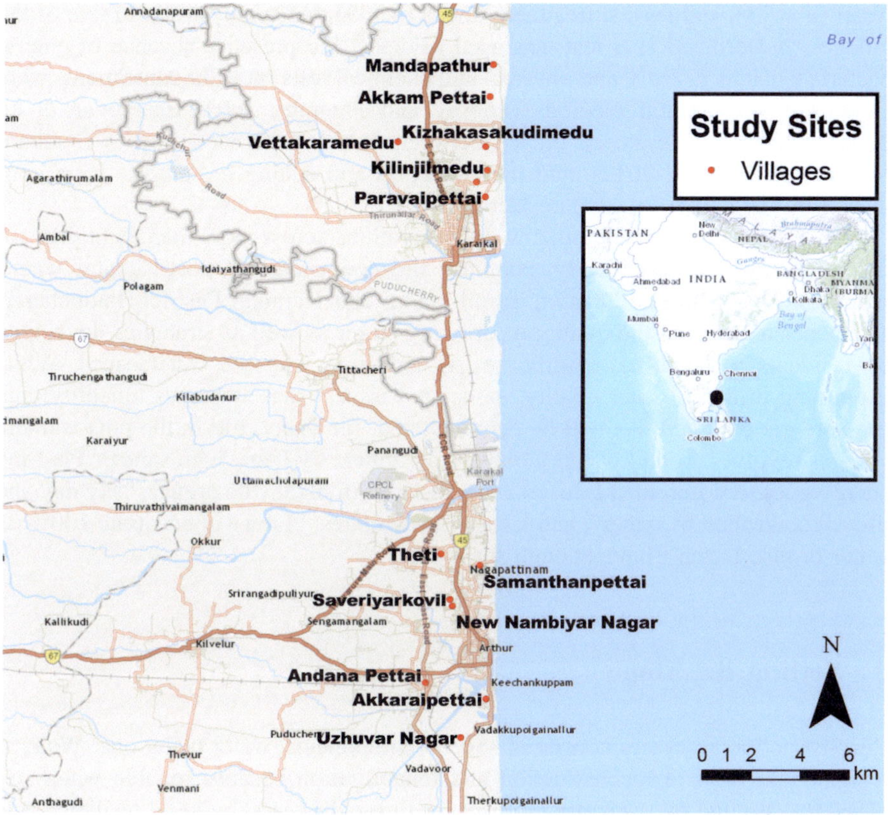

Fig. 1 The 14 study villages in Nagapattinam District, Tamil Nadu, and Karaikal District, Puducherry

comprise different religions (Hindu, Muslim, Christian), livelihoods (fishing, agriculture, daily wages), and social locations (middle income, low income, scheduled caste). Twelve of the villages were relocated, and two were reconstructed in situ (one in each district). Each village was visited at least three times, and visits encompassed both the dry (May–August) and monsoon (October–December) seasons.

Original quotations and a visual ethnography of the built environment will be used to construct a narrative from the perspective of those who live in, cope with, and manage the post-tsunami waterscapes (Geertz 1973; Smith 1981; Pile 1991; Pink 2001). Data were acquired through a mixed methods approach including: semi-structured household interviews with village residents (n = 74); key informant interviews with government officials, non-profits, and individuals with specialized knowledge on water and/or reconstruction (n = 66); and a focus group discussion in each village (n = 14) (Baxter and Eyles 1997; Chacko 2004; Longhurst 2010). Furthermore, this paper draws upon descriptive statistics generated from household water quality tests (n = 300), tests at public taps from where households retrieve water (n = 75), and tests at treatment nodes (n = 3) that supply water to five of the villages. A sterilized H_2S test was used to detect the presence-absence of enteric bacteria (such as *E. coli*), and physical and chemical tests on eight constituents were conducted via titration methods (Sobsey and Pfaender 2002; Elangovan et al. 2006). Samples were stored on ice between 2 and 8 °C according to Standard Methods (Rice et al. 2012), and the collection and testing of samples were conducted by the author in a secure laboratory.

Using these data and equipped with observations across more than eight years of fieldwork, this paper critically examines the newly created post-tsunami waterscapes in Nagapattinam District, Tamil Nadu, and Karaikal District, Puducherry. The paper focuses on the quality of drinking water as well as strategies developed by residents to adapt to, modify, resist, and cope with the waterscapes. While examining drinking water quality, associated issues such as access, quantity, and pressure are explored. As will be demonstrated, for many, life in the post-tsunami villages revolves around water, a contention upheld by Uma, who states, "First my energy goes to water, then I do my other tasks"; Ambika, who argues, "My day and life are controlled by water"; and Usha, who declares, "I must concentrate 100% on water or else I won't have enough for the day."[1]

3 Setting the Stage

Reconstruction processes consist of several components, water being one. Water is a significant facet of reconstruction and rehabilitation because reliable water services are essential for mitigating impacts of future disasters, building resilience and capacity, and addressing pre-existing socioeconomic issues. The tsunami not only prompted the largest reconstruction process in India's history, but also the capital, external assistance, and consensus support necessary for the development of improved infrastructure at such a grand scale. Within this context, Nagapattinam and Karaikal Districts reconstructed entire villages from the ground-up, water infrastructure included (see Juran 2015). The long-term outcomes of these new water-society constructs will be investigated.

In coastal southern India, the quality of surface and groundwater is generally poor as a result of both natural causes (such as iron, dissolved solids, fluoride) and anthropogenic causes (such as, nitrates, saline intrusion, sewage, and open air defecation). Furthermore, at the study villages, even piped and treated water provided by the government habitually fails to arrive safely to end consumers. For example, bacteriological tests at 75 randomly sampled public taps in the 14 villages revealed only 21 (28.0%) with an absence of faecal coliform, and a random sample of 300 households that retrieve water from those taps revealed only 61 (20.3%) with an absence of faecal coliform at the point-of-use. However, rather than probing the

[1]All names of interlocutors are anonymous except for those with official titles.

numbers, this paper explores *why* the quality of water is poor and goes on to detail associated coping mechanisms that have been developed by village residents.

Water quality at the post-tsunami villages exhibits a temporal dimension. During the monsoon, an increase in quantity supplied is matched with an increase in contaminant load, particularly bacterial flora. Continuous water supplies do not exist, so the situation is further complicated by the ability of groundwater and floodwater to seep into cracked pipes during non-supply windows. But hardships are confronted during the dry season as well. Although contaminant loads are typically less in the dry season, concentrations are typically elevated as a result of lower volumes of water resources. Furthermore, the dry season is when households, if options are available, diversify their water consumption to include more untreated sources. Thus, owing to lower quantities of piped supplies in the dry season, households are driven to consume water from shallow, untreated hand-pumps and borewells, which harness water that is less naturally filtered and notorious for harbouring biological contaminants (Black and Talbot 2005; Ferguson et al. 2011).[2] Meanwhile, groundwater is still able to infiltrate pipelines during the dry season due to cracked pipes and the high coastal water table. Ishweri of Andana Pettai captures the setting:

> The pipes are empty for 23 hours per day. Before the water comes you can hear 'poof-poof' air sounds. Then the water comes, but it is brown and murky at the beginning. We just let the water flow until it becomes clear or sometimes we use it to water plants.[3]

While the practice of not consuming front-end water supplies is beneficial, outbreaks of diarrhoea, fever, and skin rashes are a fixture of everyday village life, especially among children. Lakshmi of Vettakaramedu recounts a 2011 outbreak of diarrhoea during the monsoon:

> Many people had *joram* ['fever' in Tamil], loose motions [diarrhea], and so many health problems. When we visited the GH [government hospital], the doctor told us that water quality was the main problem. He told us to boil water and to complain to the government about the water supply. We complained to our local leaders and they contacted the PWD [Public Works Department] and the Collector's Office by letter. The government came and cleaned the tank [water tower], but said the major problem was that we broke the pipes to get water due to low pressure. He said the broken pipes allowed dirt and dirty water to enter the pipes, but we have to do that in order to get water.

---

[2]There should be a 30.5 m (100 ft) buffer between groundwater sources and sewerage infrastructure, and groundwater sources should also be placed uphill of sewerage infrastructure (Woodson 2010). At the villages, this buffer is impossible due to high housing (and therefore sewerage infrastructure) densities—which is further complicated by open air defecation. Field measurements indicate that the furthest a hand-pump or borewell is from sewerage infrastructure is 9.1 m, and the flatness of sites precludes the uphill placement of groundwater sources.

[3]The 'poof-poof' sound, due to dry pipes during non-supply windows, is common across all villages and acts as a sensory reminder of piped water being tainted as a result of no supply or pressure for a majority of the day.

As intimated by Lakshmi, water quality is dependent upon the engineered environment through which water supplies flow, which often exhibits deficiencies in terms of shoddy infrastructure, poor maintenance, and coping mechanisms by water consumers that address quantity but serve to degrade quality. Thus, it is useful to explore the multi-layered supply regime that delivers water to the villages.

4 Water Quality Woes in Post-tsunami India

First, water quality is a function of the quality of water initially supplied by the Tamil Nadu Water Supply and Drainage Board (TWAD) in Nagapattinam municipality, the Public Works Department (PWD) in Karaikal municipality, and *panchayats* (units of rural governance) in rural villages. I was permitted to test water released by three treatment nodes that supply water to a total of five study villages. Tests at Velepalayam sump in Nagapattinam municipality (operated by TWAD) not only confirmed the presence of faecal coliform, but that levels of pH, hardness, chlorides, and alkalinity exceeded desirable limits set by the Bureau of Indian Standards (BIS). Furthermore, tests at the Kunjimani Thidal and Nanavaickal treatment centres (both operated by Kottucherry Commune *Panchayat* in rural Karaikal District) revealed the presence of faecal coliform and levels of pH, hardness, chlorides, alkalinity, and iron that exceeded permissible standards of the BIS. Additionally, no tests detected a presence total residual chlorine (TRC), which is extra chlorine that imbues the capacity to eliminate biological threats encountered during the journey from treatment centres to end consumers. While I was denied access to treatment centres that supply water to the remaining villages, these data—combined with 54 of 75 (72.0%) public taps indicating a presence of faecal coliform—confirm that the initial quality of water already fails to meet codified standards upon leaving treatment centres, with quality deteriorating thereafter.

After being released from treatment centres, water must travel through shallow-dug pipes with loose couplings and cracks, or what Laltinkhuma, District Collector of Karaikal, terms "perforated pipes." Thus, water quality diminishes as groundwater and surface waterlogging, both tainted with sullage from septic tanks, leach pits, and open air defecation (the dominant method of human waste disposal in the region), seep into pipelines. It must be reiterated that an intermittent water supply and lack of pressure compound the situation. Depending on the village, water is supplied for a total of only 1–5 h across 1–3 discrete supply windows. This means that pipes are empty for a majority of the day, allowing foreign constituents to penetrate pipelines and further adulterate supplies. Additionally, the non-continuous supply fails to properly flush the system (especially near end-points), and the prevailing low in-pipe pressure even when water is flowing is insufficient to mitigate further threats: the introduction of external impurities is mitigated when in-pipe pressure is greater than pressure outside of the pipe (exfiltration), with the opposite holding true in this case (infiltration).[4] It is here that the absence of TRC must be revisited, because its deficiency further permits the degradation of quality. With the exception of one test, tests from 75 public taps and three treatment centres revealed that TRC was undetectable (should be 0.02 mg/L per BIS standards).

During its travel through inferior pipelines, water is eventually raised to local water towers for purposes of supply and pressure, from where it is finally conveyed to endpoints for retrieval by consumers. Thus, the cleanliness of water towers must be scrutinized. Water towers, which accumulate legacies of sediment, nitrate, bacterial growth, and other impurities over time, are to be cleaned by Nagapattinam municipality or the respective *panchayat* every 15 days in Nagapattinam District, and every six months by PWD or the respective *panchayat* in Karaikal District. Chandrasekharan, Commissioner of Nagapattinam municipality, describes the process:

> Tanks are cleaned every 15 days, we've been doing that for the past five years. We have two tank operators and a watchman at all tanks. They are responsible for cleaning, but others help if more labor is needed. There is a notice board on each tank that tells when it needs to be cleaned.

Similarly, Jayasankar, Lab Assistant at PWD, outlines the process in Karaikal municipality:

> Tanks are cleaned twice per year. There is no particular date to clean. There are dates painted on the pillar of the tank that state the last time it was cleaned. When it gets close to six months we tell our superior and he gives an order to clean the tank. After cleaning, we paint the date at eye level on the pillar for future reference. *Mazdoor* [multipurpose staff] workers clean the tanks along with some private hired helpers such as plumbers and people who work with water. The tanks and sumps are cleaned in one day. It takes from 8 AM until 6-7 PM with a total of about 25 workers.

Lastly, the process is detailed in rural Nagapattinam District by *panchayat* President Sankar:

> Cleaning tanks in the *panchayat* is under my authority. The tanks are cleaned every month by two workers that are hired to clean tanks, repair water lines, and fix hand-pumps provided by the *panchayat*. They do these tasks under my direction. The tanks are cleaned by first opening the drain pipe, then removing the sludge, then scrubbing the inner walls with an iron brush, and then rinsing the inside with clean water. Then, we fill the tank and add some chlorine. Cleaning takes half a day.

Therefore, the cleaning of water towers, if executed properly and according to protocol, should maintain the initial quality of inflows. However, cleaning occurs much less than what is reported by government authorities. Returning to Sankar's

[4]A Ministry of Urban Development (2011, 48) report crystallizes this contention: "In a continuously pressurized distribution system, contaminants surrounding the pipelines cannot penetrate even if there are breaks in the pipes and joints. Without continuous pressure, street run-off, drainage water, raw sewage from adjacent sewerage lines and leaky septic tanks get sucked into the water mains.".

contentions, Rajjilakshmi, a resident of Akkaraipettai who lives just 5 m from a water tower, argues that, "He [Sankar] is telling lies. The tank is cleaned only once every six months at best and it only takes a couple hours maximum." Worse yet is the case of Samanthanpettai, located in Nagapattinam municipality, as recounted by Palmani:

> The tank wasn't cleaned for six months and the water was dirty. We told the municipality to clean the tank to no avail. Then we got fed up and 10 of us cleaned the tank ourselves. We told local leaders we were going to do it and they allowed us. They just watched us clean the tank but didn't help. Since then, we've cleaned the tank five times. We clean it once every two months. There is 30 cm of sludge at the bottom of the tank each time, it's fully black!

With the cleanliness of water towers called into question, several government officials yielded from their claims, with the Commissioner of Nagapattinam municipality readily admitting:

> Even though tanks are cleaned every 15 days, sometimes they are not cleaned properly and sometimes maybe 30 days goes without cleaning—it happens. Sometimes people complain that the tank is not clean and that dirty water, insects, or worms are coming.

Adding insult to injury, Jayasankar of the PWD acknowledged that there is "no penalty" for not cleaning water towers, while Ragunath, regional Executive Engineer of TWAD in Nagapattinam District, stated, "There is no punishment, so much punishment is not there. Nobody will get sacked for not cleaning a tank." Furthermore, Murugan, Junior Engineer of Water Supply and Maintenance for Kottucherry Commune *Panchayat*, claimed, "There will be no penalty to the individual who was supposed to clean the tank, but we will take immediate action to ensure the tank is cleaned at the earliest."

Collectively, water quality is a function of several multi-layered, social, hydrological processes: (1) the initial quality of water dispensed from treatment centres; (2) travel from treatment centres through shoddy pipes compounded by intermittent supply and low pressure; (3) conveyance into poorly maintained water towers; and finally (4) distribution through additional compromised pipes to end-points. Ultimately, residents are forced to consume unsafe water, with individuals across several villages regularly observing organisms swimming in their water after two or more days of storage. Subramanian of Samanthanpettai spoke animatedly on the subject:

> *Puuch* ['Insects/fish' in Tamil]! *Puuch irukku* ['Insects/fish are there' in Tamil; said while vigorously bending his index finger, a non-verbal expression to designate small organisms]! After two days there are small bugs and animals swimming in the water! This is happening too much, and then we must put the water to wastage.

Subramanian's account, supported by field observations and testimonials from other villages, confirms the Commissioner of Nagapattinam municipality's frank admission that cases of 'dirty water' with 'insects' and 'worms' are in fact true.

Next, water encounters additional risks for deterioration once it reaches end-points. First, the spigots of most public taps are perpetually open: there is no valve to turn water on and off. Thus, taps readily allow precipitation, dirt, airborne particulate matter, and other constituents to enter. Even worse, many spigots have been equipped with rubber hoses to facilitate pressure (through siphoning) and the conveyance of water into *kodams*[5] more easily. Unfortunately, the ends of hoses come into contact with floodwater and regularly rest on the ground and in puddles of stagnant water that enclose the taps. These ephemeral, yet virtually permanent, pools that envelope taps result from a low-lying coastal topography, flat site preparation of villages, and physical depressions that have been worn in due to heavy foot traffic and queueing during water retrieval (Fig. 2). These depressions are filled with the murky water released at the beginning of water windows (in wait for the flow to become clear); *kodams* being rinsed before their filling; and water that inadvertently falls to the ground as full *kodams* are replaced a few seconds later with empty *kodams*.

Serial flooding and waterlogging during the monsoon are among the foremost issues at public taps (Fig. 3). Such water not only surrounds taps and infiltrates pipelines, but does so while inundating sewerage infrastructure and solid waste (paper, plastic bags, refuse) that dots the landscape. Furthermore, standing water is inadvertently kicked up as water seekers trudge to, from, and around the taps, their *kodams* speckled with droplets of the contaminated water along the way. Nalini, a resident of New Nambiyar Nagar, outlines these myriad risks for post-point contamination:

> It is difficult to get water in times of waterlogging. I have to walk to the standpost with my *kodams*, and I also have to hold up my sari so it doesn't get wet. In the journey, I have to be careful not to step on rocks and garbage that are under the water. When I get to the tap it is difficult to wash the *kodam* because it just gets dirty again from sitting in the dirty water and mud. I have to be careful not to get the dirty water in the clean water that I get from the tap. Also, the hose falls in the dirty water and we try our level best to rinse it off. Children play in the dirty water. They think they're having fun, but they don't realize that it's not clean and that they are splashing dirty water into the good water. Then, I have to walk back home with the *kodams*, but sometimes I can only carry one instead of two because I have to hold my sari and watch out for rocks and garbage under the water. It's a hectic situation for all the ladies.

Unfortunately, Nalini's vivid narration of water retrieval is all too common across the study villages, especially during the monsoon.

Stormwater drainage also engenders risks for post-point contamination. A lack of drainage infrastructure prevails across the 14 villages, resulting in serial flooding and unhygienic environments. However, the provision of drainage also generates water quality hazards. The location of public taps in villages without drainage is visible in Figs. 2 and 3, and taps near drainage infrastructure (which is often incapacitated by anthropogenic waste or purposefully blocked to prevent the flow of grey water in front of households) render the collection of water extremely treacherous in Fig. 4.

---

[5]*Kodams* are water collection and storage vessels. They typically hold a volume of 12–18 L.

Fig. 2 These taps in New Nambiyar Nagar are surrounded by standing water. Some taps are only accessible by jamming rubber hoses into the pipe, which inevitably fall into the water from time to time. Other taps are completely submerged, with hoses barely reaching out from the floodwater

As mentioned earlier, a discussion of pressure is required as it is inextricably linked to quality. Low pressure degrades water quality because it allows impurities to enter pipelines. Further, low pressure manifests in a weak stream at the point of retrieval, thereby impacting quantity and making the acquisition of water onerous. Pressure also exhibits temporal attributes: it is lowest at the beginning and at the end of water availability windows, and reaches its peak in the middle of windows. Additionally, measurements of flow rate indicate that pressure is lowest during the dry season and peaks in the monsoon season, when water supplies, and thus flux, are greatest.

Pressure also demonstrates spatial attributes. Public taps nearest to water towers exhibit greater pressure than taps farther from water towers and those located on the tail ends of networks. Pressure not only decreases as a function of distance from water towers, but also because taps closer to towers remove water from a finite supply that consequently has less and less water behind to induce pressure, a classic etiology of the serial and branched distribution networks at the study villages (Bhave 2003; Chinn 2009). Illustrating these points, taps in southeast Kilinjilmedu receive water before and enjoy greater pressure than those on the north and northwest fringes of the village; taps in northeast Amman Kovil Pathu dispense

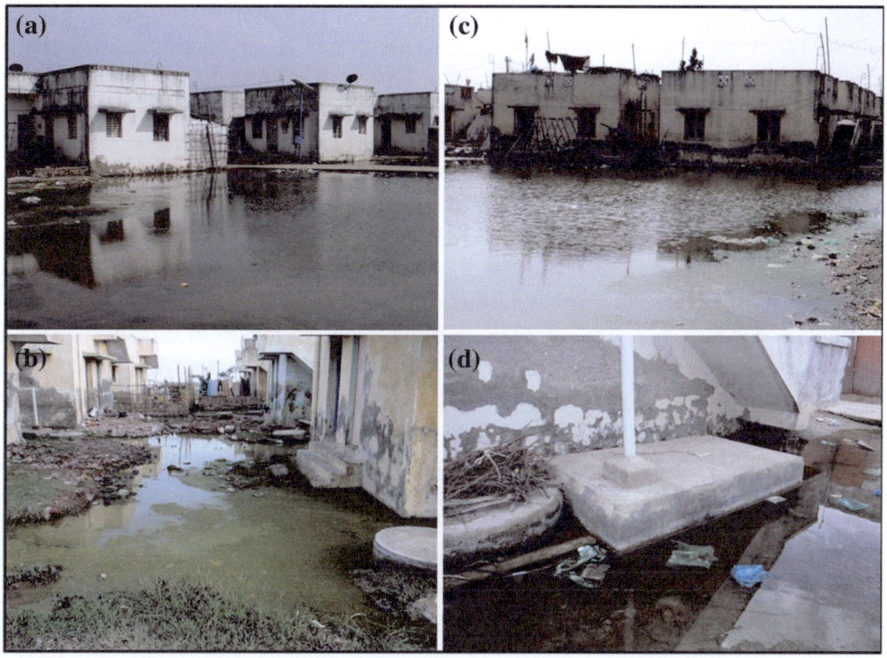

Fig. 3 Floods and waterlogging, primarily during the monsoon, represent hazards to water quality, not to mention hygiene, mobility, and quality of life. Here, standing water is visible in New Nambiyar Nagar (**a**), Saveriyarkovil (**b**), and Andana Pettai (**c**), while a septic tank is surrounded by water in New Nambiyar Nagar (**d**)

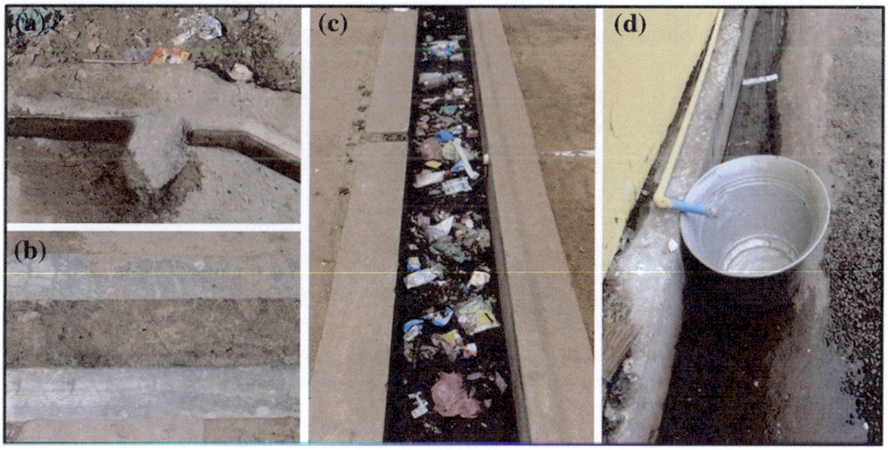

Fig. 4 Drainage ditches are purposely blocked by cement in Akkaraipettai (**a**), sand in Samanthanpettai (**b**), and solid waste in Mandapathur (**c**), while water is obtained from a tap hovering above an open drainage ditch in Kilinjilmedu (**d**)

water at a higher flow rate than those in the south and southwest; and taps in the eastern block of New Nambiyar Nagar release water faster than taps to the west. Even more extreme, taps in north and north-eastern Kizhakasakudimedu are able to dispense water while other taps remain dry, unable to eke out a single drop due to weak pressure and because those 'upstream' have already removed the precious resources, leading Prema to contend: "People on that side are lucky. They get more water than us because the pressure here is very less. We don't even get water on this side!" All of these cases are due to the former sections of the villages being located closer to their respective water tower. Collectively, these discrepancies in pressure fabricate an artificial hydrology determined by the relative location of waters tower in real space. Thus, pressure is highly variable, with the length of time to fill a 15-L *kodam* as low as 17 s (0.88 L/s) at the tap closest to the water tower in Akkaraipettai, but as appalling as 861 s (0.02 L/s) to fill the same *kodam* at the tap farthest from the water tower. Important to the focus of this paper, the outcomes materialize not only in opportunity costs, time wasted, and perturbations in social hydrology, but also in degradation of water quality.

The dominant method for addressing weak pressure is to modify the height at which water is dispensed, which often increases quantity at the cost of quality. Laws of physics dictate that a lower tap height relieves water of its travel to higher altitude, allowing it to be released from the lower height at a faster rate. This concept is not lost among residents of the study villages, with many taps having been manually lowered to augment pressure (Fig. 5). However, lowering taps to near surface level has proven futile in many cases, prompting modifications that are even more drastic. Ultimately, many residents—especially those located relatively far from their respective water tower—have dug holes up to 2 m below surface level in order to induce flow, with the flow rate after such modifications still painstakingly low. Worse yet, pressure is so low that sometimes small containers must be filled at subsurface levels and then hoisted to ground level in order to fill larger *kodams* in a one-by-one, piecemeal effort. Still, flow tapers off and frequently ceases, forcing residents to suck on the end of the hose to induce flow, essentially siphoning water from the pipeline. Yet the hardships do not cease there. The holes fill with precipitation during the monsoon season and rain events, further impacting water quality and often shutting down public taps for days or weeks at a time, as seen in Fig. 6 and attested by Vani of Kizhakasakudimedu:

> During monsoon the hole fills with rain and makes it very difficult to get water. We can't use the water because it's dirty, we can't even use it for bathing. In dry season I have to scoop water from the bottom of the hole. Sometimes I use the water for cleaning and bathing, but not for drinking. Often, I have to stop using the tap full stop and get water from a neighbor if they have water to spare. Our tap gets unusable because no pressure and too much dirty water in the hole. This is common.

Residents of the study villages have submitted complaints to government agencies in last ditch efforts to secure adequate pressure. For example, Sangeeta of Akkam Pettai complained to the *panchayat*, but was summarily rebuffed:

Fig. 5 Taps have been lowered to subsurface levels in order to access water in Saveriyarkovil (**a**) and Akkaraipettai (**b**). In Kizhakasakudimedu (**c**), a young girl fills a small container with water and subsequently pours it into a larger container near surface level

> I complained to the *panchayat* that pressure at my tap is too low and sometimes no water comes. They [*panchayat*] said there was breakage in the underground pipe when the road was built and that it would be too difficult to identify where the breakage is and too expensive to tear up the road. So, they told me to use a different tap. They don't realize that other people don't want me to come to their tap and that other taps have the same problem.

Nisha of Amman Kovil Pathu encountered a similar problem, and she did not envision the situation improving in the short-run:

> I complained one month ago that pressure is too low. PWD said a pipe was broken between the tank and the village, but they didn't fix it. MGR Nagar [a separate, adjacent post-tsunami village] has been empty, but now it is starting to fill up with tsunami people. The pressure is getting less because they use the same tank as us. Now the same water is being used by 250 houses instead of 125, so there are big pressure and quantity problems. It's not just lean amounts of water, the quality is also bad because we are getting more desperate and digging the water out of a dirty hole. It will only get worse unless the government helps.

In an effort to share a positive case of lodging complaints, Aanandi of Akkaraipettai narrates how she contacted a superior official regarding pressure:

> I was getting water from a hole below ground. I told about the pressure problems to Sankar [*panchayat* President], but the problem was fixed only after going next to K.A. Jayapal, the

Fig. 6 Access at this subsurface tap in Kizhakasakudimedu has been obstructed by monsoon rains, rendering quality poor and access impossible

> State Fisheries Minister who lives near here. He reminded Sankar to fix it and then it was fixed after a few days. Sankar had promised to fix it, but he didn't fix until I contacted Jayapal. Now I get more water, the water comes faster, and the water is more clear because it is not coming from the dirty hole.

In the end, the infrastructure and supply regime upon which pressure depends are conceived upon *ersatz* configurations to begin with, and complaints on pressure are generally destined for vain outcomes. Ultimately, the most immediate and rewarding method to increase pressure is to lower tap height, even if that means digging holes and further compromising water quality.

Storage of water in the household—made necessary by a non-continuous supply and scarce quantities—introduces additional risks for post-point contamination (Fig. 7). Storage, typically for close to 24 h and sometimes for 2–3 days when water fails to arrive, facilitates microbial growth and further deterioration of quality (Wright et al. 2004; Juran and MacDonald 2014). Furthermore, water is often stored in *kodams* that have not been washed with soap, and contain visible mildew and mineral build-up in the bottom, and drinking water is not removed via spigot but instead cup-by-cup. Not only is the cup itself a hazard, but so are the hands that are dipped, especially when those hands are attached to thirsty children returning home from school or parched adolescents who just finished a game of cricket with neighbourhood friends. At any rate, hands and cups are dipped into storage

Fig. 7 Storage of water is a fixture of post-tsunami village life. Here, water is stored in an unused bathroom (**a**), unused kitchen (**b**), and storage tank provided by non-profits (**c**)

containers every day to obtain water for consumption; each agent is potentially contaminated, serving to problematize the already poor water quality scenario.[6] In fact, water quality tests confirm these assertions of unhygienic storage and serving processes: only 16 of 38 (42.1%) randomized household samples that could be traced to bacteriologically safe taps revealed an absence of faecal coliform. Thus, even when taps dispense water free of faecal coliform, storage and serving processes (not to mention retrieval and transportation from public taps) surface as additional threats to water quality.

5 Coping with (Poor) Water Quality in Post-tsunami India

Together, water treatment and supply regimes, inadequate pressure, microgeographies adjacent to public taps, and the taps themselves, coalesce to deteriorate water quality. Furthermore, there are numerous risks for post-point contamination as a

[6]Nath (2000) found that 68% of households in rural India take water from storage vessels in the manner described here. Furthermore, Nath confirms that repeated contact with drinking water during collection, storage, and serving increases risks for post-point contamination, and this was also confirmed in the study villages (see Juran and MacDonald 2014 and Juran et al. 2016).

consequence of retrieval, transportation, storage, and serving processes. Unfortunately, this context is accompanied by limited strategies to improve water quality, let alone strategies that constitute long-term solutions. The act of lodging complaints was mentioned earlier; this strategy and others will be highlighted in this section.

One method to improve water quality is through decentralized on-site treatment centres, although this method fails to address hazards of post-point contamination. Four of the 14 study villages were provided treatment centres by non-profits. The centres at Kizhkasakudimedu and Andana Pettai remain operational, and tests reveal that they generate water free of faecal coliform that also meets physical and chemical standards of the BIS, rendering the facilities extremely valuable to the two villages. However, while the facility at Andana Pettai provides much needed potable water, access is hampered because a major highway, the East Coast Road, bisects the settlement. Due to facility location, residents in eastern Andana Pettai enjoy easy access, while residents in the western section must tote heavy vessels of water either across the bustling highway or by a long, circuitous path that avoids the highway. This hindered connectivity causes many households in western Andana Pettai to forego the safer supplies.

Even more interesting from the standpoint of humanitarian aid, the water treatment centres at Amman Kovil Pathu and Samanthanpettai have remained unutilized for nine years. The facility at Amman Kovil Pathu is unutilized because residents have been unsuccessful in convincing government authorities to connect it to both a source of water and electricity. Interviews with PWD officials establish that they refuse to connect the facility to water because (1) non-profits, not the government, introduced the facility, which leaves the PWD off the hook for maintenance; and (2) the facility's purpose of providing water is already being met by the PWD itself (that is, the PWD is already fulfilling the plant's function of providing water). Thus, the PWD argues that responsibility for the facility rests with village residents and the non-profits, not the PWD. In terms of electricity, the PWD argues that proper forms and fees have not been submitted, although electricity remains moot until an inflow of water is secured. This scenario underscores that external aid and capital were welcomed in the short-term, but that the government refused to maintain or take responsibility for such assets in the long-term, exposing serious issues of aid-coordination and project sustainability. The end result is that the facility has not treated a single drop and has instead been relegated to an artefact of humanitarian aid gone wrong. Worse yet, residents have been denied much needed potable water.

The treatment centre in Samanthanpettai remains unutilized for reasons both comparable to and different from that of Amman Kovil Pathu. While an electrical connection failed for similar reasons (forms, fees, and bureaucratic barriers), the water connection failed because plans to treat groundwater proved flawed. After construction, it was discovered that local groundwater is unfit for direct reverse osmosis (RO) treatment. High concentrations of salts, magnesium, calcium, and other dissolved solids demand that groundwater must undergo pre-treatment processes before passing through membranes. Residents faced two options: purchase and install pre-treatment mechanisms, or connect the facility to piped government water (which contains less dissolved solids and is fit for RO treatment). However, the first option was economically and technologically infeasible and the latter was rebuffed by TWAD, essentially because TWAD has no responsibility to operate or maintain infrastructure introduced by other entities (in this case, a non-profit that responded to an official state call for disaster aid). Now, the facility has fallen into such disrepair—not to mention that pipes, fittings, and valves have been stolen for repurposing and to sell for scrap—that engineering experts are required just to restore the facility's operability. Meanwhile, residents of Samanthanpettai cope with low quality provisions despite the visible presence of ameliorative infrastructure.

As for villages without treatment centres, the facilities are costly and difficult to finance in the absence of external assistance, positioning them as cost-prohibitive for the most part. The result is that households must rely on government-provided tap water and untreated groundwater gained through private means, both of which generally fail to satisfy basic water quality standards. Further, it must be stressed that the mere presence of an on-site treatment centre does not guarantee safe water, given that the water is subsequently placed into (possibly dirty) *kodams*, taken home, stored for hours or days, and consumed piecemeal by dipping cups in it. In fact, only 9 of 22 (40.9%) samples from the two operational treatment centres revealed an absence of faecal coliform when tested in the household, meaning that nearly 60% were contaminated at the point-of-use (PoU). Thus, beyond the government's responsibility to provide potable water, post-point contamination remains a formidable hazard, implying hygiene education, behaviour modification, and PoU treatment, as potential avenues for impactful intervention.

Overall, village residents are compelled to implement rudimentary strategies to manage poor water quality. For example, residents let dirty water run from taps at the beginning of water supply windows, walk gingerly to avoid splashing floodwater, lay stepping stones to walk in and out of puddles that inundate taps, and attempt to limit storage (although this is often beyond their control due to scarce supplies). PoU treatment is an effective method for improving drinking water quality. However, an analysis of PoU treatment in the study villages established that only 65 of 300 (21.7%) households practice PoU treatment, with those 65 households producing water free of faecal coliform in just 26 (40.0%) cases (Juran and MacDonald 2014). While only 40.0% of households that practiced PoU treatment were able to generate water absent of faecal coliform, households not doing so generated water free of faecal coliform in only 35 of 235 (14.9%) cases. Thus, while limitations are apparent, households engaged in PoU treatment are able to produce relatively safer water.

Another method for improving water quality is to purchase pre-packaged water, typically in 20-L jugs at an expense of about 40 Indian Rupees. Pre-packaged water is generally of better quality than government-provided water and untreated groundwater. However, households consuming pre-packaged water exhibited an absence of faecal coliform in just 3 of 7 (42.9%) cases, raising issues of unhygienic handling practices and/or poor quality to begin with, the latter complicated by the fact that the pre-packaged water market is poorly regulated both globally and in India.

A final strategy to enhance water quality is to file complaints with the government. While most water-related complaints pertain to quantity (usually when water fails to arrive for several days, and usually in the dry season), residents have complained to their governments about water quality. In fact, residents from every study village reported having lodged written or in-person complaints to *panchayat* officials, municipal officials in Nagapattinam, the PWD in Karaikal, as well as community leaders. One complaint, during a 2011 outbreak of diarrhoea in Vettakaramedu, mentioned earlier, prompted the PWD to clean the water tower that serves the village. However, the PWD swiftly deflected the outbreak upon residents, branding it a consequence of residents intercepting pipes at subsurface levels —an action residents defend as necessary to overcome low pressure. Residents of Samanthanpettai, in a case also referenced earlier, lodged complaints on the cleanliness of their water tower. Nagapattinam municipality took no action, which led residents to clean the tank themselves, a task they have since performed several times. Residents of Samanthanpettai have also pleaded with Murugaian, their Municipal Counsel representative, to connect the dysfunctional on-site treatment centre to a municipal water stream. Residents contend that several formal requests and in-person follow-ups have been neglected, but Murugaian maintains differently:

> Samanthanpettai people have never requested me to connect it [treatment center] to municipal water, but I have tried to help them anyway. I am a power man, so they must do one thing. Samanthanpettai people must give me a proper request and then I will give a proper request to the [Municipal] Chairman. This is how it must be done. Anyway, I am done with the issue.

Residents from all villages have filed complaints about water, but confronted with largely poor outcomes, residents are hesitant to raise grievances in the future, characterizing the act as tiring, unproductive, and a 'waste of time.' First, residents maintain that the complaint process is burdensome because problems are rarely rectified in the end, leading Siva of Uzhuvar Nagar to state that, "It doesn't matter if we complain, nobody will help and nobody cares about our problems." In cases when complaints actually generate results, it is often only after residents have voiced their concerns numerous times to several officials, situating the degree of work, time, and other opportunity costs incurred as extremely taxing. For example, Punita of Mandapathur claims that she submitted a complaint on water quality and pressure to the local *panchayat* President, which was then signed and forwarded to Kottucherry Commune *Panchayat*. Punita contends that "this was done 10 times" until action was finally taken. Moreover, beyond being burdensome, village residents note that government officials tend to reflexively turn the problems back on those who lodge complaints, claiming that the government is providing 'good water,' 'meeting its duties,' and that the problems have been brought on by residents themselves. Finally, some claim that they wish to share water concerns but do not know who to contact (government or non-profit), leaving them feeling disenfranchised.

More distressing, interviews uncovered that renters are hesitant to lodge complaints because they feel like outsiders and because they are technically engaged in an illegal activity. According to official government orders, all reconstructed houses built with humanitarian aid are ineligible to be sold or rented in the first 10 years. As a result of the law, several renters stated that they are not only apprehensive, but genuinely frightened to make complaints. Nagoreammal of Amman Kovil Pathu claims, "Our family rents, so I'm too scared to talk to local leaders and give work requests. I am a minority [Muslim], so if I complain I'll be ignored and it might even start trouble." Adding to the sentiment, Lakshmi, a renter at Akkam Pettai, declares:

> I complained to the [Kottucherry Commune] *Panchayat* and to PWD officials who were here fixing a broken pipe. I told them that the water is brown and not good. They said they would look into it, but never did anything. I don't want to complain again because it might put our family at risk and it might make the situation worse than it already is.

Residents of the post-tsunami villages are dissatisfied with results of the predominantly ill-fated complaint process and, adding insult to injury, Vikas of Samanthanpettai is dismayed because he pays taxes for such poor water services: "We all pay our taxes [700 Indian Rupees per annum], but we don't see any results. Water is a big problem here and the problems never get fixed even when we complain." Ultimately, the most effective action is for residents to address water problems themselves. Residents introduce temporary fixes, intercept pipelines, dig holes to access water at subsurface levels, pool funds to purchase tanker lorry water or hire plumbers, and seek water from other sources—often at a considerable distance. As Dhalia from Saveriyarkovil makes clear, "It's easier and faster to fix the problems ourselves. The government won't help, so its better we do it ourselves."

6 Conclusions

This paper critically examined the legacy of post-tsunami reconstruction in India through the lens of 'water'. A socio-geographical and visual ethnographic inquiry of 14 reconstructed villages in Nagapattinam District, Tamil Nadu, and Karaikal District, Puducherry, demonstrated that issues of water quality are widespread. Furthermore, methods developed to cope with poor water quality were detailed, and issues related to water quantity, access, and pressure were documented. An overarching theme is that the hazardous pre-disaster context was merely refabricated via reconstruction processes to fashion a hazardous waterscape of manmade, engineered proportions. While pre-tsunami settlements may have been more vulnerable to natural disasters (such as tsunamis, cyclones), it is argued that the newly constructed post-tsunami settlements are more vulnerable to unnatural disasters owing to short-sighted reconstruction processes. The outcome, couched here as a 'disaster after the disaster,' has been illustrated through the medium of water.

The ability to manage water hazards and mitigate their threats is limited, because village residents have been dealt waterscapes over which they command little power. With many factors outside residents' bounds of influence, there are few opportunities to surmount, let alone subvert, the hazardous waterscapes. Thus, residents are often forced to operate within a narrow band of functional possibilities. The result is not only constrained agency, but that subjects are driven to acclimate, acculturate, and ultimately submit to the waterscapes in which they reside. As intimated by Evangeline of Andana Pettai, "My family's approach is to simply adjust to the problems. What to do?"

Acknowledgments The author greatly thanks the research participants in Nagapattinam and Karaikal, without whom this research would not be possible. The author also thanks Annie George, BEDROC, Rajagopal Chidambaram, E. Nagarajan, N. Kalyani, and Breeanna Prince for logistical support.

References

Adams, V., T. van Hattum, and D. English. 2009. Chronic Disaster Syndrome: Displacement, Disaster Capitalism, and the Eviction of the Poor from New Orleans. *American Ethnologist* 36 (4): 615–636.

Baxter, J., and J. Eyles. 1997. Evaluating Qualitative Research in Social Geography: Establishing 'Rigour' In Interview Analysis. *Transactions of the Institute of British Geographers* 22 (4): 505–525.

Bhave, P.R. 2003. *Optimal Design of Water Distribution Networks*. UK: Alpha Science International.

Black, M., and R. Talbot. 2005. *Water: A Matter of Life and Health-Water Supply and Sanitation in Village India*. New Delhi: Oxford University Press.

Button, G., and A. Oliver-Smith. 2008. Disaster, Displacement, and Employment: Distortion of Labor Markets During Post-Katrina Reconstruction. In *Capitalizing on Catastrophe: Neoliberal Strategies in Disaster Reconstruction*, ed. N. Gunewardena and M. Schuller, 123–145. New York: AltaMira Press.

Chacko, E. 2004. Positionality and Praxis: Fieldwork Experiences in Rural India. *Singapore Journal of Tropical Geography* 25 (1): 51–63.

Chinn, T.D. 2009. Water Supply. In *Environmental Engineering: Prevention and Response to Water-, Food-, Soil-, and Airborne Diseases and Illnesses*, ed. N. Nemerow, F. Agardy, P. Sullivan, and J. Salvato, 1–131. Hoboken: Wiley.

Elangovan, R., L. Philip, and B.S. Murty. 2006. *Users Manual on Water Quality Analysis: Water Quality Test Kit-15*. Chennai: Indian Institute of Technology-Madras.

Ferguson, A., B. Mailloux, K. Ahmed, A. van Green, L. McKay, and P. Culligan. 2011. Hand-Pumps as Reservoirs for Microbial Contamination of Well Water. *Journal of Water and Health* 9 (4): 708–717.

Geertz, C. 1973. Thick Description: Toward an Interpretive Theory of Culture. In *The Interpretation of Cultures: Selected Essays*, ed. C. Geertz, 3–32. New York: Basic Books.

Juran, L. 2015. A Framework for Integrating Water Projects in Disaster Reconstruction Processes in the International Context. *Public Works Management and Policy* 20 (4): 305–321.

Juran, L., and M.C. MacDonald. 2014. An Assessment of Boiling as a Method of Household Water Treatment in South India. *Journal of Water and Health* 12 (4): 791–802.

Juran, L., M.C. MacDonald, N.B. Basu, S. Hubbard, R. Rajagopal, P. Rajagopalan, and L. Philip. 2017. Development and Application of a Multi-Scalar, Participant-Driven Water Poverty Index in Post-Tsunami India. *International Journal of Water Resources Development* 33 (6): 955–975.

Kennedy, J., J. Ashmore, E. Babister, and I. Kelman. 2008. The Meaning of "Build Back Better": Evidence from Post-Tsunami Aceh and Sri Lanka. *Journal of Contingencies and Crisis Management* 16 (1): 24–36.

Longhurst, R. 2010. Semi-Structured Interviews and Focus Groups. In *Key Methods In Geography*, 2nd ed, ed. N. Clifford, S. French, and G. Valentine, 103–115. Thousand Oaks: SAGE Publications.

Lyons, M. 2009. Building Back Better: The Large-Scale Impact of Small-Scale Approaches to Reconstruction. *World Development* 37 (2): 385–398.

Ministry of Urban Development. 2011. *Report on Indian Urban Infrastructure and Services*. New Delhi: Ministry of Urban Development, Government of India.

Nath, K.J. 2000. *Environmental Sanitation and Community Water Supply: Country Situation Analysis*. New Delhi: Independent Commission on Health in India.

Pile, S. 1991. Practising Interpretive Geography. *Transactions of the Institute of British Geographers* 16 (4): 458–469.

Pink, S. 2001. *Doing Visual Ethnography: Images, Media, and Representation in Research*. Thousand Oaks: SAGE.

Rice, E.W., R.B. Baird, A.D. Eaton, and L.S. Clesceri (eds.). 2012. *Standard Methods for the Examination of Water and Wastewater*, 22nd ed. Washington D.C.: American Public Health Association/American Water Works Association/Water Environment Federation.

Schuller, M. 2008. Deconstructing the Disaster after the Disaster: Conceptualizing Disaster Capitalism. In *Capitalizing on Catastrophe: Neoliberal Strategies in Disaster Reconstruction*, ed. N. Gunewardena and M. Schuller, 123–145. New York: AltaMira Press.

Smith, S. 1981. Humanistic Method in Contemporary Social Geography. *Area* 13 (4): 293–298.

Sobsey, M.D., and F.K. Pfaender. 2002. *Evaluation of the H2S Method for Detection of Fecal Contamination of Drinking Water*. Geneva: World Health Organization.

Virmani, S. 2008. *Disasters Within Disaster: Failed Interventions We Can Learn From*. Bhuj: Hunnarshala Foundation.

Woodson, R. 2010. *A Builder's Guide to Wells and Septic Systems*. New York: McGraw Hill.

Wright, J., S. Gundry, and R. Conroy. 2004. Household Drinking Water in Developing Countries: A Systematic Review of Microbiological Contamination between Source and Point-Of-Use. *Tropical Medicine and International Health* 9 (1): 106–117.

An Applied Anthropological Perspective on Localizing Post-disaster Aid: Lessons from Post-tsunami Aceh, Indonesia

Patrick Daly

1 Introduction

In 2006 I visited a number of coastal villages in Aceh, Indonesia that had been devastated by the 2004 Indian Ocean tsunami—the first formal research visit of what turned into a decade of field work on the post-tsunami reconstruction and its aftermath. People pointed out how high the waves were using treetops as benchmarks, and talked about the number of victims, the extent of the damage to buildings, and the disruption of local livelihoods. When I asked about their initial reactions, people recalled that they were in shock, traumatized by some combination of personal and community loss, and overwhelmed by the sheer dramatic scale of the tsunami. They also told me that within hours survivors had begun the process of tending to the injured, searching for missing persons, caring for children who had lost their parents, taking inventory of the damage, and procuring basic provisions like food and water. Such stories of localized responses are common in most post-disaster situations, and scholars have written in-depth about the resilience of communities in the face of severe hardships, and the emergence of altruism that binds people together and promotes the common good (de Vries 1995; Gibbs 1982; Gist and Lubin 1999).

However, during the main phase of the reconstruction, from 2006 to 2009, the general narrative shifted away from the successes of localized responses, to often critical commentary on the effectiveness and efficiency (or lack there of) of the massive international rebuilding effort. For all stakeholders—tsunami survivors, government officials, aid workers, and the occasional anthropologist—this period was intense, chaotic, frustrating, and overwhelming. The sudden arrival of hundreds of non-governmental organizations (NGOs), contractors, diplomats, and academics

P. Daly (✉)
Earth Observatory of Singapore, Nanyang Technological University,
Singapore, Singapore
e-mail: patrickdaly@ntu.edu.sg

S. Reddy (ed.), *The Asian Tsunami and Post-Disaster Aid*,
https://doi.org/10.1007/978-981-13-0182-7_6

fueled by billions of dollars in aid created a social and political tempest—which many referred to as the 'second tsunami'. Major evaluations, supported by more focused academic research papers, paint a picture of poor coordination between aid agencies, exclusion of local stakeholders, and the misalignment of aid and needs (Athukorala 2012; Brusset et al. 2009; Christoplos 2006; Telford and Cosgrave 2007; Daly and Brassard 2011).

These critiques are as obvious as they are curious, given that most post-disaster master plans (including those drafted in response to the 2004 tsunami) emphasize anticipating and avoiding such problems, and most assessments of post-disaster efforts list them as 'lessons learned.' Concepts of rebuilding through inclusive participatory approaches, which empower citizens and local governments, and are sensitive to local customs and traditions, are the bedrock of the main 'how to rebuild following disasters' manuals produced by leading humanitarian organizations (Doberstein and Stager 2013; Oxfam 2003; The Sphere Project 2004; UNDP 2010; Jha et al. 2010).

Given this rhetoric about community driven reconstruction and development, and the importance placed upon participatory approaches, it is puzzling that aid organizations consistently fail to connect with beneficiaries to a mutually satisfactory level. My experiences in Aceh, which generally echoes what practitioners and academics have reported from other post-disaster situations, suggests that this disconnect is largely the product of three things: (1) ignorance by many in the aid world about the nuances of local context, and how the communities that they are working with are structured and function (Daly and Rahmayati 2012; MacRae 2008; MacRae and Hodgkin 2016); (2) disregard for local structures and practices, often rooted in negative perceptions about their viability and usefulness (Daly 2015; Hagelsteen and Becker 2013; Kenny 2007); and (3) the burden and rigidity of the bureaucratic and administrative frameworks that aid agencies are compelled to work within (Castillo 2014; Edmonds 2013; Telford and Cosgrave 2007).

I have touched on these issues in other publications (Daly et al. 2012, 2016; Daly 2015; Daly and Hor 2013; Daly and Rahmayati 2012; Daly and Brassard 2011, etc.), but want to use the opportunity of this volume dedicated to long-term perspectives on the recovery from the 2004 tsunami to elaborate and provide a hopefully cohesive reflection on the relationships between the processes and structures of aid, and localized social and cultural contexts in Aceh. I draw upon a decade of experience studying community-level reconstruction in post-tsunami Aceh to highlight how recovery and reconstruction was helped, or hindered by, its alignment with local contexts. This work is strongly biased by my background as an anthropologist, but it is also tempered by years of pragmatic engagement with the processes and outcomes of aid. My goal in this paper, is to move away from critique, and outline some of the key insights we have gathered about how external aid could more fruitfully leverage upon and benefit from closer alignment with localized institutions and social practices.

2 Localizing Aid to Support Recovery and Reconstruction

Before moving on to a more empirically grounded discussion, it is important to first briefly outline the conceptual foundation of this paper. I argue that integrating aid locally is important for both recovery and reconstruction, which are two distinct but potentially overlapping things (which are, unfortunately, often conflated). Reconstruction is the process of building things—whether physical structures, institutions, livelihoods, etc. Recovery is a process through which individuals and societies respond to and cope with the impacts of a disaster, as they balance resuming their pre-disaster lives with the inevitable changes that occur. Discussions about recovery need to consider psycho-social health and community well-being, amongst other things. A core component of social recovery is the level of direct and empowered involvement disaster-affected people have with rebuilding their homes, livelihoods, and communities (Gilbert and Silvera 1996; Omer and Alon 1994; Vale and Campanella 2005). Implicit in this argument is that recovery can potentially occur without formal programs of reconstruction, and that reconstruction does not necessarily support recovery.

From this I propose that post-disaster assistance may better support both reconstruction and recovery when aid is aligned with local contexts, aspirations, needs, capacities and institutions, and allows for affected persons to play active roles throughout the rebuilding process. One way to do this is for external actors to identify how best to work within local cultural and political frameworks to minimize the foreignness of aid. Essential to this is respecting the nuanced overlapping roles that local institutions and practices play for both coping and recovery, and the practical distribution of aid.

The very capacities that people have to actually respond, and the institutions and mechanisms through which they do so, are parts of the basic pre-disaster community fabric. These include but are not limited to religions and belief systems, elements from the built environment, social structures, and cultural practices (Bolin and Stanford 1998; Daly and Rahmayati 2012; de Vries 1995; Gist and Lubin 1999). Raphael (2008) points out a number of key generic systems, such as schools, faith-based organizations, and local government, that play important roles in post-disaster situations. I want to extend this logic a bit further to include a wider range of social and cultural elements that are important parts of recovery processes. In addition to the practical roles these systems might play within a localized disaster response, the very act of using these basic and fundamental infrastructure, institutions, and practices is integral to recovery, as through doing so people are re-engaging with familiar aspects of their lives. It is questionable that members of a tightly knit community would feel that reconstruction and aid efforts are really 'inclusive' if they are not discussed and managed by the communities in venues, and through processes that the community identifies as *theirs*.

On a more practical level, working in a post-disaster situation is difficult. Conducting accurate needs assessments, establishing networks, building operational infrastructure, navigating cultural and political systems, and understanding livelihoods and local economies can pose major obstacles to external aid actors. This is especially the case for disaster responses that occur in areas where aid organizations do not have previous operational experience. While some relief and reconstruction tasks may exceed the capacities of local actors and institutions, many aspects of aid distribution and rebuilding can be handled by local actors, and/or through pre-disaster frameworks. All too often, however, the main actors in post-disaster situations are quick to establish their own institutions and frameworks —which do not always align with local contexts. This can both impede recovery and lead to less effective reconstruction outcomes. Below I use some examples from the post-tsunami reconstruction in Aceh to illustrate these ideas.

3 The 2004 Tsunami and Reconstruction in Aceh

The Aceh province in Indonesia was the 'ground zero' of the 2004 Indian Ocean tsunami, in which approximately 150,000 people lost their lives. In the days following the tsunami, graphic and horrifying images from Aceh dominated global media, including vast stretches of wreckage, debris-laden waves pushing through towns and villages, and ships displaced far inland by the powerful waves. The sheer intensity of the devastation made Aceh the center of the largest and most ambitious post-tsunami relief and reconstruction projects (Hutchison 2014; see Table 1). Funds poured in from private citizens, governments and NGO's from all over the world (Athukorala 2012; Daly et al. 2012; Clarke et al. 2010; Older, 2015; see Tables 2 and 3).

A combination of the tsunami, and the legacy of Aceh's long-standing armed separatist conflict between the Indonesian government and the *Gerakan Aceh*

Table 1 Estimates of tsunami damage in Aceh and Nias

Number of people dead	~150,000
Number of people displaced	~600,000
Number of houses destroyed	~80–110,000 in Aceh; 13–14,000 in Nias
Length of road destroyed	3000 km
Key infrastructure destroyed	14 seaports, 11 airports, 120 major bridges
Number of school buildings destroyed	2000
Number of fishing boats destroyed	5000
Amount of agricultural land affected	70,000 ha

Source BAPPENAS (2005)

Table 2 Source of funding allocated to specific projects

International donors	US$2.1 billion
NGOs	US$1.6 billion
Government of Indonesia	US$1.9 billion
Total	US$5.6 billion

Source McKeon (2008)

Table 3 Summary of funding allocations by sectors

Social sector (education; health; community, culture, and religion)	US$1.5 billion
Infrastructure (housing; transport; communications; energy; water and sanitation; flood control and irrigation works)	US$2.8 billion
Productive sectors (agriculture and livestock; fisheries; enterprises)	US$645 million
Cross sectoral (environment; governance; banking and finance)	US$630 million
Total	US$5.6 billion

Source McKeon (2008)

Merdeka (Free Aceh Movement or GAM), left Aceh poorly equipped to manage the crisis (Reid 2006). The capacity of the provincial level government was low, and relations with the Indonesian government were strained. This administrative vacuum left unprecedented space for the response to become internationalized—a move that was more or less accepted by both the Indonesians and the Acehnese (Gaillard et al. 2008; Telford and Cosgrave 2007). This differentiated the situation in Aceh from other regions affected by the tsunami such as Thailand, India and to a less degree Sri Lanka, where national and regional governments played more powerful roles.

While the initial responders, such as foreign militaries, the Red Cross/Crescent, Doctors Without Borders, etc. were well organized, skilled in emergency response, and held to strict internal mandates about the scope and level of involvement on the ground in Aceh, things rapidly descended into a total free-for-all. Countless organizations and individuals made their way to Aceh, or as close as they could get given the limited infrastructure (Brusset et al. 2009; Older 2015). The good intentions of most aside, this created a logistical nightmare. Not only was there a significant lack of coordination and basic capacity to absorb the rapid influx of people and resources, but it was impossible to evaluate the credentials, skills, experiences, and driving ideologies of people and organizations arriving (Pandya 2006; Telford and Cosgrave 2007; Zeccola 2011). For an area that suffered decades of conflict, had been virtually cut off from the outside world, and was devastated by the earthquake and tsunami, this arrival of outsiders was incredibly disorientating.

Banda Aceh, the capital of the province, was reconfigured into the hub of the humanitarian effort—with the emergence of new actors, power centers, and administrative processes. The onslaught of external actors was evident in the myriad of foreign visitors driving around in sport utility vehicles (SUVs), living in gated compounds, and socializing at the small number of establishments that sprung up to feed them (Smirl 2015). This foreign invasion became the starting point for what was meant to be (at least on paper) a community-centered reconstruction.

4 Community Infrastructure and Recovery

> A minimum degree of community integration is seen as a basis for initial steps toward recovery and rebuilding, which suggests that early positive responses to disasters should be based on greater local understanding of social and physical environments for the reduction of both short to long-term losses. (Oliver-Smith 1996, p. 306)

As I have elaborated on in other papers, I fully agree with Oliver-Smith that disaster responses can benefit from being sensitive to local environments (Daly and Brassard 2011; Daly and Rahmayati 2012). Here I discuss some of the important 'local' environments and institutions in Aceh that I feel influenced both recovery and the outcomes of reconstruction projects. In Aceh, villages, locally known as *gampungs*, are the most important unit of government for most people, as many of the decisions that affect their day-to-day lives are made at this level. In addition to their formal administrative roles, *gampungs* are dynamic and functional social entities. Social life in Aceh is heavily structured at the *gampung* level, with prescribed hierarchies, leadership structures, manners of public debate and discussion, and formal decision-making; all of which are essential for embarking upon community recovery, and the realization of 'participatory' reconstruction and development practices (Mahdi 2012). Below, I discuss some of the key elements within *gampungs* that played roles within both recovery and reconstruction—moving from culturally significant elements of the built environment to leadership structures, customary law, and community mobilization.

4.1 Cultural Spaces and Places

As I addressed in another paper, individual and community level recovery, which involves reaffirming identity, commiserating, finding solace, providing and receiving support, reflection about the nature and consequences of the disaster, and planning the way forward, is grounded in local built environments and cultural heritage (Daly and Rahmayati 2012). The physical spaces that make up *gampungs* were critical parts of both reconstruction and recovery. In much of the tsunami inundation zone, these structures and spaces were decimated, which often placed survivors lacking such cultural resources at a significant disadvantage in terms of healing and coming together to make plans and interact with external aid actors. I found that rebuilding these places were top priorities for affected communities, but that their significance was not widely recognized by external aid actors.

For the Acehnese, mosques and prayer halls provide a combination of religious and public space. The spiritual and functional important of mosques was reinforced by the higher rates of survival of these structures (often the only building left standing in a village) which has been widely interpreted within Aceh and the Muslim world as a sign of God's will (counterbalanced by more secular interpretations about design and structural integrity) (Feener and Daly 2016). It is clear from the countless personal narratives of survival and inspiration that I heard that mosques were important for people trying to cope with the psychological impact of trauma and loss. People gathered there for comfort, solace, and to perform important rituals such as burials. Additionally, in the chaos and disruption caused by the tsunami, which extended far into the reconstruction phase, regular prayer at the mosque provided an important and reassuring rhythm for people desperate for some normalcy. It is also clear that for many in Aceh, religion strongly influenced their perspective on the tsunami, and there was an increase in religious devotion following the tsunami—consistent with findings elsewhere that disasters often elicit religious responses (Bradfield et al. 1989; Gaillard and Texier 2010; von Vacano and Schwarz 2014; Fountain and McLaughlin 2016).

In addition to their spiritual and therapeutic roles, mosques and prayer halls were critical spaces in the aftermath of the tsunami for community gatherings, discussions about aid projects, and the dissemination of information. These spaces uniquely combine moral and practical elements that made them obvious and intuitive spaces for both healing and managing the pragmatic aspects of aid distribution. We found that communities benefitted from access to such spaces early in the reconstruction period, as they provided culturally suitable venues for the practical aspects of rebuilding, such as planning, meeting with representatives of donors and NGOs, and keeping the community updated on plans and progress.

In addition to these more obvious formal communal spaces, Acehnese villages contain several gendered spaces that were essential in the post-tsunami period for social bonding, community healing and discussing aid projects. For men, these were the little coffee shops that are ubiquitous throughout Aceh. It was common for these to be amongst the first make-shift structures built after the tsunami, and anyone who wanted to engage with the men in a village needed to spend time in such spaces. For women, social interactions and discussions generally occurred within the domestic sphere (Daly and Rahmayati 2012). Acehnese houses are typically structured to have spaces for women to gather and chat. While it was relatively easy to throw together temporary coffee shops, women were often left at a disadvantage as exclusive space for women was not factored into the design of the temporary barracks that many tsunami survivors lived in during the reconstruction period. Furthermore, research on post-tsunami housing suggests that the lack of gendered spaces within some of the designs for permanent housing has lead to long-term social disruption for women (Rahmayati 2016a, b).

The most readily accessible, familiar and intuitive way to make plans and distribute resources from the perspective of affected persons was through pre-existing community venues. However, it was common for respondents to say that aid organizations failed in this regard; instead relying upon methods created on the spot or brought in by the organization. People frequently mentioned how villages were fragmented during the emergency 'resettlement', and that community connections and institutions were not preserved when people were re-located to barracks or other temporary shelters. Additionally, many respondents did not feel that reconstruction and aid efforts were really 'inclusive' because they were not discussed and managed by the communities in venues, and through processes that communities identified as *theirs*. In practical terms, pre-tsunami, social processes of meeting, discussing, decision-making, consulting, distributing resources, and disseminating information were all culturally grounded within physical spaces and places in villages and neighborhoods. People often found it difficult to freely carry out such processes in the absence of the necessary infrastructure following the disaster.

It is not possible to conduct a definitive assessment of the relationship between levels of recovery and reconstruction and how quickly communities were able to access familiar social venues. However, my experiences leave me confident that identifying and providing such facilities immediately following a major destructive event can help community healing and reduce the ad hoc and chaotic nature of aid provision. No one feels at home in temporary shelters—and it is questionable how local and participatory aid processes can be in the extended absence of vital social and communal infrastructure.

Many NGOs acted out their version of participatory consultations, and most likely felt that gathering groups of men and women in a make-shift shelter was a reasonable way to engage with them. However, having spoken with many people who participated in such consultations, I came away with the impression that people were often unsettled and uncomfortable. In a subtle way, this created a distinct hierarchy between the givers and receivers of aid, as it forced tsunami-affected persons to conform to the preferred methods of aid actors.

4.2 *Localized Leadership and Decision-Making*

Good local leadership is commonly cited as one of the key variables contributing toward successful reconstruction and development projects—and enormous resources have gone into promoting 'good governance' programs by major development agencies such as the World Bank, International Monetary Fund (IMF), and the United Nations Development Programme (UNDP) (Kyamusugulwa 2013; UNDP 2010; World Bank 2000). Implicit within this premise is that local governance structures are not adequate and need to be either reformed, or replaced by external structures (especially during short-term interventions). While certainly there are issues with governance structures in most parts of the world (Boano and García 2011; Brinkerhoff and Brinkerhoff 2002), my experiences in Aceh suggest that it is not helpful to automatically discount them (Daly 2015). Like it or not, it is often impossible to get things done at a local level without some engagement with localized governance structures, irrespective of how flawed they might be. Actions that circumvent them can undermine the legitimacy of both aid efforts, and the local structures that need to be functional after the end of the reconstruction period. It can also lead to immense frustrations as alienated local leaders employ passive-aggressive techniques to resist the dictates of aid actors; drawing upon Scott's famous 'weapons of the weak' to subvert the domination of external actors (Scott 1985).

The main reconstruction plans for Aceh stressed the need to both work with local communities, and engage local leadership (BAPPENAS 2005; Government of

Indonesia 2005). The importance of *gampung* structures was mentioned repeatedly within the reconstruction Master Plan, with traditional leadership structures outlined in detail, complete with explanation of the range of mechanisms for making decisions, and passing on information to community members. The Master plan discussed how local structures of leadership and decision-making have inherent accountability features in terms of the complex sets of checks and balances offered by multiple layers of authority, diffusion of decision-making jurisdiction, and feedback channels. In Aceh, lines of political authority are reinforced with moral and social authority based upon a system in which respected elders and religious figures play important advisory roles, and can act as a counter balance to the village head, which is an elected position.

I have come across many examples in Aceh from formal evaluations, personal encounters, and anecdotal conversations with aid workers of village leadership playing strong and constructive roles within the reconstruction process. This was elaborated upon in great detail by Thorburn (2007), who showed that the strength and cohesiveness of local leadership was one of the main variables deciding the success or failure of projects at the village level. Communities with open discussions and transparent use of aid monies reported higher levels of both progress and end-user satisfaction (Bouraoui and Lizarralde 2013; Ophiyandri et al. 2013). Examples show that strong leadership, while sometimes attributed to the abilities and charisma of a single individual, is more likely the result of a village leader who worked within a broader system of local decision-making. Open discussions, free flow of information, trust, the support of advisors, moral approval, and presence of viable checks and balances were all important parts of the success of local leadership, and all key parts of traditional leadership structures (Thorburn and Rochelle 2014).

In cases where local leadership was cited as negative, it is often because donors and NGOs either side-stepped local leadership systems (Dixon and McGregor 2011), or misread traditional systems and over-invested in the village head as a way of 'ensuring' local involvement and endorsement. In Aceh, the village head is just one prominent figure embedded within a wider framework that allows for checks and balances. To fully utilize the strengths of local leadership it was necessary to also engage religious leaders such as the *imam gampung* (head of the local mosque), and other respected village elders such as the *tuha peut*, which oversee different livelihood sectors, and use of communal lands. The fact that these were not engaged at the village level by most NGOs, allowed too much of the success or failure of community-level efforts to be concentrated in the hands of the village head. The ways that many external organizations engaged with communities simplified a complex system of decision-making and left great scope for abuse by under-qualified or self-serving village heads, and very limited options for disaffected villagers to communicate issues with higher social or source of funding (donors and NGOs).

For many donors and NGOs, to make aid inclusive involved using decision-making procedures that derived from the generic 'community engagement' tool kit of the humanitarian sector. This involved using focus group discussions, 'representative' communal meetings, workshops, etc. While I have seen positive benefits from some of these supply-side efforts to increase transparency and inclusiveness, I have also seen waste and alienation. By-passing local governance structures runs the risk of delegitimizing the delicate balance of power at the village level, turning local leaders from potential partners and assets into adversaries in a contest most aid actors don't realize they are engaged in. Furthermore, some of the best examples of capacity building I have seen were the result not of NGO programs, but rather the experiences that local leaders obtained when given significant responsibility for managing reconstruction projects.

4.3 Formal and Customary Legal Systems

In addition to governance structures, external organizations in Aceh had to contend with the presence of multiple, parallel systems of legal authority. Aceh is loosely governed by a complicated network of federal (which were mostly ignored) and provincial laws, village level *adat* (traditional customary laws), and Islamic law (Shari'a), all of which were variably articulated, understood and implemented within Aceh (Daly et al. 2016; Feener 2012, 2014). This legal patchwork caused significant problems for many involved in the relief and reconstruction processes because many of the operational procedures of NGOs are contingent upon the availability of a local legal apparatus that they can easily plug into. The legal aspects of aid are often understated, but play important roles in shaping the outcomes of aid operations, as donors and organizations need to abide by international standards, laws in their home countries, and laws within host countries (Barber 2016; Daly 2016). Typically this is facilitated by normative state legal systems, which articulate the boundaries of what non-governmental organizations can do. Given the limited presence and legitimacy of the Indonesian government in Aceh because of the conflict, much had to be done at the provincial and local levels.

One of the main instruments for regulating society within Acehnese villages, and elsewhere in Indonesia is *adat*, or customary laws/practices (Reid 2004; Davidson and Henley 2007). This refers to the complicated arrangements of local customs and rules that influence behavior. Depending on the context, *adat* can carry a lot of weight, and is very influential at the village level in all facets of life. There are two things that make *adat* especially important in the discussion about reconstruction in Aceh. First, *adat* is a cultural production with extensive local legitimacy. It is both responsive to local circumstances, while at the same time implicitly references and reinforces the social order within *gampungs* and wider society. Secondly, a number of edicts, including law 44 dealing with the Special Status of the Aceh Province, have created the legal framework for viewing *adat* as a formal entity that can be implemented and enforced by provincial authorities. This leads to the duality in which *adat* represents the social fabric of traditional *gampung* structures in Aceh, factoring in a wide range of localized circumstances, and operates within broader provincial and national legal parameters. Within Aceh, a great many issues and disputes get resolved within communities based upon interpretations of *adat*, with recourse to the arbitration of other facets of the legal system being a last resort.

A potent mix of *adat* and Islamic jurisprudence shaped key issues that arose in the immediate aftermath of the disaster. Perhaps the most obvious area is land ownership and inheritance. Many lives were lost, in some cases the physical environment was fundamentally altered, and there was no uniform pre-existing land ownership and titling system. Since many organizations have strict policies that aid for housing can only be allocated when there was clear proof of land ownership, the entire rebuilding process needed some framework to operate within. While certainly not a new issue following disasters and conflicts, both the scale and the urgency of NGOs to rebuild quickly brought this into focus.

The Reconstruction of Land Administration Systems in Aceh (RALAS) project was established with funding from the UNDP to remedy this through an extensive program of community driven mapping, which served as the basis for issuing legally recognized land titles which could then be used to acquire homes and other forms of aid (Fitzpatrick 2008a, b, c, d). This ambitious program relied upon teams of Acehnese employing community level mapping to translate vernacular understandings of property into more formal, legal entities. Residents were deeply involved in articulating and demarcating properties within each community. In a discussion with one of the coordinators of the program, I was told that for the most part communities were really responsive, and once they were informed of the purpose of the program, got on with it pretty much on their own.

One of the keys to the success of this important and massive project was the simple recognition that people in communities both knew the local landscape, and had the capacities through their community leadership and social infrastructure to manage most of the complications that might have arisen involving disputes. Additionally, the project was flexible and allowed disputes to be settled locally on the basis of *adat*. Only a limited amount of outside assistance was needed in most cases to establish titling that was acceptable to both community members and external organizations. This approach, driven by people with extensive experience with community mapping in different areas of Indonesia, is an excellent example of a program that served to facilitate community recovery and reconstruction through extensive use of local laws and customs. It also illustrates a successful translation of local customs into formal legal structures.

However, I found the RALAS project to be an exception rather than the norm for large, donor driven projects. It was very common to hear complaints from Acehnese about the incongruence between the formal legal and accounting systems required by NGOs and donors, based upon codified procedures, due process, and contractual obligations, and the much more informal and vernacular mix of customary and religious practices that are flexible and contingent upon localized circumstances. Most of the external organizations were firmly bound by their own expectations and procedures; this was compounded by the fact that very few organizations understood either Shari'a or *adat*. Additionally, some NGOs had an institutional inability to deal with explicitly religious issues which prevented them from building religious structures.

In addition to institutional incompatibility with local legal systems, the diversity of the reconstruction actors led to ideological clashes. *Adat* and Shari'a are very powerful forces in Aceh, but many of the foreign aid workers that I interviewed were some combination of uncertain or uncomfortable when it came to dealing with them. A number of the western NGO workers whom I spoke with had objections to Islamic law, even though very few actually had a detailed understanding of what it entailed. Aside from objections about the impact on their personal lifestyles, many aid workers also viewed both *adat* and Sharia as backward and reactionary obstacles to development and the successful implementation of projects. Simply put, *adat* and Shari'a were powerful forces, essential for realizing community driven reconstruction, but they either were not taken seriously, or were seen as obstacles by a significant number of the aid community, including Indonesians from outside of Aceh. The lack of understanding and respect for the efficacy of such mechanisms showed in their absence in many of the project designs, and in the vast frustrations that many projects faced when their organizational approaches failed.

4.4 Community Mobilization and Volunteerism

> What has been the history of the concept of gotong royong as mutual assistance, that at first was only a way of calling up additionally labour during peak agricultural periods, and a way of helping each other when neighbors and kin were busy with feasts, deaths and disasters, and that now it has been brought together within the system of forced labor? (Koentjaraningrat 1974: 63 cited in Bowen 1986)

It can be useful to think about how people respond to a disaster in the absence of external aid or support. People have proven time and time again to have an amazing capacity to respond to horrible events (Bonanno 2004). In the absence of strong external intervention, this is mostly done through pre-existing community infrastructure, and fueled by variable notions of community support and contribution. Such examples are wide spread in post-disaster situations. These localized responses, as they extend into the reconstruction period, can take many forms, and include the organic and spontaneous formation of groups, the mobilization and expansion of pre-disaster organizations such as civil society, community organizations, and faith-based groups, to name but a few (Bolin and Stanford 1998; Cuoto 1989; Rich et al. 1995).

During one of my field trips to Aceh, I had a conversation with a project coordinator for a small NGO that collected enough resources from donors in his home country to build homes in one village. He told me that he arrived at the village about a year after the tsunami and told people there that his organization was able to provide all of the basic materials needed to reconstruct damaged housing, build basic furnishing, and restore community infrastructure. He told me, incredulously, how their response was to ask how much they—the village members—were going to get paid to rebuild their own homes and community. He was obviously perturbed by this response. Asking for pay to build ones own home and community speaks loudly of a profound disconnect in the reconstruction system. Rather than see this as an indication of greed, cynical opportunism or incredible audacity on the part of members of the village in question, it is important to question how this state of affairs came into being in a culture with commonly known and wide-spread social institutions and understandings of community volunteerism.

Many of the respondents in my fieldwork talked about supporting each other, and receiving support from outside their village. This came in the form of funds, materials, food, places to stay, etc. When talking about the communal nature of such responses, people commonly referred to *gotong royong*. Originally a Javanese expression, *gotong royong* has become widely integrated within Indonesian national dialogue since independence (Bowen 1986). It is largely taken to mean some form of mutual assistance, in which members of a community contribute their time and labor to communal projects. While this remains at the core of what the expression is about, Bowen (1986) makes it clear that it has many manifestations, and has become co-opted by the state and other levels of authority to legitimate forms of forced labor under the guise of tradition. While some scholars of Indonesia are reluctant to emphasis *gotong royong* as a progressive social force, it was explicitly mentioned by many of my respondents over the years.

For many in Aceh it is a cultural framework for understanding and implementing community support, based upon a notions of mutual and shared civic responsibility and volunteerism. In practical terms it is the spirit of how neighbors and communities come together for public projects, at critical points in the community's annual calendar, or during times of crisis. In large part the context created by incoming NGOs and governmental organizations steered people away from this default cultural setting. This was done through a number of different ways, including cash-for-work programs to inject liquidity at the household level and clear debris, the high (by Indonesian standards) sums paid out by some foreign organizations for labor, accommodation and office space, transportation and material; the abundance of often competing livelihood rehabilitation programs (where many of the participants were provided financial or other incentives to attend); and the ways through which the Acehnese were often sidelined from the reconstruction process. People were more likely than not removed from situations where they could play direct roles in reconstruction processes, often literally physically relocated to barracks or camps away from their home villages for extended periods of time. Furthermore, a range of artificial systems were installed to regulate the nature and extent of participation that people had in rebuilding their homes and communities. This undermined the basic dual premise that it is critical for recovery for people to be physically and meaningfully involved and in charge of their future, and also that there was a perfectly good cultural mechanism for motivating people to give of themselves for the public good.

The net result was the creation of an artificial 'aid' economy, and the erosion of deeply held cultural impulses to band together and cooperate in aspects of the clean-up and reconstruction that were within the capacities of many Acehnese survivors. Long after the reconstruction ended, many Acehnese still lament the degradation of community spirit and people's willingness to contribute freely to communal projects. They are quick to call their compatriots lazy and selfish, but also to point out how this was a product of the reconstruction. In tsunami affected areas, a culture of contribution has been partly replaced by a culture of expectation.

It is important to better understand how such 'cultural' understandings of participation, assistance and volunteerism factor into mobilizing and motivating citizen participation within relief and reconstruction processes. While it would be naïve to assume that it is purely a reflection of benevolence, the implicit cultural understandings of such concepts create an pact which encourages and sometimes socially coerces reluctant members of a community to get involved for the broader public good. The end result is that there are often impulses within communities to constructively be involved in helping out without promises of rewards for doing so. Participation and helping in times of crisis are not simply spontaneous, but rather often operate within implicitly understood cultural parameters which provide not only motivation and rationale, but also direction, focus, and ways to ensure that such responses strengthen rather than compete with long standing social structures. Aid organizations that tap into this can prevent potential dependency, maximize their resources, and support the empowerment that comes from people working together for their collective betterment.

5 Discussion

In this paper I have stressed the importance of aligning post-disaster assistance with local contexts, and argued that for outside agencies to operate effectively in local communities, they must understand and respect basic cultural infrastructure. In order to do so, it is essential to have detailed and practical knowledge about the basic mechanisms through which communities are structured, make decisions, and mobilize for communal purposes. It is impossible to talk meaningfully about community reconstruction and development without an appreciation for how communities are structured and function.

All too often, the external aid frameworks that come with large-scale post-disaster reconstructions marginalize local stakeholders, which both impedes recovery, and also limits the effective distribution of aid (Lyons 2009). While there was a strong emphasis during the Aceh reconstruction upon community consultation and participation, these were mostly carried out through mechanisms that were imported by external parties, such as enlisting facilitators, and holding focus discussion groups with arbitrarily selected community committees. Such approaches were specifically intended to ensure that all segments of Acehnese society were included within decision-making. However, many people I have spoken with in Aceh found the techniques used by NGOs confusing and lacking in the sustained personal contact that would have built up lasting relationships and trust with external parties. A foreign aid worker, arriving in a white SUV and holding a two hour focus group meeting with pre-selected 'cross-section' of the community did not register as a viable decision-making or information exchanging practice to many villagers in Aceh.

While it is common for external aid actors to be skeptical of local social and political infrastructure, I found that many organizations lacked the basic capacities and local knowledge to make such assessments. Many aid organizations lacked detailed knowledge of how Acehnese communities were structured and operated, relevant language abilities, and information about culture and history; all of which would have gone a long ways towards making the reconstruction more culturally sensitive, inclusive and effective.

A quick look at the majority of people staffing NGOs and governmental organizations in post-disaster situations makes it clear that there are structural flaws in team composition that limit local engagement (MacRae and Hodgkin 2016). With the increasing linkages between relief, reconstruction and development following disasters, it is becoming more common to find people with anthropological training working with NGOs. However, applied anthropological skills, and intimate knowledge of local cultures, language, and customs are far down the pecking order, buried beneath a mountain of economists, engineers, development specialists, and others deemed imperative for the technical aspects of responding to a disaster. As we saw in Aceh, it is also rare to see local actors working with external aid organizations in positions where they have the authority to influence project design. It will continue to be difficult for international humanitarian organizations to enact truly localized and participatory post-disaster reconstruction projects if they don't address the systemic imbalance between the rhetoric of bottom-up initiatives, and the practical top-down inability to understand how 'bottom-up' actually works.

Acknowledgements This paper results from countless field observations made over the course of the past decade in Aceh. During this time I benefited from collaboration with Yenny Rahmayati and the Aceh Heritage Community, as well as my more recent collaboration with the International Centre for Aceh and Indonesia Ocean Studies. Ezra Ho and Divya Hundlani at EOS assisted in the preparation of this manuscript. This research was supported in part by the National Research Foundation Singapore and the Singapore Ministry of Education under the Research Centres of Excellence initiative, and is EOS Research Contribution No. 166.

References

Athukorala, P.C. 2012. Indian Ocean Tsunami: Disaster, Generosity and Recovery. *Asian Economic Journal* 26 (3): 211–231.

BAPPENAS. 2005. *Indonesia: Preliminary Damage and Loss Assessment. The December 26, 2004 Natural Disaster*. http://www.unep.org/tsunami/reports/damage_assessment.pdf. Accessed 13 Jan 2017.

Barber, R. 2016. Humanitarian Assistance Following the 2011 Floods in Thailand and Cambodia: The Importance of Formal Invitations and Informal Relationships. In *Rebuilding Asia Following Natural Disasters: Approaches to Reconstruction in the Asia-Pacific Region*, ed. P. Daly and R.M. Feener, 315–338. Cambridge: Cambridge University Press.

Boano, C., and M. García. 2011. Lost in Translation? The Challenges of an Equitable Post-Disaster Reconstruction Process: Lessons from Chile. *Environmental Hazards* 10 (3–4): 293–309.

Bolin, R., and L. Stanford. 1998. The Northridge Earthquake: Community Based Approach to Unmet Recovery Reeds. *Disasters* 22 (1): 21–38.

Bonanno, G. 2004. Trauma, and Human Resilience: Have We Underestimated the Human Capacity to Thrive After Extremely Aversive Events? *American Psychologist* 59 (1): 20–28.

Bouraoui, D., and G. Lizarralde. 2013. Centralized Decision-Making, Users' Participation and Satisfaction in Post-Disaster Reconstruction: The Case of Tunisia. *International Journal of Disaster Resilience in the Built Environment* 4 (2): 145–167.

Bowen, J. 1986. On the Political Construction of Tradition: Gotong Royong in Indonesia. *Journal of Asian Studies* 45 (3): 545–561.

Bradfield, C., M.L. Wylie, and L.G. Echterling. 1989. After the Flood: The Response of Ministers to a Natural Disaster. *Sociology of Religion* 49 (4): 397–407.

Brinkerhoff, D.W., and J.M. Brinkerhoff. 2002. Governance Reforms and Failed States: Challenges and Implications. *International Review of Administrative Sciences* 68 (4): 511–531.

Brusset, E., M., Bhatt, K., Bjornestad, J., Cosgrave, A., Davies, Y., Deshmukh, J., Haleem, S., Hidalgo, Y., Immajati, R., Jayasundere, A., Mattsson, N., Mahaimin, R., Polastro, and T. Wu. 2009. *A Ripple in Development? Long Term Perspectives on the Response to the Indian Ocean Tsunami 2004*. Swedish International Development Cooperation Agency. https://www.oecd.org/derec/sweden/42911319.pdf. Accessed 13 Jan 2017.

Castillo, M.R.M. 2014. Development Projects from the Inside Out: Project Logic, Organizational Practices and Human Autonomy. *Journal of Human Development and Capabilities* 15 (1): 79–98.

Christoplos, I. 2006. *Links Between Relief, Rehabilitation and Development in the Tsunami Response*. London: Tsunami Evaluation Coalition.

Clarke, M., I. Fannany, and S. Kenny (eds.). 2010. *Post-Disaster Reconstruction: Lessons from Aceh*. London: Routledge.

Cuoto, R.A. 1989. Catastrophe and Community Empowerment: The Group Formulation of Aberfan's Survivors. *Journal of Community Psychology* 17: 236–248.

Daly, P., and C. Brassard. 2011. Aid Accountability and Participatory Approaches in Post-Disaster Housing Reconstruction. *Asian Journal of Social Science* 39 (4): 508–533.

Daly, P., R.M. Feener, and A.J. Reid (eds.). 2012. *From the Ground Up: Perspectives on Post-tsunami and Post-conflict Aceh*. Singapore: Institute of Southeast Asian Studies.

Daly, P., and Y. Rahmayati. 2012. Cultural Heritage and Community Recovery in Post-tsunami Aceh. In *From the Ground Up: Perspectives on Post-tsunami and Post-conflict Aceh*, ed. P. Daly, R.M. Feener, and A. Reid, 57–79. Singapore: Institute of Southeast Asian Studies.

Daly, P., and A. Hor. 2013. Similar Events, Different Disasters: A Comparative Assessment of the Aftermath of the 2004 Indian Ocean and the 2011 Japan Tsunamis. *Asian Journal of Environment and Disaster Management* 5 (1): 1–21.

Daly, P., R.M. Feener, M. Jauhola, and T. Thorburn. 2016. Blueprints for Change in Post-Tsunami Aceh, Indonesia. In *Rebuilding Asia Following Natural Disasters: Approaches to Reconstruction in the Asia-Pacific Region*, ed. P. Daly and R.M. Feener, 1–56. Cambridge: Cambridge University Press.

Daly, P. 2015. Embedded Wisdom or Rooted Problems? Aid workers' Perspectives on Local Social and Political Infrastructure in Post-Tsunami Aceh. *Disasters* 39 (2): 232–257.

Daly, P. 2016. Cycles of Destruction and Reconstruction: Responding to Disasters in Asia. In *Rebuilding Asia Following Natural Disasters: Approaches to Reconstruction in the Asia-Pacific Region*, ed. P. Daly and R.M. Feener, 1–56. Cambridge: Cambridge University Press.

Davidson, J., and D. Henley (eds.). 2007. *The Revival of Tradition in Indonesian Politics: The Deployment of Adat from Colonialism to Indigenism*. London: Routledge.

de Vries, M.W. 1995. Culture, Community and Catastrophe: Issues in Understanding Communities Under Difficult Conditions. In *Extreme Stress and Communities: Impact and Intervention*, ed. S.E. Hobfoll and M.W. de Vries, 375–393. Netherlands: Dordrecht.

Dixon, R., and A. McGregor. 2011. Grassroots Development and Upwards Accountabilities: Tensions in the Reconstruction of Aceh's Fishing Industry. *Development and Change* 42 (6): 1349–1377.

Doberstein, B., and H. Stager. 2013. Towards Guidelines for Post-Disaster Vulnerability Reduction in Informal Settlements. *Disasters* 37 (1): 28–47.

Edmonds, K. 2013. Beyond Good Intentions: The Structural Limitations of NGOs in Haiti. *Critical Sociology* 39 (3): 439–452.

Feener, R.M., and P. Daly. 2016. Religion and Reconstruction in the Wake of Disaster. *Asian Ethnology* 75 (1): 191–202.

Feener, R.M. 2014. *Shari'a and Social Engineering: The Implementation of Islamic Law in Contemporary Aceh, Indonesia*. Oxford: Oxford University Press.

Feener, R.M. 2012. Social Engineering Through Shari'a: Islamic Law and State-Directed da'wa in Contemporary Aceh. *Islamic Law and Society* 19: 275–311.

Fitzpatrick, D. 2008a. *Managing Conflict and Sustaining Recovery: Land Administration Reform in Tsunami-Affected Aceh*. ARI Working Paper 004, Asia Research Institute, National University of Singapore.

Fitzpatrick, D. 2008b. *Women's Rights to Land and Housing in Tsunami-Affected Aceh*, Indonesia, ARI Working Paper 003, Asia Research Institute, National University of Singapore.

Fitzpatrick, D. 2008c. *Access to Housing for Renters and Squatters in Tsunami-Affected Aceh*, Indonesia, ARI Working Paper 002, Asia Research Institute, National University of Singapore.

Fitzpatrick, D. 2008d. *Housing for the Landless: Resettlement in Tsunami-Affected Aceh*, Indonesia, ARI Working Paper 001, Asia Research Institute, National University of Singapore.

Fountain, P., and L. McLaughlin. 2016. Salvage and Salvation: Guest Editors' Introduction. *Asian Ethnology* 75 (1): 1–29.

Gaillard, J.C., E. Clavé, and I. Kelman. 2008. Wave of Peace? Tsunami Disaster Diplomacy in Aceh. *Indonesia. Geoforum* 39 (1): 511–526.

Gaillard, J.C., and P. Texier. 2010. Religions, Natural Hazards, and Disasters: An Introduction. *Religion* 40 (2): 81–84.

Gibbs, L. 1982. Community Response to an Emergency Situation: Psychological Destruction and the Love Canal. *American Journal of Community Psychology* 11: 116–125.

Gilbert, D., and D. Silvera. 1996. Overhelping. *Journal of Personality and Social Psychology* 70 (4): 678–690.

Gist, R., and B. Lubin. 1999. *Response to Disaster: Psychological, Community, and Ecological Approaches*. New York: Taylor and Francis.

Government of Indonesia. 2005. *Master Plan for the Rehabilitation and Reconstruction of the Regions and Communities of the Province of Nanggroe Aceh Darussalam and the Islands of Nias*, Province of North Sumatra. http://www.recoveryplatform.org/assets/submissions/200909020450_master_plan_for_reconstruction__government_of_indonesia__tsunami.pdf. Accessed 23 Jan 2017.

Hagelsteen, M., and P. Becker. 2013. Challenging Disparities in Capacity Development for Disaster Risk Reduction. *International Journal of Disaster Risk Reduction* 3: 4–13.

Hutchison, E. 2014. A global Politics of Pity? Disaster Imagery and the Emotional Construction of Solidarity After the 2004 Asian Tsunami. *International Political Sociology* 8 (1): 1–19.

Jha, A., J. Barenstein, P. Phelps, D. Pittet, and S. Sena. 2010. *Safer Homes*. Stronger Communities: A Handbook for Reconstructing after Natural Disasters. World Bank.

Kenny, S. 2007. Reconstruction in Aceh: Building Whose Capacity? *Community Development Journal* 42 (2): 206–221.

Kyamusugulwa, P.M. 2013. Local Ownership in Community-Driven Reconstruction in the Democratic Republic of Congo. *Community Development* 44 (3): 364–385.

Lyons, M. 2009. Building Back Better: The Large-Scale Impact of Small-Scale Approaches to reconstruction. *World Development* 37 (2): 385–398.

MacRae, G. 2008. Could the System Work Better? Scale and Local Knowledge in Humanitarian Relief. *Development in Practice* 18 (2): 190–200.

MacRae, G., and D. Hodgkin. 2016. Beyond the 2006 Yogyakarta Earthquake: From Sectors to Clusters in the International Humanitarian System. In *Rebuilding Asia Following Natural Disasters: Approaches to Reconstruction in the Asia–Pacific Region*, ed. P. Daly and R.M. Feener, 261–283. Cambridge: Cambridge University Press.

Mahdi, S. 2012. Factors Determining the Movements of Internally Displaced Persons (IDPs) in Aceh. In *From the Ground Up: Perspectives on Post-tsunami and Post-conflict Aceh*, ed. P. Daly, R.M. Feener, and A. Reid, 57–79. Singapore: Institute of Southeast Asian Studies.

McKeon, J. 2008. World Bank: Tracking Reconstruction Funds in Indonesia After the 2004 Earthquake and Tsunami. In *Data Against Natural Disasters: Establishing Effective Systems for Relief, Recovery, and Reconstruction*, ed. S. Amin and M. Goldstein, 185–232. Washington: The World Bank.

Older, M. 2015. When is too Much Money Worse than too Little? Giving, Aid, and Impact After the Indian Ocean Tsunami of 2004. In *Recovery from the Indian Ocean Tsunami*, ed. R. Shaw, 121–137. Tokyo: Springer.

Oliver-Smith, A. 1996. Anthropological Research on Hazards and Disasters. *Annual Review Anthropology* 25: 303–328.

Omer, H., and N. Alon. 1994. The Continuity Principle: A Unified Approach to Disaster and Trauma. *American Journal of Commonly Psychology* 22 (2): 273–287.

Ophiyandri, T., D. Amaratunga, C. Pathirage, and K. Keraminiyage. 2013. Critical Success Factors for Community-Based Post-disaster Housing Reconstruction Projects in the Pre-construction Stage in Indonesia. *International Journal of Disaster Resilience in the Built Environment* 4 (2): 236–249.

Oxfam. 2003. *Guidelines for Post Disaster Housing Reconstruction*. Oxfam UK. http://www.ifrc.org/PageFiles/95751/B.d.03.%20Guidelines%20for%20Post%20Disaster%20Housing%20%20version%201_OXFAM%20GB.pdf. Accessed 23 Jan 2017.

Pandya, C. 2006. Private Authority and Disaster Relief: The Cases of Post-Tsunami Aceh and Nias. *Critical Asian Studies* 38 (2): 298–308.

Rahmayati, Y. 2016a. Reframing 'Building Back Better' for Post-Disaster Housing Design: A Community Perspective. *International Journal of Disaster Resilience in the Built Environment* 7 (4): 1–13.

Rahmayati, Y. 2016b. Post-disaster Housing: Translating Socio-Cultural Findings into Usable Design Technical Inputs. *International Journal of Disaster Risk Reduction* 17: 173–184.

Reid, A. 2004. *An Indonesian Frontier: Acehnese and Other Histories of Sumatra*. Singapore: Singapore University Press.

Reid, A. (ed.). 2006. *Verandah of Violence: The Historical Background of the Aceh Problem*. Singapore: Singapore University Press.

Rich, R., M. Edelstein, M. Hallman, and A. Wandersman. 1995. Citizen Participation and Empowerment: The Case of Local Environmental Hazards. *American Journal of Community Psychology* 23 (5): 657–676.

Scott, J.C. 1985. *Weapons of the Weak: Everyday Forms of Resistance*. New Haven: Yale University Press.

Smirl, L. 2015. *Spaces of Aid: How Cars, Compounds and Hotels Shape Humanitarianism*. London: Zed Books.

Telford, J., and J. Cosgrave. 2007. The International Humanitarian System and the 2004 Indian Ocean Earthquake and Tsunamis. *Disasters* 31 (1): 1–28.

The Sphere Project. 2004. *Humanitarian Charter and Minimum Standards in Disaster Response*. Geneva: The Sphere Project.

Thorburn, C. 2007. *The Acehnese Gampong Three Years On: Assessing Local Capacity and Reconstruction Assistance in Post-tsunami Aceh*. Report on the Aceh Community Assistance Research Project (ACARP).

Thorburn, C., and B. Rochelle. 2014. *The Acehnese Gampong Ten Years On: A Post-post Tsunami Assessment*. Department of Foreign Affairs and Trade, Government of Australia; ICAIOS; Earth Observatory of Singapore; Monash University. http://artsonline.monash.edu.au/wp-content/arts-files/ges/Acehnese-Gampong-Ten-Years-On.pdf. Accessed 23 Jan 2017.

UNDP. 2010. *Evaluation of UNDP Contribution to Disaster Prevention and Recovery*. United Nations Development Programme. http://www.oecd.org/derec/undp/47871337.pdf. Accessed 23 Jan 2017.

Vale, L., and T. Campanella. 2005. *The Resilient City: How Modern Cities Recover from Disaster*. Oxford: Oxford University Press.

Von Vacano, M., and S. Schwarz. 2014. The Religious Dimension of Coping: The Roles of Cosmologies and Religious Practices. In *Cultural Psychology of Coping with Disasters*, ed. M. Zaumseil, M. Von Vacano, S. Schwarz, G.B. Sullivan, and J.E. Prawitasari-Hadiyono, 245–264. New York: Springer.

World Bank. 2000. *Working Together: The World Bank's Partnership with Civil Society*. http://documents.worldbank.org/curated/en/477131468767089339/pdf/multi-page.pdf. Accessed 23 Jan 2017.

Zeccola, P. 2011. Dividing Disasters in Aceh, Indonesia: Separatist Conflict and Tsunami, Human Rights and Humanitarianism. *Disasters* 35 (2): 308–328.

Land, Natural Resources and Sustainable Development in Aceh, Indonesia: Confronting Inequalities Through Post-tsunami and Post-conflict Recovery

Jane Dunlop

1 Introduction

Nanggroe Aceh Darussalam (Aceh), the westernmost province of the Indonesian archipelago, was the region most affected by the December 2004 Indian Ocean tsunami. The tsunami inspired one of the world's largest international relief efforts, with over USD7 billion of aid pledged to Aceh alone (Fan 2006).

Aceh had endured nearly 30 years of armed conflict when the tsunami devastated its coastline and up to five kilometres inland. The scale of the disaster, and the need for cooperation amongst diverse stakeholders in the peace and recovery effort, prompted a renewal of peace talks in the months following the disaster, which led to a historic Memorandum of Understanding between the Government of Indonesia (GoI) and Gerakan Aceh Merdeka (GAM: the Free Aceh Movement). The MoU sought to resolve the conflict by accommodating Aceh's desire for greater autonomy, allocating increased central government funding to support Aceh's development, and allowing the region a greater share of profits derived from the region's extensive natural resources.

Three years after the tsunami, and two and a half years after the signing of the MoU that had thus far brought peace to Aceh, this paper analyses Aceh from the perspective of both post-disaster and post-conflict recovery. The first section provides a historic overview which the next section draws on in order to analyze the effectiveness and appropriateness of post-tsunami and post-MoU efforts. It will emerge that while post-tsunami recovery efforts in Aceh have, in many cases, attempted to address existing inequalities by ensuring that as many citizens as possible benefit from the increase in available resources, in other cases, aid delivery has not been based on local needs and realities, in particular the 30-year conflict.

J. Dunlop (✉)
New Plymouth, New Zealand
e-mail: dunlopjane@gmail.com

S. Reddy (ed.), *The Asian Tsunami and Post-Disaster Aid*,
https://doi.org/10.1007/978-981-13-0182-7_7

2 A Background to Land and Natural Resource Rights in Aceh: Exploitation, Conflict and Natural Disaster

Aceh sits at the northern tip of the island of Sumatra, Indonesia. The province is endowed with natural resources, including rich and diverse forest and marine resources, oil, gas and minerals. Approximately 3.3 million hectares of Aceh's total of 5.5 million hectares is covered in forest, and oil and gas from Aceh accounts for 15–20% of Indonesia's total oil and gas earnings (Carbon Conservation and FFI 2007).

For over a century, profits from these natural resources have mostly flowed out of the region. Starting in 1874, Dutch colonial forces integrated what was then the Sultanate of Aceh—one of the major trading centres in the region with direct relations with the powerful Ottoman Empire—into the Netherlands East Indies, a new national and political structure centred on the capital Batavia/Jakarta (Brown 2005). After Indonesia's independence from the Dutch in 1949, Aceh continued to operate as a production base for the national economy. Exploitation of Indonesia's natural resources intensified greatly after President Suharto came to power (Resosudarmo 2005). An unequal share of profits derived from Aceh's natural resources coexisted with an Indonesian political economy that effectively marginalized Acehnese at all levels of government. The New Order regime was a highly centralized political and economic system backed up by strong authoritarian rule; there was little space for genuine participation by local government or civil society in public planning, decision-making, or the evaluation and implementation of government policy (Erawan 2007).

3 GAM's Fight for an Independent Aceh

Aceh had become peripheral to Indonesia's Jakarta-centred polity. This led GAM in 1967 to declare Aceh's independence from Indonesia. In its declaration of independence, GAM claimed that Aceh's revenue production from natural resources was 'used totally for the benefit of Java and the Javanese'.[1] A GAM-led rural insurgency consequently ensued, and a protracted, low-intensity conflict was fought in Aceh for almost 30 years (Schulze 2006).

Tenaga Negara Indonesia (TNI; the Indonesian Military) responded to GAM with a broad and ruthless campaign, which waxed and waned over three decades. TNI's campaign often focussed on Acehnese people as a whole due to difficulties

[1]Brown (2005), citing Aceh Sumatra National Liberation Front (ASNLF), 1976. For further reading on the background to the conflict, see also Reid (2006a, b). Barron and Clarke (2006) detail other antecedents to the conflict in Aceh, including strong Islamic faith and Aceh's connection to the world of the Indian Ocean rather than the Java Sea which influenced the development of a distinct Acehnese identity.

distinguishing GAM members from the general population and a widespread perception that Acehnese were intrinsically disloyal to the TNI (Schulze 2006). This caused a far-reaching social impact.

In 1998, the collapse of President Suharto's New Order regime was marked by increased political openness from the central government, and offered hope for a peaceful end to the conflict. Nationwide decentralization, or 'reformasi' as it is known in Indonesia, aimed to make local governments more responsive to local needs (Resosudarmo 2005; Erawan 2007). At the same time as decentralization was taking place across Indonesia, Law Number 44 of 1999 attempted to stem the conflict in Aceh—a conflict that the central government perceived as predominantly based on Islamic ideologies. This law allowed for the implementation of *sharia* (Islamic) law in social life and the education system, facilitated policies designed to preserve and empower customary institutions, and permitted the establishment of an institutionalized body of *ulema* (religious scholars) to advise on regional policy (Jemadu in Reid 2006a, b). An almost unanimous rejection in Aceh of Law Number 44 of 1999 sent a clear message that the conflict ran more deeply than the central government perceived.

Escalation of the security situation from 1999 onwards forced the central government to contemplate a more comprehensive solution. Law Number 18 of 2001 followed, granting Aceh unprecedented powers of self-governance and control over its natural resources. Ongoing security operations and a lack of political will impeded the successful implementation of Law 18/2001. After the devolution of special autonomy to Aceh, there was, in the central government's view, no longer any legitimate reason for GAM to be fighting; Aceh had been awarded fair opportunity to manage its own affairs and now had significant control over natural resources in the province (Miller 2006).[2] A Cessation of Hostilities Agreement (CoHA) in December 2001 failed to bring both sides together. In May 2003, the central government under the then President Megawati declared martial law in Aceh: 50,000 troops were deployed in the region and the military was placed in charge of most aspects of Acehnese affairs (McCulloch 2005).

The social, economic, and political impact of the military campaign and GAM's reaction was enormous. International human rights and humanitarian organizations were prevented from entering the province, over 5,000 people were killed, many arrested, and at least 125,000 people fled their homes or relocated because of the fighting (BRA 2008; also see Human Rights Watch 2003). Thousands fled to neighbouring countries, especially Malaysia (BRA 2008). The following passage articulates the social and political situation in Aceh immediately prior to the tsunami.

> [B]etween 1977 and 2004 there was not one successful attempt at addressing the primary causes of the conflict—economic and social inequalities, the feelings of exploitation, and the loss of dignity and of the space for political, cultural and social expression…These not

[2]Miller also notes that the central government was concerned that, 'the Acehnese would abuse their increased self governing powers to continue struggling for eventual secession' (2006, 304).

> only remained unresolved, but were compounded by secondary causes and grievances such as the brutality by the security forces, injustice, and the decline into poverty of many parts of Aceh. (Schulze 2006, 264)

By December 2004, nearly 30 years of conflict had resulted in 15,000 deaths and had taken a huge social toll (International Crisis Group [ICG] 2007). Reports of human rights violations by both TNI and GAM were increasing and, over two years of martial law and civil emergency had aggravated Aceh's poor material and economic conditions (ICG 2007).

4 The Impact of the Tsunami

Aceh was the region most devastated by the 26 December 2004 Indian Ocean tsunami, which was triggered by an earthquake measuring 9.0 on the Richter scale with a shallow epicentre only 150 km from Aceh's shoreline (Bappenas 2005). The tsunami swept debris and seawater up to 5 km inland across 800 km of Aceh's coastline (Bappenas 2005; BRR 2006). 130,000 people were killed, 93,000 declared missing, and over 500,000 displaced (Fitzpatrick 2007a, b, c).

Total damage and loss has been estimated at more than USD4.6 billion. More than 2,000 school buildings, eight hospitals and 114 health centres were damaged or destroyed, and 3,000 km of road was made impassable; 14 of 19 seaports were badly damaged, 8 of 10 airports damaged, and 120 arterial bridges destroyed; 120,000 new houses needed to be rebuilt, and 85,000 required rehabilitation, as they were rendered inhabitable by the tsunami (BRR 2006).

The International Labour Organization (ILO) estimates that more than 600,000 people lost their jobs as a result of the disaster—about a quarter of the employed population (ILO 2006). The fisheries sector, which accounted for over 130,000 jobs and provides a livelihood for almost 70% of the coastal population, suffered the greatest impact. Large employment loss also occurred in agriculture, with about one-fourth of cash crops and rice fields damaged (ILO 2006).

The tsunami also had a devastating effect on land and land-related institutions in Aceh. In summary, the tsunami:

- destroyed at least 7000 ha of land (15,000 land parcels) due to an inundation of sand, mud, and salt, or gauging;
- obscured or obliterated a large numbers of boundary markers;
- ruined most buildings of the Badan Pertanahan Nasional (BPN; the National Land Office) (Fitzpatrick 2007a, b, c);
- caused widespread loss of personal identity documents;
- damaged or destroyed almost all land-related records, including those held by individuals, local authorities and the BPN; (Fitzpatrick 2005) and
- affected the social memory of land rights due to widespread mortality and the deaths of local officials and leaders, including up to 30% of BPN staff (Oxfam 2007a).

5 The Memorandum of Understanding (MoU)

In the months immediately following the tsunami, the GoI and GAM entered into serious negotiations to resolve the conflict. The negotiations were in part brought about by the scale of the disaster and a pressing need for cooperation amongst different actors during reconstruction and recovery. The political situation in Jakarta had also evolved, with a new democratically elected government coming to power in September 2004 led by President Susilo Bambang Yudhoyono (SBY) and Vice President Kalla. The result of the peace negotiations was a historic Memorandum of Understanding (MoU) between the Government of Indonesia and the Free Aceh Movement signed in Helsinki on 15 August 2005. The MoU was described by Edward Aspinall, an expert on the history and conflict in Aceh, as "the best chance for a negotiated peace since the insurgency began" (Aspinall 2005).

6 Post-tsunami and Post-conflict—Rehabilitation and Reconstruction

In the aftermath of the severe destruction caused by the tsunami, the global community pledged an unprecedented USD7 billion to support Aceh's reconstruction. Organizations from over 40 countries committed to the recovery effort, including the United Nations, Red Cross and Red Crescent societies, development banks, intergovernmental agencies, governments, and non-government organizations (BRR 2006). In order to manage the financial assistance pledged to Aceh, and in light of the weakened capacity of the provincial Aceh government both after the disaster and the conflict, Badan Rehabilitasidan Rekonstruksi (BRR; the Reconstruction and Rehabilitation Agency) was established by presidential decree and commenced operations in Banda Aceh in May 2005. BRR's main tasks were to restore livelihoods and infrastructure and to strengthen communities by directing a coordinated, community-driven reconstruction and development programme (BRR 2006).

To facilitate the implementation of programmes aimed at post-conflict reconciliation and reintegration following the signing of the MoU, Badan Reintegrasi-Damai Aceh (BRA; the Aceh Peace Reintegration Agency) was established in February 2005. BRA's mandate is to support and coordinate post-conflict recovery and peace-building work that will foster the reintegration of ex-combatants, non-combatants and amnestied prisoners into civil society, and to undertake integrated programmes aimed at generating employment and improving livelihoods of victims (BRA 2008).

In December 2006, direct local elections were held for the posts of governor and district heads and the caretaker government stepped down. Aceh elected Irwandy Yusuf as governor, previously a member of GAM, and put 17 ex-GAM members in place as district heads (ICG 2007).

This section analyses the effectiveness of several post-conflict and post-tsunami initiatives relating to land, natural resources, and economic development. It examines the extent to which various actors have supported the capacity of the individuals and communities of Aceh to recover from the socio-economic costs imposed by both the tsunami and conflict.

6.1 Land Tenure Security

Given the severe destruction the tsunami caused to land and land-related institutions, one of the immediate priorities for recovery was to ensure land tenure security for those who were affected by the tsunami (Bappenas 2005). Secure access to land and natural resources is central to achieving both immediate and long-term, sustained, recovery after the conflict and the natural disaster. At the household level, secure access to land allows family labour to be put to productive use, and protects against hunger and poverty by meeting subsistence needs and providing opportunities for income generation. In a post-disaster or post-conflict situation where community support mechanisms have often weakened or collapsed, secure land and resource rights also play a critical role in protecting vulnerable groups, especially women and those burdened by poverty, from exploitation and abuse.[3] Secure land rights lay a foundation for long-term sustainable development and poverty reduction as collateral for loans or credit. Where land or other property rights are protected, investment in the reconstruction of homes and livelihoods is possible without risk of these improvements being later destroyed or forcibly removed (Deininger 2003). Land tenure security was important for house construction to commence with a sufficient degree of certainty.

The Reconstruction of Aceh Land Administration Systems (RALAS) project has been the primary process for confirming land ownership in Aceh after the tsunami. Designed and administered by the World Bank and implemented by BPN, RALAS is a USD28.5 million systematic land-titling project, which aimed to issue 600,000 land title certificates in tsunami-affected and adjacent areas by the end of 2008. Most estimates are that, prior to the tsunami, registered land titles only covered 20% of affected land parcels, with the remaining 80% held under *adat* (customary) ownership and subject to local systems of documentation.[4] Under RALAS, BPN

[3]See Agarwal and Panda (2005); correlating immovable property ownership and immovable property (land or a house) with both physical and psychological violence based on household surveys in Kerala (India), the paper concludes that women owning immovable property are found to face a significantly lower risk of marital violence than property less women; and UNCHS (Habitat) 1999: access to land, housing and property is one of the principle factors determining the economic and social well-being of women, especially in situations of conflict and reconstruction, when their rights are violated on a mass scale.

[4]Other estimates are much lower, for example, Bappenas (2005) estimates that only 9% of landowners in Aceh had formal land certificates before the tsunami.

provides land title certificates to land owners free of charge, irrespective of whether the landowner had a BPN land title certificate before the tsunami or not.[5] This policy was a clear attempt to implement the principle of 'building back better' in Aceh after the tsunami.

RALAS involves three main stages. The first and most successful stage involves a process of community-driven adjudication (CDA), where communities come to agreement over land boundaries in order to produce a map of their village. This bottom-up process puts Acehnese communities at the centre of the land-titling process by actively involving landowners in the demarcation of boundaries and determination of land ownership. Each landowner signs a statement of ownership that is witnessed by neighbours and approved by the *Keucik* (village head) (BPN 2005). Landowners also mark out their own land boundaries by placing monuments provided by BPN. According to an Oxfam land rights policy paper:

> While it took on a myriad of forms at the local level, community-driven adjudication generated sufficient documentation and certainty for housing providers to commence reconstruction, and avoided substantial delays by foregoing 'top-down' mechanisms for re-establishing tenurial certainty. In practice, the first document generated by community-driven adjudication—the signed statement of ownership—provided sufficient 'legal' authority for most forms of house reconstruction…While the extraordinary circumstances of disaster meant that land rights and certainty did delay reconstruction, there is no doubt that delays and disputes would have been much worse if the designers of RALAS had not incorporated community-based agreements as the basic mechanism for adjudicating land rights. (2007c, 12)[6]

The success of CDA can be attributed to high community involvement and exclusion of bureaucratic involvement, which can often slow down the process of determining land ownership, and, therefore, discourage landowners from being involved.

While CDA has been an effective process for determining land ownership at the village level, the second and third steps of RALAS have been less effective. Under the second stage of RALAS, BPN staff survey land ownership boundaries based on community land maps in order to produce an official map delineating village boundaries. Finally, land parcels are registered and ownership certificates are issued by BPN in line with the surveyed map. The processes involved in surveying and issuing land certificates has been slow. By September 2007, only 100,000 land title certificates had been distributed under RALAS, and there was no possibility that the original target of 600,000 titles would be reached by the end of 2008.[7] This would affect the land tenure security of many housing beneficiaries who would not receive

[5]'Building Back Better' has been a regular theme in discussions about Aceh's reconstruction since former President Bill Clinton, UN Special Envoy for Tsunami Recovery, presented this challenge on his first visit to Aceh in 2005.

[6]Note however, that BPN later refused to use the CDA process due to concerns over the legal validity of this process.

[7]For an explanation of why the issue of land title certificates has been slow, see Oxfam (2007c).

land ownership certificates before their housing constructor left Aceh.[8] It is important for stakeholders in Aceh to support the completion of RALAS so that all landowners are given equal opportunity to receive land title certificates free of charge.

6.2 Women's Rights to Land

Concerns have emerged over the RALAS project due to a low percentage of land certificates recorded in the name of women. Preliminary data obtained from research conducted by Oxfam in 2006 and 2007 reveals that around 68% of land title certificates issued under RALAS were recorded in men's names only, 27% in women's names only, and less than 5% in both men's and women's names (Fitzpatrick 2007a, b, c). While an absence of baseline data—a serious issue in itself—makes it impossible to determine the precise significance of these figures, a number of factors indicate that these figures may not reflect women's actual land ownership in Aceh.

Women in Aceh, as in the rest of Indonesia, enjoy full rights to own land, either in their own name or in joint ownership with one or more persons. This is confirmed by every element of Aceh's plural legal system: national legislation, *sharia* law as articulated in the Koran, and *hukumadat* (customary law). Indonesian law also recognizes the concept of joint matrimonial property, or *hartabersama*, as property acquired by a husband and wife during the course of their marriage (Harper 2006). According to research conducted by the International Development Law Organization (IDLO):

> Joint matrimonial property may consist of tangible assets or entitlements. Indonesian law prevents such property from being sold, transferred or used as collateral without the consent of both husband and wife…In the absence of a pre-nuptial agreement, Indonesian law requires that joint matrimonial property be divided equally between husband and wife, whether they separate through divorce or death. Where one spouse dies, therefore, 50% of the joint matrimonial property will be transferred to the surviving spouse, and the remaining 50% will be added to the deceased's *hartabawaan* and will be distributed among that spouse's heirs (including the surviving spouse (Harper 2006).

It is well-recognized that land titling processes can deny women their legitimate rights to land (Article 97, KHI). According to the World Bank:

> Land administration projects can influence the initial gender distribution of land rights at watershed transitions in property systems. Failure to provide wives with land rights can affect their standing in the community, their bargaining power within their households, and their sense of self-worth. It can also result in real landlessness and poverty when women are widowed, divorced, or abandoned (Bruce 2006).

[8]Three years after the tsunami, many housing providers had already left Aceh or were planning to conclude their activities in 2008. BRR planned to finish all activities before April 2009.

Research conducted in Indonesia suggests that married women in particular were at risk of losing their rights to land through processes of systematic land titling (Lindsey and Philips 2005, cited in Harper 2006). Where land rights are formalized, land that is a woman's independent property is often incorrectly recorded in the name of both husband and wife, or the husband's name only; and land that is jointly owned by the husband and wife is often recorded in the husband's name only. This inaccurate reflection of land ownership on the formal land ownership certificate places women at risk of losing rights during inheritance distribution or upon divorce.

The low number of certificates issued in both male and female names is likely to have been affected by Acehnese social norms and understandings (Lindsey and Philips 2005, cited in Harper 2006).[9] The sociocultural norm in Aceh is that men normally handle the paperwork and carry out public affairs on behalf of the family: it is considered respectful to register property in the name of the male income earner (Harper 2006). There is also a widespread misunderstanding within communities and amongst village leaders regarding the possibility of recording land in joint ownership of the husband and wife and the associated benefits of doing so (Harper 2006). In the Acehnese context, it is, therefore, clear that placing the onus on individuals and families to jointly register matrimonial property creates a significant risk that such property will be registered only in the name of a husband (Harper 2006, citing Lindsey and Philips 2005).[10]

By failing to accurately record women's rights to land during the process of formalizing land ownership, RALAS potentially denies some Acehnese women their right to own land. This is particularly significant in Aceh, where social boundaries are being reshaped after a major disaster and over 30 years of conflict. Women, an already vulnerable group, face additional risks during recovery due to a breakdown of usual community support structures (UNCHS [Habitat] 1999). Measures must be put in place to ensure that programmes aimed at increasing land tenure security do not, in fact, decrease land tenure security for a particular group, in this case, women.

Where cultural norms work against the issue of land title certificates to women who have a legal right to own the land, community legal awareness-raising and empowerment projects can be implemented in conjunction with the systematic land titling programme. The World Bank recommends '[p]ublic education, gender awareness training at all levels of the project, and using local approaches for gender inclusion…[to] encourage joint titling of land to men and women, especially where it is culturally acceptable' (World Bank 2005). As of November 2009, agencies such as the World Bank, the United Nations Development Programme (UNDP) and IDLO were implementing community legal education and awareness-raising

[9]For further analysis on the influence of custom on women's rights to land, see Agarwal (1994).

[10]Research conducted in other parts of Indonesia suggests that married women are at particular risk of losing rights to land through processes of systematic land titling.

projects in Aceh, focussing on issues relating to land, inheritance, and guardianship.[11] An opportunity exists to connect community legal awareness-raising with systematic land-titling projects in order to educate communities about their rights, protect women's rights to land during the process of land certification, and ensure the official land register of ownership accurately reflects land ownership realities on the ground.

6.3 Landless Survivors of the Tsunami

While RALAS addresses the issue of land ownership documentation, the project does not offer a policy response for landless tsunami survivors. Landless tsunami survivors can be divided into two groups: the first group is the 13,000 families requiring resettlement because the land they owned prior to the tsunami has been damaged or destroyed due to an inundation of mud, sand or salt, or because it was gouged away by the sea; the second group is of those who did not own land before the tsunami but rented it from a third party, lived with permission on other people's land (usually extended family), or illegally squatted on government land. Those of the latter group are known as 'renters and squatters' in post-tsunami Aceh.

Resettlement has been an ongoing policy issue since immediately after the tsunami. As early as April 2005, the Indonesian government's Official Master Plan for Rehabilitation and Reconstruction of Aceh and Nias (Republic of Indonesia 2005), a foundation document which aimed to guide government and non-government agencies involved in reconstruction of Aceh and Nias, stated that the government would make land available for those "whose land can no longer be used for living as a result of the disaster". The government promised anyone falling within this category a 200-m^2 plot of land with a 36-m^2 house. While the decision to move to a new location would be "completely decided by the people concerned" (that is, there would be no forced resettlement), if people did decide to move, the land would be provided "at a location stipulated by the government" (Republic of Indonesia 2005, 6–1).

After long delays, BRR's resettlement policy was finally released in May 2006, 17 months after the disaster: Regulation 20 (BRR 2006). The policy states that beneficiaries will make the final decision for resettlement site selection: however, in practice, this has not occurred. In many cases, the policy came too late as most land for resettlement was acquired during the policy vacuum that preceded the regulation. The urgent need for land to build housing also meant there was little time for effective community consultation and participation. District government or BRR

[11]In Aceh, for example, the World Bank was implementing a Mediation and Community Legal Empowerment (MCLE) programme, the UNDP had a Legal Awareness Raising Campaign (LARC) in Aceh, and the International Development Law Organization (IDLO) had two projects: Community Mediation and Legal Awareness Raising, and Raising Awareness of Women's Rights through Film.

officials have selected most resettlement sites (Fitzpatrick 2007a, b, c). According to Oxfam research:

> While some households have received land for housing from family or friends, or have purchased land themselves, sometimes with the assistance of a local or national NGO, most have been relocated to the 850 hectares of land acquired by the government for the purposes of resettlement. (Fitzpatrick 2007a, b, c)

Despite concerns over the suitability of some of the sites allocated for resettlement, three and a half years after the tsunami most beneficiaries have moved into new houses constructed on land provided by BRR. The Indonesian government deserves credit for issuing a clear policy response soon after the disaster. The policy expressed in the Master Plan (Republic of Indonesia 2005) sent a clear message that those whose land had been destroyed had rights to receive land and housing as part of the recovery effort.

Addressing the shelter needs of the second group of landless tsunami survivors (families who rented or squatted on land before the tsunami) has presented additional challenges. A restitutionary principle which predicated housing assistance on land ownership before the tsunami has been one of the major hurdles for this group of tsunami survivors, whether or not they owned a house that was destroyed by the tsunami.[12] In May 2007, Oxfam policy paper identifies renters and squatters as:

> Renters and squatters are among the poorest and most vulnerable victims of the tsunami disaster. They are disproportionately represented in the barracks. They are among the last in line for housing assistance. Yet, for over a year, they were not identified as a separate group requiring protection. No data was collected as to their needs for land, housing and livelihoods. Their land rights were not recorded or restored through community land mapping and titling programs. Until the policy efforts of recent months, they were the forgotten people in programs of return and rehousing for displaced victims. And when policymakers did turn their attention to their plight, Regulation 21/2006 gave them substantially less assistance than that made available to other victims of the disaster. (Oxfam 2007a)

Regulation 21 of 2006 provided that pre-tsunami renters and squatters who still required housing would receive a cash payment equal to 40% of the cost of constructing a basic 36-m^2 house. This payment could only be used for the purposes of land and housing.

In September 2006, two major demonstrations by some 2,000 barrack residents were held outside the BRR offices in Banda Aceh. The demonstrators demanded that BRR pay more attention to the conditions of people living in barracks and argued that Regulation 21 was not an adequate solution for the housing needs of renters and squatters (Fan 2007). With at least 70,000 people still in the barracks at the time of the protests, including some 20,000 renters and squatters, BRR publicly acknowledged that the political situation was 'far from easy' and presented a 'challenging issue' for BRR (Fan 2006).

In January 2007, after significant advocacy efforts led by the Land Task Force—made up of agencies such as Oxfam, UN-Habitat, UNDP, and Aus-Aid—BRR

---

[12]Land and housing can be owned separately under Indonesian law.

formally announced the revision of its renters and squatters policy. The new regulation gave renters and squatters clear status as beneficiaries, and guaranteed land from BRR, and the provision of housing from either BRR or an NGO or international agency. The land provided by BRR would be between 60 and 100 m^2 and would be owned by the beneficiary with a land certificate issued in their own name. A BRR house would be 21 m^2, and an NGO or international agency house would be 36 m^2

In November 2009, three years after the tsunami, over 2,000 households had been provided land and housing under the renters and squatters policy.[13] While 4,000 families still remained in the barracks, most of them renters and squatters, these families now had a certain future. Implementation of the renters and squatters policy would provide a significant improvement to the lives of this group of landless tsunami survivors, who would be given land ownership despite their not owning land before the tsunami. Not only was the policy a practical solution for those who would be left homeless when the barracks closed, it also increased equity in the division of post-tsunami resources in Aceh and guarantees land—a resource which is at the heart of reconstruction and recovery—to families who are most in need.

6.4 Grievances Over Land

A number of land disputes have surfaced since the MoU was signed. These disputes arose during previous administrations, but are emerging only now due to the safer political environment. They are a result of processes used to exploit Indonesia's natural resources during the Suharto period, when rights to resources were mainly granted to industries or companies that were close to Suharto and that played a role in supporting his regime (Resosudarmo 2005). For many years, while local resources and local land was being exploited by these companies, local communities were receiving little or no benefit from these activities (Resosudarmo 2005).

One prominent dispute that emerged in Aceh related to an oil palm and rubber concession in the early 1990s. In July 2007, thousands of residents from each of six villages in Aceh, with support from Lembaga Bantuan Hukum (LBH; the Legal Aid Foundation), protested in front of the East Aceh Parliament claiming that the company PT Bumi Flora, jointly owned by Indonesian and international investors, had wrongfully taken their land. The villagers demanded that their lands be returned and Bumi Flora be put on trial. Bumi Flora denied having seized the land, saying they paid fair compensation (World Bank 2007a, b, c).

While the company held the deeds for 8,300 ha of land to be used for palm oil and rubber plantations until 2024, the villagers claimed that 3,400 ha of this land was theirs, and was taken from them by force. According to the villagers, after they

[13]Personal correspondence with BRR, 1 April 2008.

refused the token amount of compensation being offered by PT Bumi Flora, the company used the TNI to intimidate them into accepting. If they had protested against the offer, they were at risk of being accused of being rebels (and, thus, enemy combatants), with potentially fatal implications. When a farmer's organization resisted land seizures in 1991, three leaders were murdered (World Bank 2007a, b, c).

In August 2007, eight members of LBH were named by the police as criminal suspects for distributing pamphlets about the Bumi Flora case (World Bank 2007a, b, c). This led to strong protests from civil society, human rights NGOs, conflict victims' associations, and Komite Peralihan Aceh (KPA; Committee for the Transition of Aceh), GAM's current political structure established to help in the transition of former combatants from GAM's military to a civilian life. These groups claimed that the police had "reverted to a pre-MoU repressive approach by criminalizing the advocacy of basic human rights—and by doing so they intended to divert public attention away from the land seizure case itself" (World Bank 2007a, b, c). PT Bumi Flora subsequently dropped the civil charges against LBH staff and proposed that the case be resolved through mediation and negotiation (World Bank 2007a, b, c).

In response to a request by the community, the governor of Aceh called a meeting with the affected parties, including the *Bupati* (regent) of East Aceh, and representatives from the community and PT Bumi Flora. The outcome of the meeting was that PT Bumi Flora agreed to return two hectares of land to each of 965 villagers, a total of 1,930 ha. The company also promised housing and roads to the community to replace those destroyed during the conflict.

This land dispute illustrates that it was not only tsunami-related issues that would have to be tackled in the coming years, but also those remaining from the conflict and years of authoritarian rule where citizens of Aceh were denied access to justice. The formal legal system remained very weak and the case outlined above reveals that the police were accustomed to using Suharto-era tactics that denied people's social and political rights (UNDP 2007). The challenge in coming years would be to support processes that ensure that disputes are resolved openly and transparently with respect to the rule of law. There is a legacy of mistrust in the community as a result of what they endured during the conflict (UNDP 2007).

7 Economic Recovery and Long-Term Development

In 2007, the World Bank and the Asian Development Bank (ADB) revealed that Aceh was experiencing an artificial economic bubble fuelled by post-tsunami reconstruction (World Bank 2007a, b, c; ADB 2007). They warned than once post-tsunami reconstruction is over, the region would face a risk of a structural downturn, with serious implications for employment and poverty alleviation (World Bank 2007a, b, c; ADB 2007). While tsunami reconstruction efforts have generated short-term employment opportunities, especially in the construction and service sectors, neither the USD7 billion dollars pledged to Aceh nor the improved security situation since the signing of the MoU had made a noticeable difference to employment opportunities in other sectors (World Bank 2007a, b, c; ADB 2007). The post-tsunami aid had, at best, restored people's economic well-being to pre-tsunami levels (World Bank 2007a, b, c; ADB 2007). Reconstruction efforts had only a minor impact on the agriculture sector, Aceh's main source of employment, and poverty remained widespread, with remote villages and rural areas in the interior of Aceh experiencing the highest poverty levels (World Bank 2009). According to ADB figures, the number of farmers living in poverty was increasing "at an alarming rate" (ADB 2007, 1). Over 30% of rural households lived below the poverty line (World Bank 2009). In addition to rural households, high poverty levels were also associated with large families, female-headed households, households with lower education levels, and those that work in the agricultural sector (World Bank 2009).

Areas affected only by the conflict have received little attention from the international community compared to those affected by the tsunami.[14] This artificial bifurcation of post-conflict and post-tsunami aid has resulted in the large majority of Aceh's aid resources being concentrated in areas affected by the tsunami, despite evident needs in other areas. This perpetuates disparities present before the tsunami and could be a potential source of new inequities and grievances. For example, contention between provincial and local government was high in late 2007 with a number of districts in central Aceh seeking political support to break away from Aceh and form their own province.

A central limitation of post-disaster economic development and livelihoods programmes in Aceh was their short-term planning focus and inability to commit funding for more than one year (ADB 2007). ADB has recently highlighted that, "[t]here is no long-term development plan to guide or allocate donor funds and to ensure that this expenditure is part of a long-term growth strategy that will result in a substantial increase in farmers' income and promote economic growth in the rural communities" (ADB 2007, 1). While the urgent need to ensure people's survival in the immediate aftermath of the tsunami does provide some justification, in hindsight, too much attention was concentrated on people's immediate needs for far too long. This was at the detriment of engaging in long-term planning.

Now that the initial pressure on post-tsunami rehabilitation needs was tapering off, there was greater attention being focused on issues surrounding long-term sustainable growth. Although much of the aid resources had already been utilized,

[14]A number of reasons have been given for this division. In some cases, international organizations were obligated (or at least felt obliged) to spend money in post-tsunami areas as most donations to them were received from private individuals on the basis of the tsunami campaign. It is also possible that because the conflict remained active, at least in the initial period after the tsunami, most international agencies were not permitted to work in areas that were not affected by the tsunami. Although this did not remain the case, it is likely to have made agencies wary.

Aceh's special autonomy status offered significant opportunities to tackle the province's economic needs. The MoU contemplated the promulgation of a new Law on Governing Aceh (LoGA), which was passed by the Indonesian House of Representatives on 11 July 2006 and signed by President SBY on 1 August 2006. Both the MoU and LoGA addressed one of the root causes of the conflict by establishing mechanisms to ensure that a significant share of Aceh's natural resource revenues will remain within Aceh. Under LoGA, Aceh would also receive an additional allocation from the central government's Dana Alokasi Umum (DAU; General Allocation Fund), a discretionary grant from the central government aimed at equalizing fiscal capacity of regional governments throughout Indonesia (Barron and Clarke 2006). The LoGA specified that Aceh would receive an additional 2% of the DAU for 15 years, starting in 2008, followed by an extra 1% for a further five years. The additional resources from the DAU have been estimated at IDR 4 trillion per year (over USD421 million), equal to about one-third of Aceh's annual revenue (Barron and Clarke 2006, citing World Bank 2006a). The money will be paid into a Special Autonomy Fund to be managed by the Acehnese provincial government. According to LoGA, this fund must "be used to pay for development, especially construction and maintenance of infrastructure, community economic empowerment, poverty eradication, and funding for education, social programs and health".[15] The MoU also obliges the government to provide suitable farmland, employment, or social security to ex-combatants, pardoned political prisoners, and other civilians who have suffered a demonstrable loss due to the conflict.[16]

The availability of DAU funds for 20 years provided the Government of Aceh with resources to address obligations under the MoU as well as to develop long-term plans which reflect the needs of Aceh's citizens. With the enormous injection of fiscal resources from the DAU, the provincial government had the funding needed to increase employment, decrease poverty, and improve public service delivery in the province. It was now the responsibility of the Acehnese government to develop a long-term plan targeted at the interdependent objectives of sustainable peace and poverty alleviation. Years of conflict compounded by strong rule from Jakarta, however, meant that, while state functions have been devolved, the provincial Acehnese government and local authorities had little experience in leading, or implementing, development plans.[17] The World Bank identified a number of areas where the Government of Aceh was particularly weak, including: planning and budgeting, accounting and reporting, undertaking external audits, regulation, and the management of public debts and investment (Barron and Clarke 2006; World Bank 2006a). Capacity weaknesses are one area in which the international community needs to target their support.

[15] Article 183.

[16] MoU, Art. 3.

[17]. See, for example, Erawan (2007) (discussing that while decentralization was aimed at making local governments more responsive to local needs, decentralization has had varying success throughout Indonesia; this is explained by the differing capacity of local actors to respond to political, societal and institutional conditions in their localities).

The Government of Aceh began to articulate its vision for Aceh's recovery and development. In May 2007, Governor Irwand Yusuf publicly announced a land distribution project that aims to decrease rural poverty by distributing 140,000 ha of land to poor families in Aceh. These families would be supported by the government to establish crops such as palm oil, rubber, and cocoa on their land. A month later, in June 2007, the governor declared a moratorium on logging activities in the province to protect Aceh's remaining 1.8 million hectares of natural forest. To explore new business, technology, and financing partnerships, the governor also led teams on several trade and investment missions to Malaysia, India, Europe, China, South America, and the United States.

The governor expanded on his vision to alleviate rural poverty and conserve Aceh's forests by drawing on the principles of environmental conservation and pro-poor economic development through an integrated ecosystems management approach. His vision was articulated in the Green Economic Development and Investment Strategy for Aceh Province, Indonesia ('Aceh Green'), which was officially launched during the United Nations Framework Convention on Climate Change (UNFCCC) 13th Conference of the Parties, held in Bali, December 2007. The introduction to the Aceh Green Strategy stated that: For Aceh's reconstruction efforts and the peace process to succeed, the Governor knows his policies and programs must soon generate both employment and income opportunities for all Acehnese, with a particular focus on the poor and ex-combatants, while at the same time ensuring the protection and preservation of Aceh's natural resources—particularly its extensive inland forests, watersheds and marine reserves—as key resources for sustainable economic development available for future generations.

Aceh Green integrated the themes of conservation and sustainable land use management, community development, commerce and renewable energy through the following six activities.

- Primary forest protection and management.
- Reforestation and forest restoration.
- Community forestry and agro-forestry development.
- Small-holder estate crop development.
- Public infrastructure development.
- Geothermal energy and micro-hydro energy.

Much was riding on the ability of the new government to generate employment and increase service delivery in the province. The success of Aceh Green and other development strategies required a transparent leadership that was accountable to all Acehnese people and that was capable of planning and implementing inclusive, socially responsible, and environmentally conscious policies in contrast to the inequities of the past. The governor's commitment to the principles of poverty alleviation and sustainable development, as articulated in Aceh Green, were progressive: however, there were concerns over whether Aceh Green would be planned and implemented in an equitable manner. Indonesia's negative track record in environmental protection raises questions about the capacity of the government to adhere to its own ‘green’ codes.

In essence, Aceh Green provided a platform from which different levels of government and other stakeholders, including NGOs, donors and private investors, could focus their support of Aceh’s development potential in a sustainable manner. A number of NGOs and international agencies, including Flora and Flora International (FFI), Carbon Conservation, Oxfam, and the International Finance Corporation (IFC) committed to support various aspects of Aceh Green, including: conservation, forest protection, livelihoods, climate change, strategic planning, financing, and facilitating broad stakeholder discussions. International investment bank Merrill Lynch also expressed interest in a carbon agreement (Gunther 2008). The challenge was to move on from the post-tsunami rehabilitation phase into a more holistic, long-term development framework that met the needs of all Acehnese, not just those affected by the tsunami. Poverty remained a real problem in Aceh, and the disparity of aid delivery between tsunami-affected areas and those areas only affected by the conflict could hamper the sustainability of peace.

8 Conclusion

Three years after the tsunami, Aceh was working through a complex post-conflict and post-disaster period. With the signing of a Memorandum of Understanding between the Government of Indonesia and GAM that ended 30 years of civil war, post-tsunami reconstruction of Aceh occurred alongside a historic peace process. At the heart of the conflict were disputes over rights to land and other natural resources, the emergence of a political economy that effectively marginalized Acehnese at all levels of government, and escalating impoverishment of large sections of the population (Brown 2005).

Communities in Aceh were aware that the fulfilment of promises under the MoU is crucial to their lives in the next few years (World Bank 2006a, b). With greater autonomy and an enormous increase in fiscal resources, the Acehnese government had the opportunity to alleviate poverty and improve service delivery in the region. Much depended on the ability of the new Acehnese government and other stakeholders to support long-term recovery in Aceh.

In a region where unequal access to land and other benefits from the region’s natural resources was the central motive for a secessionist conflict that lasted nearly 30 years, the attention of all actors in Aceh should be on creating and implementing inclusive development policies. A number of post-tsunami recovery efforts attempted to address inequalities in Aceh and ensure that as many citizens as possible would benefit from the increase in fiscal resources and greater control over land and natural resources in the province. For the most part however, an absence of understanding about Aceh’s conflict history and an artificial bifurcation of aid between post-tsunami and post-conflict regions threatened the sustainability of the achievements and presented immediate challenges to be policy makers and development practitioners in the province.

References

Agarwal, B. 1994. *A Field of One's Own: Gender and Land Rights in East Asia*. Cambridge: Cambridge University Press.

Agarwal, B., and P. Panda. 2005. Marital Violence, Human Development and Women's Property Status in India. *World Development* 33 (5): 823–850.

Asian Development Bank (ADB). 2007. *Aceh Agriculture Development Project 2008 to 2012: Project Proposal, April 2007, Mimeo*. Banda Aceh: ADB.

Aspinall, E. 2005. The Helsinki Agreement: A More Promising Basis for Peace in Aceh? *Policy Studies* 20: 1 (Washington, D.C.: East-West Center).

Badan Rehabilitasi and Reintegrasi Aceh Nias (BRR, Aceh and Nias Reconstruction Agency). 2006. *Aceh and Nias Two Years After the Tsunami*. Banda Aceh: BRR.

Badan Reintegrasi-Damai Aceh (BRA; Aceh Peace Reintegration Agency). 2008. *BRA Role and Strategy in Building Sustainable Peace in Aceh, Mimeo*. Banda Aceh: BRA.

Barron, P., and S. Clarke. 2006. *Decentralizing Inequality? Centre-Periphery Relations, Local Governance and Conflict in Aceh, Mimeo*. Washington, D.C.: World Bank.

Brown, G. 2005. *Horizontal Inequalities, Ethnic Separatism, and Violent Conflict: The Case of Aceh, Indonesia, Mimeo*. Jakarta: UNDP Human Development Report Office.

Bruce, J.W. 2006. Reform of Land Law in the Context of World Bank Lending. In *Land Law Reform: Achieving Development Policy Objectives*, ed. J.W. Bruce, et al. Washington, D.C.: World Bank.

Carbon Conservation and Flora and Fauna International (FFI). 2007. *Reducing Carbon Emissions from Deforestation in the Ulu Masen Ecosystem, Aceh, Indonesia: A Triple-Benefit Project Design Note for CCBA Audit, Mimeo*. Banda Aceh: Carbon Conservation and FFI.

Deininger, K. 2003. *Land Policies for Growth and Poverty Reduction: A World Bank Research Report*, 2003. Washington, D.C.: World Bank and Oxford University Press.

Erawan, I.K.T. 2007. Tracing the Progress of Local Governments Since Decentralisation. In *Indonesia: Democracy and the Promise of Good Governance*, ed. R.H. McLeod and A. MacIntyre. Singapore: Institute of South Asian Studies.

Fan, L. 2006. *The Struggle for Land Rights in Post-tsunami and Post-conflict Aceh, Indonesia*. The World Bank, Land Policies and Legal Empowerment of the Poor, November 2–3.

Fan, L. 2007. Protecting Land Rights in Post-tsunami and Post-conflict Aceh, Indonesia. In *Reducing Global Poverty: The Case for Asset Accumulation*, ed. O.N. Caroline. Moser, Washington, D.C.: Brookings Institution Press.

Fitzpatrick, D. 2005. *Restoring and Confirming Rights to Land in Tsunami-Affected Aceh, Mimeo*. Canberra: UNDP–Oxfam.

Fitzpatrick, D. 2007a. *Oxfam International Policy Paper: Access to Housing for Renters and Squatters in Tsunami-Affected Indonesia*. Singapore: Asian Research Institute.

Fitzpatrick, D. 2007b. *Oxfam International Policy Paper: Housing for the Landless: Resettlement in Tsunami-Affected Aceh, Indonesia*. Singapore: Asian Research Institute.

Fitzpatrick, D. 2007c. *Oxfam International Policy Paper: Managing Conflict and Sustaining Recovery: Land Administration Reform in Tsunami-Affected Aceh*. Singapore: Asian Research Institute.

Gunther, M, (2008). *Merrill Lynch's Carbon Bet: Why a Wall Street Firm Wants to Save a Forest in Sumatra*. Fortune Magazine, April 18.

Harper, E. 2006. *Guardianship, Inheritance and Land Law in Post-tsunami Aceh*. Banda Aceh: International Development Law Organisation.
Human Rights Watch. 2003. *Aceh Under Marital Law: Inside the Secret War, Mimeo*. New York: Human Rights Watch.
International Crisis Group, Indonesia. 2007. *How GAM Won in Aceh, Mimeo*. Jakarta/Brussels: International Crisis Group.
International Labour Organisation. 2006. *One Year in Aceh and Nias, February 2005–February 2006 Progress Report, Mimeo*. Banda Aceh/Jakarta: ILO.
Jemadu, A. 2006. Democratisation, the Indonesian Armed Forces and the Resolving of the Aceh Conflict. In *Verandah of Violence: The Background to the Aceh Problem*, ed. A. Reid. Singapore: Singapore University Press, NUS Publishing.
McCulloch, L. 2005. *Aceh: Then and Now, Mimeo*. London: Minority Rights Group International.
Miller, M.A. 2006. What's Special about Special Autonomy in Aceh? In *Verandah of Violence: The Background to the Aceh Problem*, ed. Anthony Reid. Singapore: Singapore University Press.
Reid, A. 2006a. Introduction. In *Verandah of Violence: The Background to the Aceh Problem*, ed. A. Reid, Singapore: Singapore University Press, NUS Publishing.
Reid, A. 2006b. *Verandah of Violence: The Background to the Aceh Problem*. Singapore: Singapore University Press, NUS Publishing.
Republic of Indonesia. 2005. *Master Plan for Rehabilitation and Reconstruction for the Regions and Communities of the Province of Nanggroe Aceh Darussalam and the Islands of Nias, Province of North Sumatera, Mimeo*. Jakarta: Bappenas.
Republic of Indonesia-Bappenas. 2005. *Indonesia: Preliminary Damage and Loss Assessment, The December 26, 2004 Natural Disaster, Mimeo*. Jakarta: Bappenas.
Republic of Indonesia–BRR and BPN. 2005. *Manual of Land Registration in the Affected Tsunami Areas at Nanggroe Aceh Darussalam and Sumatra Utara: Reconstruction of Aceh Land Administration System (RALAS), Mimeo*. Jakarta: BRR and BPN.
Resosudarmo, B.P. 2005. Introduction. In *The Politics and Economics of Indonesia's Natural Resources*, ed. B.P. Resosudarmo. Singapore: Institute of Southeast Asian Studies.
Schulze, K.E. 2006. Insurgency and Counter-Insurgency: Strategy and the Aceh Conflict, October 1976–May 2004. In *Verandah of Violence: The Background to the Aceh Problem*, ed. A. Reid. Singapore: Singapore University Press, NUS Publishing.
United Nations Centre for Human Settlements (UNCHS [Habitat]). 1999. *Women's Rights to Land, Housing and Property in Post-Conflict Situations and During Reconstruction, Mimeo*. UNCHS (Habitat): Nairobi.
United Nations Development Programme (UNDP). 2007. *Access to Justice in Aceh: Making the Transition to Sustainable Peace and Development in Aceh*. Jakarta: UNDP.
World Bank. 2005. *Gender and Best Practices in Land Administration Projects: A Synthesis Report*. Washington, D.C.: World Bank.
World Bank. 2006a. *Aceh Public Expenditure Analysis: Spending for Reconstruction and Poverty Reduction: Preliminary Report*. Banda Aceh/Jakarta: World Bank and BRR.
World Bank. 2006b. *Village Survey in Aceh: An Assessment of Village Infrastructure and Social Conditions*. World Bank: Banda Aceh/Jakarta.
World Bank. 2007a. *Aceh Conflict Monitoring Update, 1–30 July 2007, Mimeo*, 2007. Banda Aceh: World Bank.
World Bank. 2007b. *Aceh Economic Update, November 2007, Mimeo*, 2007. Banda Aceh/Jakarta: World Bank.
World Bank. 2007c. *Public Financial Management in Aceh: Measuring Financial Management Performance in Aceh's Local Government*. Banda Aceh/Jakarta: World Bank.
World Bank. 2009. *Aceh Poverty Assessment 2008: The Impact of the Conflict, the Tsunami and Reconstruction on Poverty in Aceh*, 2009. Jakarta/Washington, D.C.: World Bank.

Part II
Build Back Better? Relief Materials, Livelihoods, Shelters and Discontentment

The Sustainability of Humanitarian Aid: The Nicobar Islands as a Case of 'Complex Disaster'

Simron Jit Singh, Marina Fischer-Kowalski and Willi Haas

1 Introduction

"Natural"[1] disasters have been an inalienable part of human existence and are known to cause untold human suffering in a myriad of ways, whether it is the loss of lives, devastation of property, spread of disease and physical deformations, or in the decimation of entire civilizations (Fagan 1999). They occur on nearly all places on this globe, with varying magnitudes and with increasing frequency. "What is a disaster" is an ongoing debate and subject of several publications by sociologists, anthropologists, and environmental geographers (Quarantelli 1998; Oliver-Smith and Hoffmann 1999; Oliver-Smith 2002; Perry and Quarantelli 2005; Smith 2006). Since the purpose of this contribution is not to visit this debate, we adopt the UN definition that defines a disaster as, "a serious disruption of the functioning of society, causing widespread human, material and/or environmental losses which exceed the capacity of the affected society to cope using only its own resources" (UN 1992). Thus, not all adverse events may be classified as a disaster, only those

This chapter is based on a paper presented at the conference "Re-examining disaster, recovery and reconstruction: Social science perspectives on the tsunami" held in New Delhi, 14–15 January 2008. Versions of this chapter have been published in Singh (2007, 2009) and Singh and Haas (2016).

[1]There is a strong leaning in the social sciences to suggest that "there is no such thing as a natural disaster", since who lives and who dies is to a greater or lesser extent a "social calculus" (Smith 2006) and exposure to socio-economic vulnerability (Bankoff et al. 2004).

S. J. Singh (✉)
Waterloo, Canada
e-mail: simron.singh@uwaterloo.ca

M. Fischer-Kowalski · W. Haas
Institute of Social Ecology, University of Natural Resources and Life Sciences, Vienna, Austria

S. Reddy (ed.), *The Asian Tsunami and Post-Disaster Aid*,
https://doi.org/10.1007/978-981-13-0182-7_8

that affect humans, overwhelm response capacity, and legitimize the need for external help.

This chapter deals not so much with the causes, losses, and sufferings caused by the disasters themselves, but with the inappropriateness of existing structures and mechanisms organized to respond to a disaster in its aftermath. We argue that, while existing disaster response structures do well to alleviate suffering in the period immediately following a disaster (that is, rescue and immediate relief phases), they seem highly problematic when they translate to the rehabilitation (also referred to as 'recovery') and reconstruction (also referred to as 'development') phases, in effect, leading to what we term 'complex disasters' (Singh and Haas 2016).

Using the case of the Nicobar Islands after the 2004 tsunami, the paper attempts to illustrate the notion of 'complex disaster' and describe how disaster response on these islands undermined the resilience of the socio-ecological system and thereby the social, economic, and ecological sustainability. We begin by introducing the socio-ecological system of the Nicobar Islands as it was before the tsunami, followed by a more elaborate narrative on the human response in the aftermath. The fourth section provides a generic discussion on the notion of complex disaster, while the final section illustrates the same using the case study.

2 The Pre-tsunami Socioecological System of the Nicobar Islands

The Nicobar Islands are located some 1,200 km off the east coast of India, and are part of the larger Andaman and Nicobar archipelago that runs from north to south in an arc in the Bay of Bengal (Fig. 1). The 24 tropical islands of the Nicobars are not only home to an outstanding terrestrial and marine biodiversity—nearly one-tenth of which are endemic—they are also inhabited by 23,500 Nicobarese, an indigenous community (Census of India 2011).[2] Since 1956, the Government of India has afforded protection to the Nicobarese through a special legislation, the Andaman and Nicobar Protection of Aboriginal Tribes Regulation, which regulates entry to these islands. As a consequence, before the tsunami, contacts with the Nicobarese had been limited, such as with government employees and their contractors, immigrant traders, and occasional researchers. Despite the protection, the Indian government introduced and ran several welfare programs such as primary health and education, inter-island ferry service, and electrification of some villages (Singh 2003).

The Nicobarese are of mongoloid origin with their language rooted in the Austro-Asiatic family. We do not know exactly when they came to occupy these islands. Some scholars trace their origins (at least those of the northern islands), as immigrants from the Malay-Burma coast before the Christian era. Among the Nicobarese themselves, diverse myths and legends concerning their origin exist.

[2]Prior to the tsunami of 2004, the population of the Nicobarese was 26,565 (Census of India 2001).

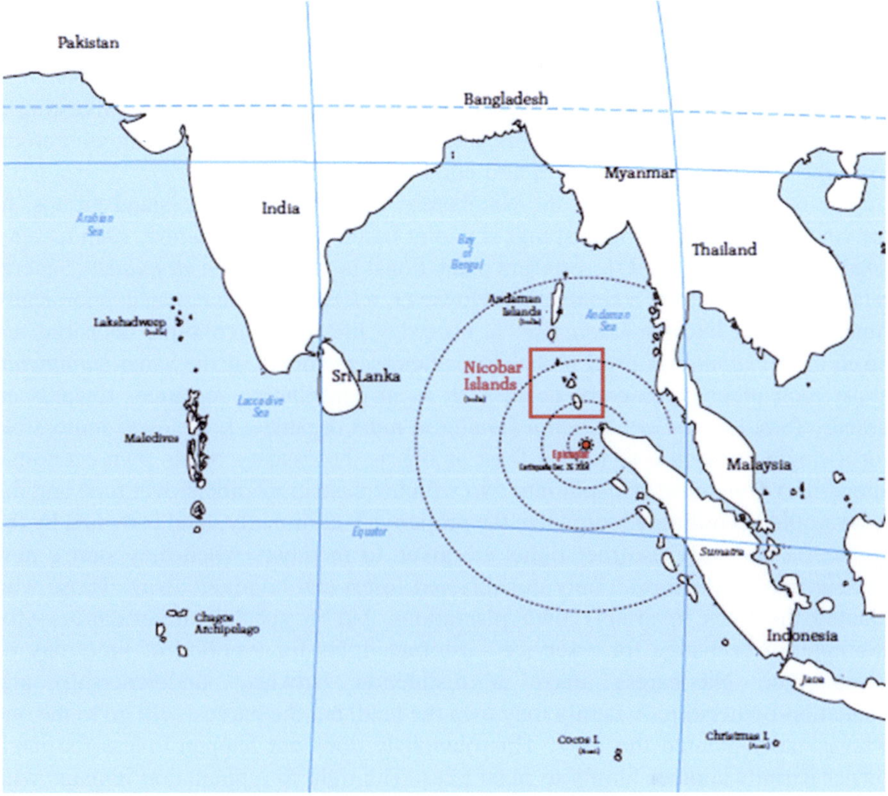

Fig. 1 Location of the Nicobar Islands *Courtesy* Ulrich Schueler

Each island has its own particular narrative and some have even more than one. Culturally, the Nicobarese show affinities to Southeast Asian cultures. But owing to the long period of isolation from the land of origin, suggested by the distinctiveness in language, one may find only traces as such. Cultural characteristics include layered cosmologies, spirit-mediumship, carved figures to attract or ward off spirits, and secondary burial of the dead. However, one finds huge variations in cultural expressions between the various islands, and equally so in the way the Nicobarese language is spoken (Singh 2003).

Prior to the tsunami, all Nicobarese villages were located along the coast, usually sheltered behind mangroves or within a bay. A typical Nicobarese dwelling is perched on stilts facing the sea with coconut palms in the background. Outrigger canoes provided for easy access to villages located either along their own coastline or on another island. The Nicobarese are largely subsistent with an economic portfolio comprising of fishing, hunting-and-gathering, pig-rearing, and producing and selling copra in lieu of rice, sugar, cloth, kerosene, and other necessities. Although their dependency on the market has increased considerably, capital accumulation was still largely absent. *Copra*[3] is made only when there is a requirement of food or other commodities from the market. A few Nicobarese maintain food gardens for growing an assortment of crops such as papaya, bananas, sugarcane, yams, pineapples, oranges, lemons, and jackfruit, while harvesting a large variety of edible leaves, tubers, and fruits from the forest, as well as seafood from the surrounding mangroves and coral reefs (Singh 2003).

The social organization of the Nicobarese varies between the island groups. In the central islands, the smallest unit is a joint family called *kamuanse*, each under a head. Ownership of land (forests and plantations) is vested in the *kamuanse*. Several related *kamuanse* form a *kamunchia*. However, a *kamunchia* is not an economically functional unit like the *kamuanse*. In everyday life, socio-economic decisions are taken at the *kamuanse* level alone. Nonetheless, members of the same *kamunchia* assist each other in times of need, such as in contributing resources towards an ossuary feast (*kinruaka*), which a *kamuanse* must organize. It is nearly impossible for a single *kamuanse* to hold a feast as big as the ossuary on its own economic strength. A *kamuanse* can split into two with increase in members over time and the head would accordingly reorganize the property. Traditionally, land is owned by the *kamuanse* and only usufruct rights are given to members when they start a new nuclear family. As a rule, only uncultivated forest can be given away. Those who planted the trees invariably own plantations, but in special circumstances—for example, in exchange for a service—usufruct rights for a period of time may be given. The Nicobarese make a distinction between landownership and plantation-ownership. A family may own the land, but the harvest will go to the one who actually planted the trees. This obviously does not happen unless the land-owner permits another family to plant trees. The right to a plantation remains with the planter until the trees are old and cannot bear fruit any more. Severe conflict may arise if someone attempts to plant coconut trees on another's land without permission. If a guilty verdict is passed, the trees have to be removed immediately. If the trees are allowed to stand there, it will be the planter who has the right to harvest the produce and not the owner of the land (Singh 2003; Singh 2006; Ramanujam et al. 2012).

On an average, economic activities put together took about an hour of a person's time each day (excluding food preparation time), leaving plenty of time for leisure and elaborate rituals and festivities, some of them lasting months. In describing the economics of hunters-and-gatherers, Sahlins (1972) and Gowdy (1997) argue the under-utilization of land, labour, and technology is a strategy to ensure an ample supply of time for leisure and social relations essential for the stability of the society. Calling them the 'original affluent society', Sahlins (1972) argues that hunter-gatherers had a very efficient system of satisfying their limited wants with extremely low time-investments leaving them with plenty of time for leisure and

[3]Copra is dried coconut meat that is used as a raw material for the extraction of coconut oil.

festivities.[4] Time-use studies among the Nicobarese has shown similar trends (Singh et al. 2001; Ringhofer et al. 2014). A good example of this would be the Ossuary feast—*Kinruaka*—celebrated for months on end to carry out the secondary burial of the deceased. Only by providing secondary burial can an heir claim legitimate ownership to the property once owned by his deceased ancestor. In doing so, the inheritor expresses gratitude to the deceased for having planted the coconuts on which they now can live. Strictly speaking, until the *Kinruaka*, the inheritor does not have the right to harvest a single coconut from the plantation of the deceased, except for the purpose of organizing the ceremony. Organizing a *Kinruaka*, hence, is no trivial task. It requires years of accumulation and social support to be able to do so. It reveals the strength of the new owner, both in economic as well as social terms, the latter manifested in his will to uphold the traditions. Upon success, the organizer is held in high esteem; his views are honoured, and he is spoken of favourably in the society. On the contrary, those who have not yet been successful may face ridicule, especially in times of dispute (Singh 2006).

Traditionally, village elders, some of who were also elected as village chiefs called 'captains', handled conflicts in Nicobarese society. These village captains together form the Tribal Council of an island(s) with a chairperson. In this sense, the Tribal Council is a political body, a parliament of elected village chiefs, presided over by a Chairperson. Like a local government, the Tribal Council represents the island(s) and negotiates with external institutions (such as the Indian administration and, of late, NGOs) on the various welfare/development programmes to be realized on the islands. Only those programmes/schemes approved by the Tribal Council are eligible for implementation.[5] In the aftermath of the tsunami, the Tribal Councils were responsible for overseeing the distribution of aid and supplies. Given such an important role, a major limitation of the Tribal Council had generally been its inability to handle external negotiations, as these elders lacked educational qualifications and the language to communicate with officials, and were often not aware of the advantages or disadvantages of government schemes proposed to them. The need was felt for a Tribal Council that would be comprised of educated, articulate, and enterprising young leaders who could lead the islands towards effective development by taking advantage of some of the benefits being offered by the government. Consequently, in the elections of 2003, younger leaders, who were articulate and at least partially educated, replaced most of the older leaders. While this enhanced the performance of the captain's role as communicators and negotiators with the outside world, it did not come without cost. In effect, this meant electing leaders who lacked experience in traditional knowledge, wisdom, and the

[4]One example is that of the! Kung Bushmen where Lee (1969) estimates their working time to be 2 h 9 min per capita/day, for food appropriation and preparation, and the remaining time is spent in other pursuits such as resting in camp or visiting other camps.

[5]The Tribal Council has the same status as that of Panchayati Raj Institutions (PRI) of local self-governance, across India. The PRIs are empowered to decide on the use of allocated budget for their villages.

history of land ownership, a limitation that proved extremely problematic when dealing with some of the post-tsunami challenges.

Thus, we see that the various segments of the Nicobar socio-ecological system are inextricably linked with, and dependent upon, each other. Social organization is closely related to property ownership (economy) and access to resources (ecology), as are rituals, ceremonies, norms, and values (culture). Together they serve to reproduce the system in terms of power-relations, hierarchies, social functions, economic stability, ecological sustainability, and general wellbeing of the population. Any higher-level intervention or changes in the framework conditions will trigger changes in one or more aspects of the Nicobarese society, the consequence of which could vary from a slight irritation to a breakdown of the entire socio-ecological system, depending upon the system's capacity to resist change.

3 The Tsunami and the Humanitarian Response in the Nicobar Islands

Many will remember the 2004 tsunami for its enormous devastation to life and property. For the Nicobarese, the disaster distinguishes their sense of time from the 'then' and 'now'. It is clear that life will never resemble how it used to be before the tsunami in any of its sociocultural and economic complexities. The islands being so close to the epicentre, large-scale destruction was inevitable. For several minutes, the earthquake shook the islands, followed by eight consecutive tsunami waves that caused the most devastation. Being topographically flat, the islands and the people were easy victim to the waves. In some places, the tsunami swept the islands from one end to the other. When the disaster subsided, some of the islands had sunk almost two metres from their original level with very little signs of human settlement. Fallen trees covered the sea around, while the coast was clogged with smashed corals and debris from houses and other infrastructure. Where the villages once stood, there was only water. Large parts of the islands were not only deformed but, in some cases, were also reduced in size by a third. Trinket Island was broken into three parts with the ocean cutting through from one end to the other at low-lying places.

In a matter of minutes, the tsunami had destroyed entire villages and cultural artefacts, with 3,449 Nicobarese either dead or missing. In addition, there was tremendous loss to material property, including more than 120,000 livestock and 6,000 ha of coconut plantations (Sekhsaria 2009), and cultural artefacts (some of them hundreds of years old). The destruction of mangrove and coral ecosystems, which had been the main source of (protein-rich) seafood, were lost to the ocean, creating a new coastline, making navigation difficult. In short, life had changed entirely for the Nicobarese in a few minutes.

The national and international response to the tsunami was overwhelming. Approximately ten billion dollars were donated or pledged in what is termed as the world's largest fund-raising exercise (Jayasuriya and McCawley 2008). People donated generously in cash and kind. In a matter of weeks, the Nicobarese were swamped with a variety of goods, some of which they could use, while others they had never seen before or had no use for. Besides the wide distribution of 'relief' material, the government ensured a constant food supply to the several relief camps on each island.

Confronted for the first time with the idea of aid and development, the Nicobarese found it difficult to grasp the dynamics. Under the Protection of Aboriginal Tribes Regulation of 1956, entry into the Nicobars had been highly regulated, and so was the Nicobarese' interaction with the outside world. Now, for the first time, they were approached by large donor organizations with large volumes of aid packages. At first, the Nicobarese were not impressed. Unable to work and rebuild their lives, they were extremely agitated and suffocated in the relief camps that were set up for them. "Leave us alone. We can manage on our own. We don't need biscuits and chips. We need to make our homes and plant our gardens. Give us tools, if you wish to help us", is what some had begun to say. Some even believed that outside interference and non-indigenous settlers had caused the tsunami. "This is our land. Please leave us alone. Otherwise we are sure to die" was the remark of a leader from Katchal Island.

But aid continued. The event was much too big and the international media attention high to leave the Nicobarese alone. A series of temporary and permanent rehabilitation programmes were launched by the government to bring relief to the affected population. These included cash compensation for loss of lives to next of kin, damage to property, intensification of social welfare programmes, cash for work schemes in agriculture and rebuilding of villages, loans for entrepreneurs, intensification of welfare programmes, free food supply, measures to revive the local economy, and long-term infrastructure development. The question of location of intermediate and permanent shelters became a crucial one, as was the question of design. Following coastal security regulations, the new settlements were to be built on higher hinterlands, in several cases, a couple of kilometres from the shore. This induced discussion with landowners as well as discontent over being located far from the shore. The sea had always been an integral part of the Nicobarese's being in this world, in all its social, cultural, and economic complexity. But the law took its course, and for security against future calamities, land surveys and construction programmes were initiated.

Following the rescue operations, all inhabitants of Chowra and Bompooka Islands were moved to the nearby island of Teressa in relief camps, as these locations were rendered unsafe. Having spent several unhappy months at the camps, the inmates were eager to return to their island and start a new life. Kephus, the Chief of Bompooka Island, was unable to understand why he could not return to his island despite the fact that he and his people had clear plans upon returning. After building their shelters, they would establish horticultural gardens with bananas, pineapples, jackfruits, and various other fruits and vegetables. Unfortunately, what stopped Kephus and his people was the lack of boats and the permission to leave the relief camp on Teressa Island. The same intention was true of the people of

Chowra. Yet, in the 18 months of 'exile' on Teressa, the persistent Chowrites built about a hundred small canoes and ten festive ones. Further, the men made regular voyages to Chowra in these canoes to plant their orchards well before the first monsoons. Within a month of their return, the Chowrites had repaired their houses and cleaned the debris. In fact, it was the only island that actually resembled its past condition. The anxiety to begin a new life and fend for themselves, despite the recent trauma, reflects the resilience of the Nicobarese in the face of tragedy, along with the ideology that life must go on, with aid or no aid.

Chowra was certainly an exception. In general, the scramble for relief and exotic goods, the lack of work, filled with anxiety to begin a new life, and stress in the cramped environment of the relief camps triggered new dynamics among the inmates. Inter-generational conflicts, questions of leadership (several old leaders had died), the redistribution of land and resources, and a restructuring of the former joint family system started to take place. The situation got even more complicated when the tribal councils were confronted with differences in opinions among the various indigenous leaders on the issue of which path to opt for in the wake of opportunities offered by the government and various donor organizations. While the elders preferred those options that were close to the traditional lifestyles, the younger leaders opted for new alternatives. The discussion was endless: whether they should give up sea transport altogether and take to roads; concrete housing instead of traditional huts; land-based economy versus off-land labour and employment; traditional joint family unit or nuclear families for receiving compensation.

The stakes were high, and directed at the very foundation of their sociocultural framework. While, on the one hand, modern options were seen as a sign of development, economic sustainability of these options was doubted. Aid money would not flow incessantly, and eventually the local economy was to maintain it all. With most coconut trees destroyed, and, thereby, the economy, the question of their subsistence was still open. At the same time, rapid cultural transformation was seen as hazardous. The disintegration of the traditional joint family system would have implications on the entire socio-economic structure, as it was directly linked to land and resource distribution, and, thereby, social control (Ramanujam et al. 2012).

3.1 *The Role of Aid Organizations*

Dozens of local, national, and international aid organizations had established their offices in Port Blair, the capital city of the region. With no permission to go to the Nicobar Islands, most catered to the few relief camps in town. Despite the fact that the government was providing food to the tsunami victims, the aid organizations contributed with add-ons such as cheese, soups, noodles, chicken, chocolates and cola (Abid 2006; Abid, this volume). Competition among the various NGOs was evident with each wanting to give more than the other. With such good care and no work to engage in, a number of residents took to alcohol, triggering conflicts and harassment of women in the camps. It was reported that not even the female police constables were spared. Local people noticed the huge volumes of food and commodities the aid organizations purchased at high prices (not all without a hint for commissions), the wastefulness of food supplies rotting in warehouses due to mismanagement, the exorbitant salaries of the staff from the mainland, and the extravagant lifestyle of aid workers in the best local hotels. It did not take long for the tsunami victims and the local population to receive an impression that aid organizations had enormous amounts of money, and their biggest problem was how to get rid of it. In Car Nicobar, it was widely reported that NGOs liberally distributed presents to the indigenous leaders to be able to remain on the island and work there. Word went around that the needs of an aid organization were no less than those of the victims, when it came to spending. Large volumes of purchases by aid organizations caused scarcity of essential goods on the islands, as well as raised the prices of most commodities in the market, causing tremendous difficulties to the inhabitants of Port Blair not affected by the tsunami.

In the first few months, aid organizations in the Nicobars were mainly engaged in the distribution of relief materials, and in organizing a few training/capacity-building workshops. Among these were several products for which there was little use, and if they were of use, they only created conflicts for want of possession of the limited exotics: radios run with mechanical power, woollen blankets (unfit for tropical climate), saris (which the women did not know how to wear), ceiling fans (when there was no provision for electricity), bicycles (without pathways to ride them), and a variety of other consumer goods and food products alien to the local culture and conditions. Some of the interventions were even absurd. For example, an international organization imported several galvanized water tanks from Australia, each with a capacity of a million litres and installed it at the highest points in Car Nicobar, Teressa, Chowra, Katchal, Kamorta, and Nancowry. The effort took nearly a year but nobody questioned where the water would come from. Currently, these large tanks are relics of a self-fulfilling act with little consideration of the needs of the affected population or of the donor's money.

In general, most of the interventions in the Nicobar Islands were state-driven, and aid organizations were not permitted to work directly, except through or in cooperation with the government. This happened mainly in the form of contribution to housing and infrastructure costs, or by undertaking projects planned by the government. There were a few exceptions though, such as Oxfam India, which was able to strike a direct partnership with the Tribal council of the Central Nicobars. However, after a year of working together, the Tribal Council refused to renew the contract that terminated at the end of 2005. According to the Council, working with Oxfam revealed the hypocrisy of the aid organizations; that, while propagating the ideologies of participation, transparency, and accountability, they did not actually practice these themselves. Despite several reminders, only 22 of the 44 NGOs working after the relief phase provided information on their budgets and expenditures to the government and public. A Nicobarese leader once remarked: "They [NGOs] are fooling both the donors and the community. They are great actors who come, create temporary sets for shooting, like in a film, and once the shooting is over, they leave, and the sets disintegrate. The toilets made by UNICEF fell down in no time, but look at the photo of the toilets in their annual report, they look so beautiful." Abid (2006) notes: "The toilets introduced by UNICEF-OXFAM have been a virtual disaster since they did not take into account prevalent practices and water resource availability."

In most instances, it was clear that projects were 'supply-driven' rather than 'need-driven'. The large volumes of money that had been collected had to be spent, no matter how, and tangible results reported back to the donors. This 'one-size-fits-all' approach has also been criticized by Abid (2006): "These organizations are driven by their own agendas and they have heedlessly introduced new concepts, ideas, schemes, and projects without taking into account the socio-cultural milieu of the district."

With little leverage to operate freely on the Nicobar Islands, several NGOs served as watchdogs of government activities and to act as a pressure group together with the media. While important, often this role was misused by exaggerating the shortcomings of the government. Headings like 'Deadly Administration', 'India Islands relief is denied', 'Anger over two rupee tsunami aid', were splashed in the national and international media, thus raising eyebrows against the ruling government. To circumvent the ensuing assault from opposing political parties and the risk of negative public opinion, the government was placed under much pressure to overemphasize its own role in rehabilitation. Thus, large amounts of money were allocated under various schemes to infrastructure, housing, and as compensation to all victims.

3.2 The Impact of State Interventions

Soon after the tsunami, the government announced an immediate relief of Rupees 2,000 per family.[6] When local officials realized that Nicobarese live in large extended families, they suggested (in good faith indeed) that they organize themselves into nuclear units so each could be entitled to the sanctioned amount. It took quite a while to educate the Nicobarese on the concept of a nuclear family. The list, when finally ready, became the blueprint for all compensation packages that followed. Bank accounts were opened for all heads of nuclear families for the issue of cheques related to several forms of compensation. This was the beginning of the disintegration of the extended family system and of future conflicts.

In compliance with the national policy, the government announced a package of cash compensation to the next of kin for each person missing or dead in the tsunami. Another package that was offered was to compensate land and crop loss per hectare. Put together, most families received amounts up to several hundred thousands of rupees. Conflicts arose in both forms of compensation, since traditional rules did not match the Indian legal framework. For example, in the Central

[6]Approximately 70 Indian Rupees (INR) makes one US $.

Nicobars, it is the norm that the husband must go and live with his wife as *ungrung* (slave). This being so, he has no right over the wealth of his wife or her family. Now according to Indian law, the next of kin in case of the death of the wife is the husband. Without due consideration of the traditional system, cheques were issued in the husbands' names, creating conflicts. Indeed, the possibility of receiving large sums of money further spawned greed and jealousy, visible in the conflicts over who was next of kin for those dead. What's more, compensation for land and crop loss (also payable to nuclear families alone) caused the splitting of land, which was previously jointly held, thus leading to conflicts in several households. In this new situation where land had to be translated into cash compensation, the confusion between ownership and usufructs arose quite often.

Last, the generosity of the government in compensating losses with cash was a predicament in itself. Never before had the Nicobarese so much money at a time. Pre-tsunami, whenever something was required from the market, copra was made and sold or exchanged immediately for rice, sugar, cloth, or other necessities. In other words, capital accumulation among the Nicobarese was rare, if not unknown. At the same time, the concept of investment and savings for future needs is alien to the common world of the Nicobarese. Where large amounts of cash are made available at once to a society in which the concept of time is compressed to the present, there certainly were some repercussions to be expected. What happened immediately was an increase in the purchase of consumer goods such as motorbikes, TVs, DVD players, mobile phones, music systems, and junk food. The Nicobarese have a special liking for 'red alcohol' (whisky and rum), and so substantial amounts of money was used to purchase cheap 'red alcohol' (at exorbitant prices due to the fact that these bottles had to be brought into the islands illegally). Besides the damage to health, the money had been a burning hole in the tribal pocket, as the Nicobarese ended up paying two to three times the going price to immigrant traders for all commodities.

One of the schemes of the government was that of promoting community plantations (coconuts, cashew nuts, fruits, and vegetables) under the 'cash-for-work' scheme. The agriculture department requested each village to sanction a portion of land from the 'village commons' for establishing new plantations. The Nicobarese were not only provided with seeds and saplings, but also paid daily labour wages. How the harvest would be shared was not clear to anyone. Since this programme was initiated top-down—which under mainland conditions may have been successful—it was hard for Nicobarese to explain to the agriculture department that the concept of village commons does not exist on these islands. In effect, some family or the other owns every piece of land. But the plea fell on deaf ears. Eventually, the captains sanctioned some land to this programme, and in at least one case, the captain had his personal land planted at the government's cost. To the Nicobarese participating in the scheme it was only the daily wages that mattered, whosoever owned the land, as they were unable to comprehend how usufructs can be divided. This is because, as mentioned before, rights to harvest goes to one who planted the trees. In this case, there were several Nicobarese who contributed to the work at different times, and at different stages.

There was yet another problem. The availability of sufficient money provided little incentive to most Nicobarese to engage in economic activities in general. Thus, the government's 'cash-for-work' scheme suffered heavily. Of the coconut and cashew saplings that arrived, not even half were planted before the monsoons. At the end of one year, hardly a quarter of the budget allocated to revival of agriculture was used, due to the simple fact that manpower was not available. At a time when food rations were still being provided free of cost and money was lying in the bank, the willingness of the Nicobarese to work remained low.

Humanitarian aid and interventions in general had a detrimental effect on the instruments of social containment. The death of a large number of elders in the community was a great social loss. Families that received less than others expressed discontent, and began to stir up turbulences in the society. In particular, discontent by younger people who had got into the habit of alcohol was even more problematic. They realized that it was easy to make a quick buck with all the aid organizations around and the free distribution of cash by the government. Village captains were under pressure to contain this, but at times, the captains themselves were young and incompetent. Some villages, after the death of old and experienced captains, replaced their leaders with temporary captains (cynically referred to as 'tsunami captains') who were poorly informed of the history of land-use and family structures. Nevertheless, their importance had increased since they were the only ones authorized to deal with the aid organizations and the government. During land surveys, it was usually the captain who accompanied the officer and arrived at agreements. Several instances were reported when the captain demarcated parcels of land with the government, for development projects and revival of infrastructure such as school, hospital, and shelters, without consulting the community or the actual landowners. Arbitrary decisions by such leaders only aggravated the problem, especially in cases where loss of land had to be compensated with cash. Claims to land were not always fixed on the basis of actual ownership or usufruct rights, but subjectively fixed for either personal gain, or to settle an old dispute, or to favour friends and family. The temptations were simply too high and social containment too weak (Ramanujam et al. 2012).

4 What Is a 'Complex Disaster'?

In this section, we return to our original thesis briefly mentioned in the introduction, namely the inability to reconcile the logic of humanitarian aid with that of sustainable development, in effect, leading to what we term 'complex disasters'. That is to say, it is the inherent inconsistencies in the system of meaning, goals, structures, and approach that makes humanitarian aid and the goal of sustainable development incompatible.

Ever since the publication of the Brundtland Commission report (WCED 1987), 'sustainable development' has been high on the political and scientific agenda, defined as "development that meets the needs of the present without compromising the ability of future generations to meet their own needs" (WCED 1987). While such a vague definition of sustainable development allows for a broad political consensus, it drove scientists on a vigorous path to operationalize the concept over the next decades. The massive amount of scientific research that followed came to warrant the label of "sustainability science" (Kates et al. 2001), offering a variety of interpretations and approaches to sustainability by the natural and social sciences alike (Atkinson et al. 2007). Despite the lack of consensus on the understanding of sustainable development and how to go about it (see Daly 1990, 1996; Sachs 1992; Fischer-Kowalski and Haberl 1997, Kates et al. 2005), there is sufficient evidence that environmental problems are a consequence of the way humans interact with their environment. Since the 1970s, despite improvements in technologies, global material extraction has tripled while per capita material consumption has nearly doubled, with most of the increase in the non-renewable categories (UNEP 2016). In effect, a transition towards sustainability would require a fundamental shift in consumption patterns from those typical in industrialized nations to a modest level (Fischer-Kowalski and Haberl 2007; Wackernagel and Rees 1996; Haberl et al. 2016).

While such a paradigm shift has yet to occur, we find much of the developing world at close heels to their industrial forerunners. The model of "third-world" development and industrialization promoted by global neo-liberal policies remains a powerful force in reinforcing the growth of capitalism based on high consumption, a logic counter to sustainable development. This process, popularly known as 'globalization', is further invigorated by an ever-increasing number of development programmes worldwide. It is argued that hardly any part on this planet is untouched by development aid that presently totals $100 billion annually (Riddell 2007). At the same time, scholars from the field of development studies, ecological economics, and political ecology have argued that the process of development where the north aids the south has met with little success. On the contrary, it has mostly served to reinforce global inequalities via a global division of labour (both in terms of production and consumption) contrary to the goals of sustainable development (Ramalingam 2013).

Indeed, humanitarian and development aid workers are a significant global force contributing to the globalization of the local by opening up new markets, bringing about changes in the mode of production, consumption patterns, and lifestyle that has a direct bearing on the environment. Research has revealed an explicit link between increased resource flows in the course of industrialization, social development, and changes in the natural environment (Fischer-Kowalski and Haberl 2007; Haberl et al. 2016) often leading to distributional/social conflicts (Martinez-Alier et al. 2010; Muradian et al. 2012). Thus, interventions caused by development programmes are powerful enough to be considered within the discourse on global environmental change if we time and again invoke the mantra of sustainable development.

Another trend that has been observed is the increasing synergism between development aid and the agencies engaged in disaster response. At present there are about 400 disasters reported each year, a 20-fold increase in the number of disasters since the 1950s (EM-DAT).[7] Disaster affected areas are now increasingly viewed as locations where long-term development programmes could be initiated. This has two reasons. First, during a humanitarian crisis, and its aftermath, funds are not a limiting factor. The second reason relates to the phenomenon termed 'disaster capitalism' where the issue is not reconstruction, but of reshaping everything (Klein 2007). Already in the 1980s, associations between disasters and social vulnerability began to be proposed. It began to be argued that the magnitude of impact by a natural event depends very much on the social characteristics of the population being affected (Blaikie et al. 1994; Hoffman and Oliver-Smith 2002; Bankoff et al. 2004). In other words, disaster began to be seen not as an accidental event, but a consequence of structural factors such as poverty, social organization, and processes that contribute to vulnerability. International disaster management agencies and scholars agree that the key to reducing vulnerability to natural disasters is to focus on prevention, mitigation, and preparedness. There is growing consensus among practitioners that the root cause of vulnerability is poverty and underdevelopment (Hewitt 1983; Gilbert 1998; Guha-Sapir et al. 2004).[8] As a remedy, development programmes with a long-term focus on structural changes and economic growth are proposed, which in its current form remains a major driver of 'complex disasters'.

But first, what is a 'simple' (or first order) disaster? The immediate effect of the tsunami was characterized by massive destruction of the 'physical' attributes of the Nicobarese socio-ecological system: the loss of life and property (housing, cultural artefacts, livestock); the halting of the local economy due to the destruction of most coconut palms; the breakdown of communication and transport infrastructure; the loss of land caused by the sinking of the islands; and the consequent damage to mangrove and coral ecosystems that had been a main source of seafood. In short, the Nicobarese socio-ecological system was physically damaged in terms of demographic changes, loss of livestock, property, subsistence/market economy, and the local ecology. This we would refer to call as the effects of a 'simple' or first order disaster.

A complex disaster is also an external event, on top of a (simple) disaster, that affects the social system's ability to regenerate, to govern its own recovery, by interfering with its cultural, economic and political regulation. Thus, a 'complex disaster' is a state more vulnerable than what was caused by the disaster itself, and is a consequence of inappropriate human interventions by: (a) creating dependency on higher systems for continuous resource flows for its survival; (b) profoundly undermining the capacities of the system at self-recovery, or loss of community

[7]These figures must be read with caution since much of this rise can be attributed to the increase in the variety of sources used (such as, insurance companies, WHO, World Food Programme) as well as more people reporting disasters, however small they may be, due to improved communication technologies, and want of humanitarian aid. However, in the last 30 years, with acute monitoring and improved reporting mechanisms there still has been a four-fold increase in the number of recorded disasters (Guha-Sapir et al. 2004).

[8]For a review on evolving paradigms and approaches to disasters, see Khuswaha (2007, 2008).

resilience; and (c) causing fundamental changes in the environmental relations of the affected society. These changes may challenge medium and long-term ecological sustainability. Thus, in describing the notion of complex disasters we identify three interdependent aspects affecting the socio-ecological system in distress that enhances system vulnerability.

(a) **Increasing dependency on higher systems**: Since the 1980s, as part of the process of globalization and neoliberalism, large parts of the agrarian 'developing world' are increasingly being integrated within a global division of labour and the world market. Under the rubric of development, nation states have devised programmes to expedite this process by introducing a variety of services (education, medical, legal), transport infrastructure, subsidies, and fossil fuel based technologies in agriculture. While they indeed improve the quality of life to some extent, by providing access to clean water, health care, legal rights, and so on, these interventions require heavy inputs of resources from the outside to sustain these (Fischer-Kowalski et al. 2011). In other words, these economies—still largely unchanged and quintessentially retaining an agrarian mode of production—are not able to generate an income to pay for the quality of life based on increased resource flows without subsidies from outside. Over time, these societies become dependent on constant supplies, subsidies, and services to meet their needs. Humanitarian aid, if inappropriately organized, may guide the system into a similar system of dependency and vulnerability. Dependency refers to a state where the social system cannot sustain itself without the continuous flow of resources from outside the affected system's resource boundaries, while unable to generate the necessary income to pay for these resource inflows. Such a state of affairs is usually accompanied by an eventual integration into the global division of labour that might inherently be unequal, based on the unsustainable exploitation and export of natural resources (Nayak 2004).

(b) **Loss of community resilience**: Magis (2010) defines community resilience as the "existence, development, and engagement of community resources by community members to thrive in an environment characterized by change, uncertainty, unpredictability, and surprise." In other words, it is the social system's capacity to collaborate to achieve a common goal. Berkes and Ross (2013) offer an integrated concept of community resilience, combining complex adaptive system from socio-ecological resilience framework (Walker et al. 2004; Gunderson and Holling 2002; Folke et al. 2010; Folke 2006) to identify place-based social strengths and connections, triggered by agency and self-organizing from psychology of development and mental health literature (Paton and Johnston 2001; Luthar 2006). In the context of hazards and disasters, societies are able to overcome the physical losses caused by such events through community resilience, maintaining their pre-disaster social fabric, or through accepting marginal or larger change in order to survive (Gaillard 2006). The capacity of societies to reorganize themselves is seated in their existing institutional structure. Breakdown of institutions as a consequence of inappropriate interventions may result in the loss of these inherent attributes to restore and reorganize, indicating a breakdown of community resilience. A complex disaster occurs when the non-material attributes (embodied in formal and informal institutions) that normally allow any society to self-organize and stand on its own feet are compromised.

(c) **Changes in society-nature interactions**: The third variable central to the notion of complex disasters relates to *changes in society–nature interactions*, fundamental to sustainable development. It is argued that existing environmental problems are anthropogenic, owing to the way humans interact with their natural environment (Fischer-Kowalski and Weisz 1999). Here, the emphasis is on the biophysical flows that a society organizes with its environment in terms of the material and energy exchanges necessary for the reproduction and maintenance of a society. This is termed 'society's metabolism', meaning that societies organize (similar to organisms) material and energy flows with their natural environment: they extract primary resources and use them for food, machines, buildings, infrastructure, heating and many other products and finally return them, with more or less delay, in the form of wastes and emissions to their environments. The quantity and structure of matter and energy a society draws from its environment largely depends on their mode of subsistence and lifestyle, which in turn is related to technology. The higher the throughput of matter and energy per capita, the more is the impact on the environment in two ways: resource scarcity on the input side, and on the output side with respect to the capacity of the ecosystem to absorb wastes and emissions. Changes in society-nature interactions accompany changes in the social metabolism, that is, the quantity and quality of throughput required to sustain the social system.

5 Relief for the Nicobar Islands: A Case of Complex Disaster?

Let us now return to our case to see the extent to which the concept of complex disaster can be illustrated. The word 'aftermath' has its roots in agriculture, meaning 'a second mowing' of grass or crop. Indeed, the aftermath of the tsunami is characterized by a second mowing of what has survived the disaster itself. The Nicobarese are least aware that they are not only subjects of a catastrophe, but of an aftermath, a phase dominated by the aid industry that thrives on it (Klein 2007). The issue here is not reconstruction, but *reshaping* everything, using the desperation and fear caused by a catastrophe, to engage in radical social and economic engineering. The major concern is not so much the material remains, such as buildings and streets, but the immaterial attributes inherent in nearly all societies, that is the ability to reorganize and rebuild one's future. Unfortunately, the logic of aid not only underutilizes these attributes, but contributes to their systematic elimination. Instead of providing the Nicobarese with tools to reconstruct their homes as they always had done, they had been made to wait for the fulfilment of the promise of an expensive modern house, each in settlements away from the sea. The former anxiety of taking control over their lives and of rebuilding their future is now hardly

Table 1 Medium term consequences of disasters and complex disasters

Medium term outcome of A	A: disaster	B: complex disaster	Medium term outcome of B
Demographic regeneration, new leaders	Loss of lives, particularly of elders (heads of extended families)	Financial compensation of the dead to nuclear family	Introduction of nuclear family pattern, extended family dissolved
Reconstruction of traditional housing, building boats, Creating new gear	Loss of housing and equipment (tools, boats, fishing gear…)	Supply of new type of ready-made housing, centralized infrastructure (roads, water tanks, electricity…)	Creating a dependence on permanent external supplies (cement, fossil fuels, motorized transport, electrical grid…)
Gradual regrowth of coconut palms, regeneration of fish, growing vegetables for food supply	Loss of nutrition sources: land, animals and coconut palms	Supplying food and monetary income for ample access to consumer goods (from aid sources)	Creating a completely new sociometabolic pattern but no lasting means for its supply
Reconstruction of low-labour, ecologically well adapted subsistence society, gradually modernizing by education, solar electricity…		Enforced rapid sociometabolic transformation to an industrial mode without resource base—high risk of future collapse and anomie	

visible among the Nicobarese. Instead, the focus has shifted on acquiring more and more through aid, and in the continued experience of reckless spending.

We defined a complex disaster as a state more vulnerable than what it had been after a disaster, characterized by the loss of self-organizing capacities of the affected community to re-establish itself, increasing dependency on higher systems as well as changes in society–nature interactions (Table 1). Would the Nicobar Islands qualify as a case of complex disaster? In the first case, we take it that the community resilience in reorganizing itself is embodied in the various socio-political institutions that have been developed and maintained over time. In the Nicobars, the two important institutions that had contributed to the stability of the system had been the Tribal Council and the joint family system—both maintained and regulated by a set of rules, norms, and conventions—to effectively perform their function of law and order, and regulate access to land and resources. As we have seen above, both these institutions have been severely affected by post-tsunami interventions. The Tribal Council has not only lost its credibility as a legitimate political representative of the people, but also failed to contain conflicts arising from unequal distribution of food aid and cash compensation. On the contrary, members have been lured into corruption for personal gains. The joint family has lost its authority as provider of food and basic necessities for its members, since most aid benefits are now accrued to nuclear families. At the same time, most families have experienced conflicts and breakdown of kinship ties in a bid to maximize cash benefits offered by the government as compensation for those who died in the tsunami and for the loss of land.

In short, we find a breakdown of important institutional structures, together with existing norms and conventions that governed them, that had provided stability to the socio-ecological system as a whole.[9]

Second, the Nicobarese began to depend on a substantial flow of resources from outside to meet their present needs.[10] In the relief phase of the aftermath, which lasted for half a year, every Nicobarese family received clothes, tools, and an elaborate packet of household goods and consumables. Electronics, motorbikes, junk food, and alcohol were purchased from the cash compensation received by the Nicobarese for losses incurred. Free food was distributed in abundance for 5 years, so much so that substantial quantities were either found rotting in the warehouses or in the villages, or sold to traders in the black market by the village captains. The state provided housing —intermediary shelters and permanent houses—with most of the materials for construction imported from the neighbouring Andaman Islands, an estimated 200,000 MT, an eight times increase in the amount of built stocks per capita as compared to pre-tsunami figures.[11] Needless to say, the maintenance of these houses requires a ceaseless flow of materials at the expense of the Nicobarese. Already in 2009, the use of water, energy, and materials had gone up substantially from pre-tsunami figures. According to the Andaman Public Works Department (APWD), water consumption increased to 70–100 L per capita per day (as compared to the Indian average of 40 L). In calculating the Domestic Material Consumption (DMC), an indicator of material use, we found a six-fold increase: from 3.4 tons/capita/year to 24t/cap/year. Meanwhile, the Domestic Energy Consumption (DEC) doubled from 30 to 54 GJ/cap/year, with a 20-fold increase in the consumption of fossil fuels as compared to pre-tsunami figures, much of it used to produce electricity, and the remaining for motor-bikes, cooking gas, and boats (Fig. 2).

Five years of aid flow led the Nicobarese to adopt a new way of living based on a much higher consumption than before. The issue here is that hardly any of this is locally produced but must be obtained from outside as aid, subsidy, or trade. Change in lifestyle is thus accompanied by an increased dependency on resources from outside the local system. High levels of stress, together with hypertension, high blood pressure and diabeties, have often been reported among the Nicobarese, caused in large part to their inability to raise enough income, pay electricity bills, or repair their homes (Yusuf, per comm. 2016). Many have fallen back on loans and credits from migrant traders, a practice known from the times prior to the tsunami that had nearly all families indebted for years into the future (Singh 2003).

---

[9]The changing role and impacts on these institutions—the Tribal Council and the joint family system—due to the tsunami has been discussed elsewhere (Ramanujam et al. 2012).

[10]The Material and Energy flows (MEFA) in the Central Nicobars have been studied in great detail before the tsunami (see, Singh et al. 2001; Singh 2003; Singh and Grünbühel 2003; Singh and Schandl 2003).

[11]Violent protests were undertaken by inhabitants of the Andaman Islands against indiscriminate quarrying of stones for construction of permanent houses. Quarrying and stone crushing activities have led to severe air pollution, causing scarcity of water and disturbance due to the use of dynamite (Raju 2007; Ahmed 2007; Giles 2008).

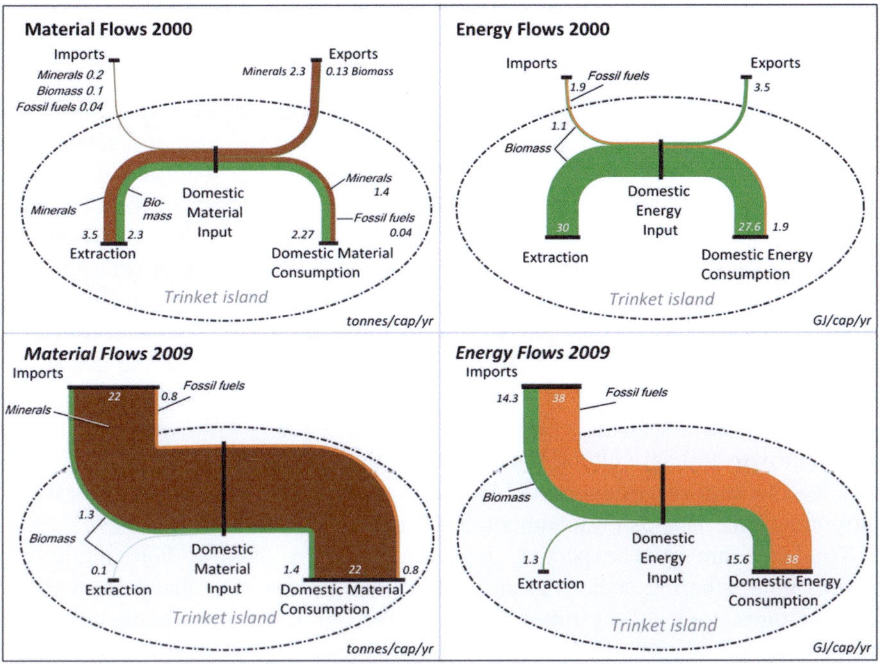

Fig. 2 Comparing material and energy flows (MEFA) before and after the tsunami Adapted from Singh and Haas (2016)

Dependency on aid and the new affluent lifestyle have consequences on the society and the environment. As a hunting-and-gathering society, producing copra for the market when there was a need for commodities did not require a disciplined investment of working time. Coconut trees, once planted, provided fruit for nearly a hundred years all the year round without much maintenance and without having to worry about seasons. The pigs scavenged the forest for three-quarters of their diet, and hunting and fishing were combined with leisure. So how will the Nicobarese sustain a newly adopted lifestyle once the effects of aid have diminished? Presumably, the only means of livelihood that is readily accessible to the Nicobarese is growing and selling a variety of vegetables, fruits, and fish to the local market. Unfortunately, very few know how to grow vegetables and fruits. Efforts were indeed made by the government to teach the Nicobarese agriculture. However, this was not very successful in most cases. It required not only the knowledge of farming, but also working with seasons and a higher investment of labour time to equal almost 6 h of work each day, a five-fold increase from pre-tsunami time-use in economic activities. Another problem is the market size. Local consumption (by non-Nicobarese) can potentially absorb only 1,000 kg of vegetables and 500 kg of fish per day. Assuming the Nicobarese do produce all of that and manage to sell it, it would still meet only 40% of the total household income required to keep present consumption levels up (Wildenberg and Singh 2012). The rest will need to come from selling copra and wage labour in construction activities.

Finally, how does a changed lifestyle affect the local environment? Modelling results indicate that to meet present household demand (of cash and subsistence) requires between 3,500 and 4,000 ha of land on Kamorta Island alone for coconut plantations, notwithstanding area needed for vegetable production. Consequently, forest and grassland would reduce by 15 and 10% respectively over the next 30 years with a high level of forest fragmentation. The combined effect of this will be a negative impact on drinking water quality and quantity, on soil erosion leading to lower productivity, on the availability of forest products and the conservation status of some of the endemic fauna and flora elements found on the island. The water situation could get even more critical if we consider the decline in water availability, with a scenario where water demand is likely to increase due to population growth and agriculture, if any. Finally, the erosion of the top soils might not only lead to land degradation but also have an undesired effect on the coral reefs surrounding the islands (Wildenberg and Singh 2012).

What we are then expecting is a sudden shift from a horticulture cum hunting-and-gathering society to an agrarian one, a transition that historically, in other cultures, took a long time. While, on the one hand, this entails changes in land-use (large amounts of land are required for coconut plantations and agriculture), at the same time, on the other hand, are serious cultural constraints in terms of willingness to work. In other words, the only way to maintain a higher consumption lifestyle would be to establish new patterns of society–nature interactions that can provide higher productivity (from land and sea) by introducing new technology, establishment of a functioning market, while at the same time coping with the 'willingness to work' constraint at the psychological and physical levels. To what extent this is possible or not is yet to be seen. In any case, the original affluence of the Nicobarese with their 'limited wants and unlimited means' seems to have been replaced by a condition of unlimited wants and insufficient means; from abundance to scarcity. Seemingly, the Nicobarese are now on a ceaseless quest to overcome a state of permanent scarcity, the starting point of all modern economic activity. For the moment we can only observe the hopelessness of the Nicobarese with their 'bourgeois impulses and Palaeolithic tools' as they attempt to pull themselves out of a complex disaster.

Acknowledgements The authors are grateful to the Austrian Science Fund (FWF) for financial support for the research in the aftermath of the tsunami (project RECOVER). The first author is particularly grateful to the Andaman and Nicobar Administration for granting research permission in the Nicobar Islands between 1999 and 2009, and for recurrent logistical support. The first author also thanks Chandra Bhushan Kumar, Ankita Mishra, Narendra Bundela, Mohammed Abid, and Ashok Meena, officers from the Indian Civil Services, for sharing their insights and discussions during the relief and rehabilitation process in the aftermath of the tsunami, although the opinions presented here are entirely that of the authors.

References

Abid, M. 2006. *Relief and Rehabilitation: Role of Civil Society Organizations*. Report submitted to the Andaman and Nicobar Administration.

Ahmed, Z. 2007. Agitation against Polluting Quarries Gain Momentum. *Light of Andamans*, 27, 9 July 2007.

Atkinson, G., Dietz, S., and E. Neumayer, (eds.). 2007. *Handbook of Sustainable Development*. Edward Elgar.

Bankoff, G., G. Frerks, and D. Hilhorst. 2004. *Mapping Vulnerability. Disasters, Development and People.*. London: Earthscan.

Berkes, F., and H. Ross. 2013. Community Resilience: Toward an Integrated Approach. *Society and Natural Resources: An International Journal* 26 (1): 5–20.

Blaikie, P., T. Cannon, and Wisner B. DavisI. 1994. *At Risk: Natural Hazards, People's Vulnerability, and Disasters*. London: Taylor and Francis.

Daly, H. 1990. Sustainable Growth: An Impossibility Theorem. *Development* 3 (4): 45–47.

Daly, H. 1996. *Beyond Growth: The Economics of Sustainable Development*. Boston: Beacon Press.

EMDAT—*The International Emergency Disasters Database. Centre for Research on the Epidemiology of Disasters*. http://www.emdat.be/.

Fagan, B. 1999. *Floods, Famines, and Empires*. New York: Basic Books.

Fischer-Kowalski, M., and H. Haberl. 1997. Tons, Joules, and Money: Modes of Production and Their Sustainability Problems. *Society and Natural Resources* 10: 61–85.

Fischer-Kowalski, Marina, and Helga Weisz. 1999. Society as Hybrid between Material and Symbolic Realms: Towards a Theoretical Framework of Society-Nature Interaction. *Advances in Human Ecology* 8: 215–251.

Fischer-Kowalski, M., and H. Haberl. 2007. *Socioecological Transitions and Global Change, Trajectories of Social Metabolism and Land Use*. Cheltenham: Edward Elgar.

Fischer-Kowalski, M., Singh, S. J., Ringhofer, E., Lauk, C., and A. Remesch. 2011. *Socio-metabolic Transitions in Indigenous Communities and the Crucial Role of Working Time. A Comparison of Case Studies*. In preparation for *Human Ecology*.

Folke, C. 2006. Resilience: The Emergence of Perspective for Social-Ecological Systems Analyses. *Global Environmental Change* 16 (3): 253–267.

Folke, C., S.R. Carpenter, B. Walker, M. Scheffer, T. Chapin, and J. Rockstrom. 2010. Resilience Thinking: Integrating Resilience, Adaptability and Transformability. *Ecology and Society* 15 (4): 20.

Gaillard, Jean-Christophe. 2006. Traditional Societies in the Face of Natural Hazards: The 1991 Mt. Pinatubo Eruption and the Aetas of the Philippines. *International Journal of Mass Emergencies and Disasters* 24 (1): 5–43.

Giles, D. 2008. Quarry and Crusher Operations vs. Human Lives. *Andaman Chronicle*, 10 July 2008.

Gilbert, C. 1998. Studying Disasters: Changes in the Main Conceptual Tools. In *What is a Disaster? Perspective on the Question*, ed. E.L. Quarantelli, 11–18. London and New York: Routledge.

Gowdy, J. 1997. *Limited Wants*. Unlimited Means.: Island Press.

Guha-Sapir, D., Hargitt, D., and P. Hoyois. 2004. *Thirty Years of Natural Disasters 1974–2003: The Numbers*. Louvain: UCL Presses Universitaires De Louvain.

Gunderson, L.H., and C.S. Holling (eds.). 2002. *Panarchy: Understanding Transformations in Human And Natural Systems*. Washington, DC: Island Press.

Haberl, H., M. Fischer-Kowalski, F. Krausmann, and V. Winiwarter (eds.). 2016. *Social Ecology: Society-Nature Relations across Time and Space*. New York: Springer.

Hewitt, K. 1983. The Idea Of Calamity In A Technocratic Age. In *Interpretations of Calamity From The Viewpoint of Human Ecology*, ed. K. Hewitt, 3–29. London and Sydney: Allen and Unwin Inc.

Hoffman, S., and A. Oliver-Smith. 2002. *Catastrophe and Culture: The Anthropology of Disaster*. Santa Fe: School for American Research Advanced Seminar Series.

Jayasuriya, S., and P. McCawley. 2008. Economics of Reconstruction: Lessons from the Post-Tsunami Experience of Indonesia, Sri Lanka and Thailand. In *Paper prepared for the International Conference on Re-Examining Disaster, Recovery and Reconstruction: Social Science Perspectives on the Tsunami*, New Delhi: Jawaharlal Nehru University, 14–15 Jan 2008.

Kates, R.W., W.C. Clark, R. Corell, J.M. Hall, et al. 2001. Sustainability Science. *Science* 292 (2001): 641–642.

Kates, R.W., T.M. Parris, and A. Leiserowitz. 2005. What is Sustainable Development? Goals, Indicators, Values, and Practice. *Environment: Science and Policy for Sustainable Development* 47 (3): 8–21.

Khuswaha, P.K. 2007. *Vulnerability and the Paradigm Shift: Debating India's Disaster Management Framework (2005)*. Unpublished Dissertation in M. Phil., Centre for Studies in Science and Policy, School of Social Science, Jawahar Lal University.

Khuswaha, P.K. 2008. *Disaster Studies and Development: A Critical Review of India's Disaster Management Framework*. JTCDM Working paper, no. 5. Mumbai: Tata Institute for Social Sciences.

Klein, N. 2007. *The Shock Doctrine: The Rise of Disaster Capitalism*. London: Allen Lane, Penguin Books.

Lee, R. 1969. Kung Bushmen Subsistence: An Input-Output Analysis. In *Environment and Culture Behaviour*, ed. A. Vayda. Garden City New York: Natural History Press.

Luthar, S. S. 2006. Resilience in Development: A Synthesis of Research across Five Decades. In *Developmental Psychopathology*, vol. 3, *Risk, Disorder, and Adaptation*, 2nd ed., ed. D. Cicchetti, and D.J. Cohen, 739–795. Hoboken, NJ: Wiley.

Magis, K. 2010. Community Resilience: An Indicator of Social Sustainability. *Society and Natural Resources* 12 (5).

Martinez-Alier, J., G. Kallis, S. Veuthey, M. Walter, and L. Temper. 2010. Social Metabolism, Ecological Distribution Conflicts, and Valuation Languages. *Ecological Economics* 70 (2): 153–158.

Muradian, R., M. Walter, and J. Martinez-Alier. 2012. Global Transformations, Social Metabolism and the Dynamics of Socio-Environmental Conflicts. *Global Environmental Change* 22 (3): 577–587.

Nayak, P.K. 2004. The Chilika Lagoon Social-Ecological System: An Historical Analysis. *Ecology and Society* 19 (1): 1.

Oliver-Smith, A., and S. Hoffmann. 1999. *The Angry Earth: Disaster in Anthropological Perspective*. London: Routledge.

Oliver-Smith, A. 2002. Theorizing Disasters. Nature, Power and Culture. In *Catastrophe and Culture. The Anthropology of Disaster.* ed. S. Hoffman, and A. Oliver-Smith. Santa Fe: School for American Research Advanced Seminar Series.

Paton, D., and D. Johnston. 2001. Disasters and Communities: Vulnerability, Resilience and Preparedness. *Disaster Prevent. Manage.* 10: 270–277.

Perry, R.W., and E.L. Quarantelli. 2005. *What is a Disaster? New Answers to Old Questions*. Bloomington, IN: International Research Committee on Disasters.

Quarantelli, E.L. 1998. *What is a Disaster? Perspectives on the Question*. London and New York: Routledge.

Ramanujam, R.V., S.J. Singh, and A. Vatn. 2012. From The Ashes Into The Fire: Institutional Change In The Post-Tsunami Central Nicobar Islands. *Society and Natural Resources* 25 (11): 1152–1166.

Ramalingam, B. 2013. *Aid on the Edge of Chaos: Rethinking International Cooperation in a Complex World*. Oxford: Oxford University Press.

Raju, G. 2007. Quarry Issue Exposes People's Messiahs. In *Light of Andamans*, 27, 9 July 2007.

Riddell, R.C. 2007. *Does Foreign Aid Really Work?*. New York: Oxford University Press.

Ringhofer, L., Singh, S. J., and M. Fischer-Kowalski. 2014. Beyond Boserup: The Role of Human Time in Agricultural Development. In *Ester Boserup's Legacy on Sustainability*, ed. Marina Fischer-Kowalski, Anette Reenberg, Anke Schaffartzik, and Andreas Mayer, vol. 4. Human-Environment Interactions Series, Springer.

Sachs, W. (ed.). 1992. *The Development Dictionary*. London: Zed Books.

Sahlins, M. 1972. *Stone-Age Economics*. Chicago: Aldine.

Sekhsaria, P. 2009. When Chanos Chanos Became Tsunami Macchi: The Post-December 2004 Scenario in the Andaman & Nicobar Islands. *Journal of the Bombay Natural History Society* 106 (3): 256–262.

Singh, S. J. 2003. *In the Sea of Influence: A World System Perspective of the Nicobar Islands*. Lund Studies in Human Ecology 6. Lund: Lund University Press.

Singh, S.J. 2006. *The Nicobar Islands: Cultural Choices in the Aftermath of the Tsunami*. Vienna: Oliver Lehmann, Czernin Verlag.

Singh, S.J. 2007. The Political Economy of Aid. A Comment on Post-Tsunami Nicobars. *Light of Andamans*, vol. 32, no. 9–10, 10 March 2007.

Singh, S. 2009. Complex Disasters: The Nicobar Islands in the Grip of Humanitarian Aid. *Geographisce Rundschau International Edition* 5 (3): 48–56.

Singh, Simron J., and Clemens M. Grünbühel. 2003. Environmental Relations and Biophysical Transitions: The Case of Trinket Island. *Geografiska Annaler* 85B (4): 187–204.

Singh, Simron J., Clemens M. Grünbühel, Heinz Schandl, and Niels B. Schulz. 2001. Social Metabolism and Labour in a Local Context: Changing Environmental Relations on Trinket Island. *Population and Environment* 23 (1): 71–104.

Singh, S.J., and W. Haas. 2016. Complex Disasters on the Nicobar Islands. In *Social Ecology: Society-Nature Relations across Time and Space*, ed. H. Haberl, M. Fischer-Kowalski, F. Krausmann, and V. Winiwarter, 523–38. New York: Springer.

Singh, Simron J., and H. Schandl. 2003. Socio-Economic Metabolism in the Nicobar Islands: Empirical Research in Society-Nature Interactions. In *Exploitation and Over-Exploitation in Societies Past and Present*, ed. B. Benzing and B. Herrmann, 169–183. Münster: LIT.

Smith, N. 2006. *There Is No Such Thing As A Natural Disaster*. Social Science Research Council–SSRC. Retrieved on 5 Feb 2017. http://blogs.ubc.ca/naturalhazards/files/2016/03/Smith-There%E2%80%99s-No-Such-Thing-as-a-Natural-Disaster.pdf.

United Nations. 1992. *Department of Humanitarian Affairs, Internationally Agreed Glossary of Basic Terms Related to Disaster Management (DNA/93/36)*. Geneva: United Nations.

UNEP. 2016. Global Material Flows and Resource Productivity. In *A Report of the Working Group on Decoupling of the International Resource Panel*. UNEP.

Wackernagel, M., and W. Rees. 1996. *Our Ecological Footprint, Reducing Human Impact on the Earth, Gabriola Island*. Philadelphia: New Society Publishers.

Walker, B., C. S., Holling, S. R., Carpenter, and A. Kinzig. 2004. Resilience, Adaptability and Transformability in Social–Ecological Systems. *Ecology and Society* 9(2):5. http://www.ecologyandsociety.org/vol9/iss2/art5/.

Wildenberg, M., and S.J. Singh. 2012. Integrated Modelling and Scenario Building for the Nicobar Islands in the Aftermath of the Tsunami. In *Human/Nature Interaction in the Anthropocene, Potentials for Socio-Ecological Systems Analysis*, ed. Glaser, B. et al. Routledge.

World Commission on Environment and Development (WCED). 1987. *World Commission on Environment and Development: Our Common Future*. New York: Oxford University Press.

The Tsunami and Its Aftermath: Resilience and Rupture of the Social Fabric Among the Nicobarese

Mohammed A. Abid

1 Introduction

The tsunami that hit the coastal belts and island territories of several countries in the Indian Ocean on 26 December 2004 caused widespread devastation in the Indian states of Tamil Nadu, Puducherry, Kerala, and Andhra Pradesh. But the worst affected were the Nicobar group of islands, located more than one thousand kilometres from the Indian mainland. These islands are inhabited by the Nicobarese, an indigenous people of Mongoloid origin (Singh 2003), who live on the 13 inhabited islands of the Nicobar Archipelago. These far-flung islands are a part of the Union Territory of Andaman and Nicobar Islands, which is directly administered by the Government of India, through the federal Ministry of Home Affairs.

The Nicobarese are isolated, pre-agricultural island communities, who have been cut off from the cultural influences of the Indian subcontinent by the treacherous Ten-Degree Channel, which separates their islands from those of the Andaman group and the mainland. However, as inhabitants of islands which bestride an ancient international sea trade route, they have not been immune to external influences. They carried on barter trade with traders and accepted new forms of clothing, ornaments, and diet (Justin 1990). These changes have, however, been voluntary and gradual, and the trend varies from island to island, depending on the extent of exposure to the 'outside world'.

After 1947, the Government of India adopted a policy framework which was aimed at allowing indigenous people to 'develop along the lines of their own genius' (Singh 1989). A number of steps were taken to protect the interests of the indigenous people of the Andaman and Nicobar Islands. The Nicobarese were placed in the category of 'Scheduled Tribes', which conferred special rights upon

M. A. Abid (✉)
School of Human Resource Management, Tata Institute of Social Sciences, Hyderabad, India
e-mail: mohammeds4@gmail.com

S. Reddy (ed.), *The Asian Tsunami and Post-Disaster Aid*,
https://doi.org/10.1007/978-981-13-0182-7_9

Table 1 Demographic impact of the disaster on the Nicobarese

S. No.	Particulars	Figures
1	Total population of A&N Islands (2001 census)	3,56,152
2	Indigenous population (2001 census)	29,469
2	Total number of people dead/missing	3,513
4	Total number of Nicobarese dead/missing	2,955

Source Basic statistics 2002–2003 directorate of economics and statistics, A&N administration; post tsunami livelihood measures by Janak Digal development commissioner A&N administration

them. The Andaman and Nicobar Protection of Aboriginal Tribes Regulation (ANPATR) was enforced in 1956, restricting the entry of non-tribals into the islands and preventing them from carrying on trade or owning property there. The government also encouraged cooperative movements among the Nicobarese and initiated a variety of development programmes, some of which saw enthusiastic participation by the local people (Singh 2003).

Admittedly, state-led development often failed to produce adequate results due to lack of 'sensitivity' on the part of government functionaries. Moreover, some of the state interventions tended to exacerbate existing inter-island inequities and created imbalances (Singh 2003). But on the whole, the Nicobarese benefited from the protective policies of the state and managed to retain their social and cultural integrity. They had full freedom to manage their own affairs, and their leadership and institutions of local self-governance received de facto recognition from the state and were consulted in all administrative matters. In fact, the Nicobarese emerged arguably as one of the few examples of the success of the 'tribal policy' pursued by the Indian State (Dhingra 2005).

The tsunami disaster had a devastating impact on the Nicobarese (see Table 1) and unleashed a host of challenges and dilemmas for this community. This chapter seeks to examine the impact of the disaster and of post-disaster interventions on the social fabric of the Nicobarese of Katchal Island. It also endeavours to understand the strategies adopted by these people to cope with the unprecedented changes in their lives. An attempt is also made to outline the issues, concerns, and lessons for state and non-state players engaged in disaster management.

2 The Island of Katchal

Katchal is an elevated, bean-shaped coral island, in the central Nicobars, spread over an area of 174 km^2. As per the 2001 Census, the island had a population of 5,312 people, about half of whom were indigenous Nicobarese. The non-tribals on

Table 2 Nicobarese population in Katchal

S. No.	Particulars	Figures
1	Total population (2001 census)	5,312
2	Nicobarese population (2001 census)	2,636
3	Nicobarese population after tsunami	1,407

Source Basic statistics 2002–2003 directorate of economics and statistics, A&N administration

the island included Sri Lankan Tamil settlers, Ranchis, Telugus, and Malayalees.[1] In the entire Nicobar group, Katchal was the worst affected island, not only in terms of loss of life and livelihoods, but also in terms of the destruction of social capital and cultural heritage of the community. The Nicobarese in Katchal lost approximately 46.6% of their population and were effectively reduced to the status of a minority group on the island in the immediate aftermath of the disaster (Table 2).

The Nicobarese population was clustered in six main settlements along the eastern, western, and northern coastlines of the island. These settlements comprise 33 Nicobarese villages as per the 2001 Census. Five of the six main settlements were completely washed away by the Tsunami waves. The 2011 census, therefore, records the area and population of these 33 villages as zero. The settlement of Upper Katchal, (East Bay) being located on elevated ground near the coast, escaped the fury of the tsunami, though it did suffer some loss of life and property. The island also suffered severe damage to its ecosystem and natural resources, resulting in complete loss of livelihoods for the surviving Nicobarese.

The pre-tsunami economy of the Nicobarese was subsistence-based and revolved around coconut plantations, which were cultivated for personal consumption as well as for the market. Areca nut plantations were also maintained for commercial export. Livelihood activities also included fishing, pig rearing and horticulture, which were practised purely for subsistence. Some of the Nicobarese held government jobs. But not a single Nicobarese was engaged in shop-keeping or industrial activity. The destruction of about 700 ha (M. S. Swaminathan research foundation 2005) of coconut plantations in Katchal deprived the Nicobarese of their only major source of income (Tables 3 and 4).

A peculiar feature of the social organization of the Nicobarese in Katchal (as on other islands) is the prevalence of the joint family system. The joint family is called the *kamuanse*, which is headed by an elder and owns all land and other resources. It is the functional unit for taking socio-economic decisions. Women enjoy a high status in this family system, and the eldest daughter of the family usually owns the

[1]As a part of the Indo–Sri Lankan agreement of 1965, 48 Tamil speaking families of Indian origin, working in plantations in Sri Lanka, were brought to the island to work on a rubber plantation set up by the government. Their population has gradually increased and was close to 600 at the time of the disaster. Ranchis is the local name given to *adivasi* people brought to the islands as forest labour before and after Independence. Many of them came to Katchal between the 1960s and 1980s to work on the rubber plantation and elsewhere.

Table 3 Damage to plantation crops in Katchal

S. No.	Particulars	Figures
1	Submerged areas in hectares	331.25
2	Reclaimable area in hectares	1297.25
3	Total Damaged area in hectares	1628.50

Source Post tsunami livelihood measures by Janak Digal development commissioner A&N administration

Table 4 Changes in land cover features in Katchal due to the tsunami

S. No.	Particulars	Area in hectares	
		Year 2004	Year 2005
1	Coconut	930	195
2	Other Vegetation	14,511	12,240
3	Mangrove	547	1
4	Sand	323	760
5	Degraded	0	1,473
6	Total	16,311	14,384

Source 'Action Plan for Development of Post-Tsunami New Andaman,' M. S. Swaminathan research foundation (2005)

property and the livestock (Singh 2003). The joint families work as a social safety net for the vulnerable—the aged, orphans, and widows. This social institution has now come under tremendous stress.

The Katchalese also suffered a huge loss of social capital. The 'chief captain' of the island along with five of the six 'first captains' of villages, who together comprised the tribal council, died in the tsunami (see Table 5). As a result, the tribal council—the island's key political institution for local governance—ceased to function. Several village elders, educated youth, and government servants—who commanded respect and influence in the community due to their traditional position, or their ability to negotiate with the 'outside world'—also lost their lives. In the vacuum thus created, the Nicobarese of Katchal had virtually no leaders who could represent them or speak on their behalf.

Table 5 Tribal council of Katchal as on 26 December 2004

S. No.	Village	First captain
1	Kapanga	Samuel Alben
2	Jhoola	Nicolas Patel[a]
3	Jhansin	Noah Jacob
4	West Bay Katchal	Mariam Suresh
5	Ponda	Lawrence
6	East Bay Katchal	Daniel

Harry Saul, chief captain and chairman, tribal council
[a]Only first captain who survived the tsunami (*Source* SRO Katchal)

All the tribal habitations, except those in Upper Katchal, were wiped out without a trace, leaving behind only the debris of trees and concrete structures, which mostly housed government infrastructure. For instance, only one stilted wooden structure (the traditional Nicobarese dwelling hut) has survived on the island. The community's public places and monuments, the churches, and mausoleums have all been destroyed. None of their artefacts—the outrigged canoes, the *kareavous* (carved wooden idols of men, animals or mythical creatures) or the *hentakoi* (carved and painted wooden figures used for rituals)—have survived. There is no documentary or visual representation of cultural events like the Ossuary feast or *Anuchoilu* (the nocturnal fishing festival), which can link the younger Nicobarese to their cultural heritage (Singh 2006). The death of a large number of elders who were repositories of traditional knowledge, skills, and oral history makes the disconnection even more severe.

3 Initial Response of the Community

Entire families were carried away by the tsunami and the survivors gradually trekked their way to the high altitude Mildera area in the late afternoon of 26 December 2004. Several of the survivors had very little clothing left on their bodies. Many of them were injured and had no clue what had happened to their near and dear ones who were still missing. They were shell-shocked, but being a quiet and rather shy people by nature, the Nicobarese did not give vent to their grief in verbal and dramatic ways. They occupied three extant school premises, which later became relief camps. The Nicobarese were provided food and succour by some unaffected non-tribal people of Mildera and Beachdera—mostly lower-level government functionaries and shopkeepers. External help of any kind was not available from any quarter for the first five days. During this period, the Nicobarese remained calm and passive recipients of help from others.

The relief camps and other arrangements were streamlined with the arrival of senior civil servants in Katchal in the first week of January. As the Special Relief Officer (SRO), the author held the first meeting with the leaders of the Nicobarese on 6 January 2005, which was attended by some captains and other leaders. They appeared completely lost and remained silent throughout the meeting, and my repeated queries about their needs and requirements elicited no response. However, I began interacting with the Nicobarese during my morning and evening inspections of the relief camps every day. Finally, in a meeting convened on 10 January 2005, which was attended by more leaders including one of the senior most surviving Captains, the Nicobarese voiced their concerns and 'demands' for the first time. Unlike the non-tribal populace, they had no complaints about the supply of food items, tents or clothing, though these arrangements were inadequate. Instead, they wanted medical attention, a temporary space for the church and some cloth for stitching garments for their priests. They also felt that the locations of their earlier habitations along the coast were unsafe, and, therefore, wanted to now live in the elevated Mildera area.

By the third week of January, the Nicobarese had started asking for tools and implements (*dao* and *batu*) to clear the land and construct houses and lay out gardens or *bagichas*. It was obvious that the community had now come out of the initial trauma, and, in fact, appeared ready with its own plan of action. They were looking for minimal support from the state to get on with their lives. In the absence of tools, the Nicobarese felt helpless and frustrated, but these could not be provided to them for the next few months.

4 Aid and Assistance by the State

Apart from setting up relief camps and providing food items and articles of daily use, the government provided basic medical facilities in the camps. Moreover, the government declared payment of various types of ex-gratia compensation to the affected persons for losses suffered due to the tsunami disaster (see Table 6). The following features characterized this package:

- The 'ex-gratia' compensation was paid in cash, which had implications for monetization of the economy and also for social relationships.
- Monetary values were attached not only to movable and immovable property, but also to human lives, the value varying according to the relationship with the survivors.
- The payments were made on the basis of the definition of the family as consisting of spouses and children.

Table 6 Types of ex gratia payments

S. No.	Category	Amount in rupees (INR)
1	LG relief fund	2,000
2	Identified dead/missing	1,00,000
3	Serious injury	5,000
4	Household articles	10,000
5	Temporary relief	3,000
6	Relief to small businesses	10,000
7	Orphans below 18 year	2,00,000
8	Widows	1,00,000
9	Unmarried girls	1,00,000
10	Payment for plantation crop loss per hectare	94,465

Source 'Living with hope…life in Nicobar post tsunami.' Vivek Powal DC Nicobar

- As per the extant CRF (Calamity Relief Fund) norms, compensation was payable to owners of property who lost the same and not to those who had lost livelihoods.

These compensation packages were announced at a time when no demand for any kind of compensation had been raised by any individual or group among the Nicobarese.

The process began with the disbursement of an amount of INR 2,000 as interim payment to each family out of the Lieutenant Governor's Relief Fund. The disbursement of this amount posed serious problems as the extended joint families of the Nicobarese had to be split into smaller nuclear units for the purpose of making payments in accordance with the norms.

In order to provide some livelihood for the affected persons, the government started engaging the local youth, on payment-basis, in various tasks associated with the relief operations, like loading and unloading of relief material, clearance of debris, and so on. In accordance with the usual procedure followed in cash-for-work programmes, muster rolls were to be maintained and payments were to be handed out directly to the workers at the end of the day. This practice also proved to be problematic in the context of the socio-economic organization of the local people.

When the youth started receiving the payments, the village elders and captains objected to the practice and demanded that the payments should instead be made to the captain or the head of the village, who would, in turn, distribute the same among all the people in accordance with their needs. This problem has to be understood in the context of the fact that in Nicobarese society, property is owned and enjoyed jointly and not individually. Nicobarese have large joint families constituting a single household and the possibility of payment to heads of households could have been explored but death of many elderly persons and financial rules made this difficult. Second, the practice of payment for relief work, helped the Nicobarese to understand that the state was obliged to compensate them for relief work, which they had hitherto considered to be their own collective responsibility. The disaster thus brought the insulated Nicobarese in direct and extensive contact with outside agencies, with concepts and norms which had not influenced their lives in a significant manner earlier. The powerful and pervasive role of the State, and modern concepts of citizen's entitlements, rights, and obligations, began to gradually seep into the consciousness of the Nicobarese.

Damage assessment was undertaken by the government in various sectors. This exercise was not devoid of complexities. The people were not in the habit of quantifying their assets and were prone to giving inaccurate details. They also tended to undervalue some of them in keeping with the traditional practice of camouflaging their prosperity. In case of many families in Katchal, it was very difficult to ascertain the extent of damage suffered because most or all members of the family had perished and there were not enough knowledgeable people among the surviving villagers. However, the community used its residual social resources and weakened networks effectively to assist the official machinery in the task of damage assessment. The fact that only two cases of disputed compensation payment came up in Katchal is an indication of this collaborative effort.

The announcement of these packages, and the disbursal of cash payments, affected the Nicobarese society in diverse ways. All of a sudden, a huge dose of capital was injected into the local economy, which operated largely on the basis of barter arrangements. The disbursals, which were driven by timelines fixed by the central government, could not be preceded or accompanied by any kind of counselling about judicious use of this wealth. As a result, there was a marked increase in consumerist tendencies. After the tsunami, many young people who were suffering from the psycho-social impact of the disaster due to the death of family members used the compensation money to indulge in alcohol and substance abuse. This affected their involvement in the process of livelihood recovery. In many cases, local non-tribal traders have exploited them, and much of their compensation has passed into the hands of non-tribals. Moreover, individuals who were now 'empowered' by cash were less susceptible to the social control of the traditional socio-political system, and, in many cases, intent on charting their own course.

The dependency-inducing nature of some of the assistance and rehabilitation support provided to the Nicobarese, tends to affect real recovery and weakens the resilience of the community. The government continued to provide free food in the form of rations to the Nicobarese for a long period after the disaster (till April 2007), The elders of the community feel that availability of these doles made the youngsters lazy, and that they neither took initiative to do something of their own nor enrolled themselves in government livelihood schemes. The nutritional status of the Nicobarese, as reflected in nutritional anthropometric data collected both before and after the disaster, has been found to be, on the whole, very satisfactory, and comparatively better than the non-tribal population (Tyagi and Prasad 2002; Brahmam et al. 2005). The Nicobarese still have adequate access to their items of traditional diet like coconut, pandanus, tubers, and marine fauna. The regeneration of livelihoods of any kind is possible only if the people have adequate incentive to work.

Similarly, it was felt that the provision of gas stoves and one-time supply of free LPG (liquefied petroleum gas) cylinders could also have long-term consequences. This has introduced the Nicobarese to a facility whose regular supply in the islands is uncertain. Cooking gas is always in short supply, even in Port Blair. In the initial period after the tsunami, many Nicobarese were buying the same at exorbitant rates since they were cash-rich at this stage. The apprehension was that, before long, they might become accustomed to the facility, which might create difficulties later. However, the supply situation appears to have improved and most families in Katchal are now reportedly using LPG supplied in the island at the actual price.

5 Role of Non-state Players

The entry of non-government organizations (NGOs) into the Nicobars remained restricted in the initial days of the post-tsunami relief operations, but soon, the administration adopted a liberal policy and allowed several NGOs to work in the district. The inaccessibility of the remotely located islands, and their rich, but complex socio-cultural contexts posed challenges to the NGOs. Most of them were either unable or unwilling to reach out to the worst-affected communities of Nicobarese on the most inaccessible islands like Katchal, Chaura, and Teresa, and tended to concentrate their efforts in district and tehsil headquarters. Moreover, many of these organizations came with fixed programmes, approaches and agendas, which were not suited to the socio-cultural milieu and could not respond to the emerging needs of the affected community.

A cult-based social service society insisted on setting up a *langar* (community kitchen) in the culturally composite relief camps of Katchal, and had to be prevented from such dependency-creating initiatives by the SRO. Some other NGOs gathered livelihood data by using survey methodology, which is not appropriate for pre-agricultural societies. Organizations which espouse community-based rehabilitation, instead of developing creative mechanisms to support traditional institutions, like the joint families, started talking of 'neglect' and 'discrimination' of orphans in these families without giving verifiable details of the same (Abid 2006).

Some organizations responded positively to the infrastructural and capacity-building needs of the community. A team from the Tata Institute of Social Sciences (TISS) stayed with the affected Nicobarese during the initial relief period, created detailed profile data of the different social groups, and trained volunteer teachers from the community. A certificate course was also started by the Institute to develop the perspectives of the Nicobarese youth and prepare them for informed and meaningful participation in the development process.

Similarly, *Save the Children* designed and constructed culturally sensitive *aanganwadis* (child development centres), which would help in the rejuvenation of the traditionally-inclusive social system. But overall, the work of non-government agencies did not have any significant impact on the socio-economic system of the Nicobarese in Katchal. While the NGOs provided a great deal of initial support to the community, their outreach remained limited due to constraints of location, budgets, timelines and understanding of the traditional milieu of the Nicobarese.

6 Housing and Habitat

The representatives of the Nicobarese along with the local officers had identified three sites for the resettlement of the Nicobarese by mid-February 2005, on the basis of the parameters of safety, accessibility, availability of land, and proximity to water

Table 7 Temporary Shelter sites in Katchal which later became Permanent Settlements

S. No.	Name of place	Altitude (approx ft.)	Minimum distance from sea (km)	Distance from main road (km)	Available area (ha)	Number of families
1	Sallo Tikry	+70	1.5	1.3	1.44	122
2	Sallo Tikry (Wimco Block)	+60	2.00	0.75	0.83	35
3	Mildera (Japan Tikry)	+30	5.00	0.76	3.00	259

Source SRO Katchal

sources (see Table 7). Though these sites were suitable for habitation, relocation to these sites would affect the entire pattern of life of the Nicobarese. It may not be possible to maintain the linkage between the habitat and economic activities as the plantation-based economy has been destroyed and may take long time to revive.

Earlier, their houses had been located in clusters of 8–10 households. The houses themselves consisted of a single, large room to accommodate the joint family. They also used to have additional houses close to their plantations. The traditional stilted circular structures made of wood and thatch was common, but some also used modern pucca houses. The Nicobarese of Katchal now wanted houses which were a blend of tradition and modernity. They were keen on building their own houses in accordance with the needs of different family units with some support from the government (TISS 2005). The government, however, decided to provide houses to all tsunami-affected people entirely at its own cost. But the manner in which some of these decisions were taken left the Nicobarese quite exasperated.

When the sites were selected, the people had little idea that these were meant for locating intermediate shelters only. After much confusion, it was decided that the present sites of intermediate shelters would be used for construction of permanent houses too. However, the area of these housing units (450 ft^2) was too small to accommodate the extended joint families. This has resulted in the break-up of the joint families, which will have long-term implications for Nicobarese society.

The size and design of these houses (which can only accommodate nuclear families and have no provision for storage of fuel wood like the traditional houses) has influenced the community's energy-use patterns. Cooking in separate kitchens for small families using LPG and extensive use of electric power for lighting and electronic gadgets has increased the use of fossil fuels. The pre-agricultural society is also becoming increasingly mechanized, which will also gradually affect the social structure. The interior electrical wiring for the housing units has been problematic and incidents of fires have been reported from the areas of E Wall and Japan Tikry in the last few years (based on the author's discussions with Tribal leaders in 2015–2016).

Moreover, the tsunami had left single women, children, and elderly persons, as the sole survivors in many families. These individuals would have joined other families as part of a natural process in the cohesive Nicobarese society. But the arbitrary determination of the number of housing units (315) by the local officials and the small size of the units has disrupted this process. The numbers of units are lesser than the number of families and two families are using one unit in many cases. However, using their creative skills, the people of Katchal have improvised the structures, making them better ventilated and comfortable to suit their needs.

It is imperative to make adequate land available at the housing sites for community facilities like churches and for accommodating the organic growth of the habitations in future. Approach roads to the main roads and to livelihood points like fishing coasts and plantations also need to be provided, where they do not exist.

7 Social Change

The Nicobarese society is undergoing rapid changes. My conversations with officials and Captains from Katchal 2015 and June 2016 revealed that the Nicobarese joint family continues to be the functional socio-economic unit of the community, but it is an institution which is under tremendous stress today. Traditional economic activities, like preparing the ground, planting, maintaining, and harvesting the plantations, maintaining the horticulture gardens and copra-making, require combined efforts by members of joint families. These activities also engender a spirit of fraternity, cooperation, and unity in the family. The control exercised by the head of the household ensured need-based consumption, care, and protection of the vulnerable. The destruction of the plantation-based economy, which played a critical role in sustaining the joint family, has weakened the latter institution. Many joint families in Katchal have lost a large number of their members. The nature of some interventions like ex-gratia payments, housing, provision of amenities, and promotion of non-plantation livelihoods have encouraged patterns of living and consumption which accelerate the trend towards monetization of the economy, nuclearization of families, and stratification of the society.

Compensation payments have been used to purchase consumer durables and electronic goods in large numbers and these include TV sets, DVD players, refrigerators, motorbikes, gold ornaments, satellite phones, and even public address systems. For instance, in 2006, 18 motorbikes had been purchased by youth in 'E Wall' settlement alone, often at exorbitant prices from non-tribals. Petrol was being sold at the rate of INR 60 per litre by the government store. The Nicobarese fancy for 'novelty', coupled with the availability of surplus cash, has encouraged this trend, but this could also lead to acquisitive tendencies, which may affect the joint family system.

The consumption of liquor is a part of traditional Nicobarese culture, though the Church and educated Nicobarese have been concerned about growing alcohol abuse among the youth and have been trying to promote temperance with mixed results. In the pre-tsunami period too, influential tribal figures, like Mr. Wilson the Principal of the Sr. Secondary school in Katchal, were actively involved in anti-liquor campaigns. A disturbing feature of alcohol consumption in the post-disaster period was the rising price of liquor and large-scale availability of illicit liquor. A bottle of 750 mL. of liquor (IMFL; Indian-Made Foreign Liquor), which costs about INR 200 was being sold in Katchal for INR 800 in 2007. The illicit liquor trade was in the hands of non-tribals. The tribal leaders of Katchal informed the author earlier this year that a women's movement against alcohol abuse gained ground across the islands in 2014, leading to raids against illicit liquor trade, illegal brewing, and even closure of Andaman and Nicobar Islands Industrial Development Corporation (ANIIDCO) liquor shops in some islands.

The successful reorganization of the socio-political leadership (discussed below) notwithstanding, the loss of social control is apparent. The chief captain complained that many of the *badmash* (rogue) boys had been idling and not doing any work on the plantations or horticulture and other schemes floated by the government. He lamented that they carelessly spent all their money and were taking loans from non-tribals, which would be repaid later by encashing the FDs (fixed deposits created in banks for parking some ex-gratia payments like plantation loss compensation). The chief captain's concerns were not misplaced. Random discussions with Nicobarese youth and examination of their bank passbooks revealed a tendency for reckless spending.[2]

The indigenous leadership is concerned that much of the compensation money is gradually passing into the hands of non-tribal traders, who have exploited the Nicobarese and dominated economic activity on the island.

The extended joint family system, particularly its mechanisms for providing care and protection to the vulnerable individuals like the aged, orphans, and the disabled, have remained functional and indeed vibrant in the immediate aftermath of the disaster. The well-known Nicobarese tradition of adopting children ensured the integration of 32 orphaned children and 34 single-parent children into extended joint families in Katchal. In fact, not a single child (out of the 133 orphans and 193 single-parent Nicobarese children) were admitted to orphanages or handed over to government care providers. However, as a part of the larger national welfare template, special schemes are being implemented by the administration for these children, with support from the Rajeev Gandhi Foundation and the Prime Minister's Relief Fund. But the most important supportive role is that of the guardians themselves, who provide a family atmosphere in familiar cultural surroundings, in spite of their resource constraints. The example of Lona,[3] an unmarried Nicobarese schoolteacher is indeed an inspiring story of hope and courage. Having lost 27 members of her own

---

[2]For instance: Vespasian, an 18-year-old orphan from Jhansin village had originally received INR 1 million as compensation. He had only INR 100,000 left in his account. Robert, a 25-year-old from West Bay, received INR 600,000 as compensation but had only INR 200,000 left. Japtha, 40 years, gave INR 130,000 to his son to 'celebrate' his vacations in Mumbai (earlier Bombay). His passbook showed a withdrawal trend of INR 30,000 every month. Mary used the compensation money to buy a jeep for her son from a local trader for INR 150,000. The vehicle broke down within a week, and the trader demanded INR 10,000 to fix it.

[3]See Zubair Ahmed, 'Our Mother Lona,' *Light of Andamans*, 3 December 2007, 8.

family, she was herself in need of support. But she decided to take care of 13 children in the age group of 3–21 years, some of whom she had adopted before the disaster. She undertook this responsibility, attempting to manage her 'family' with the salary she receives as a teacher in a government school. She lives in rented accommodation and still ensures schooling—and even extra tuitions—for the children.

While the community continues to show these signs of resilience, as mentioned earlier, in the long run, the monetization of the economy will result in the accumulation of individual assets and accelerate nuclearization of families—and this will affect the community's ability to respond to the needs of the vulnerable groups. The willing enrolment for the widow- and old-age-pension schemes of the government for the first time is an indication of these needs.

Similarly, in terms of gender, the society continues to be egalitarian. The continuing trend of the election of women as 'First Captains' and 'Second Captains' in the reorganized tribal council of Katchal is reassuring. The modernization of the economy and change in skill sets may eventually affect the status of women adversely (Porwal n.d.). These consequences can be minimized to some extent by focused, culturally sensitive, assistance and welfare measures by the state, which place a premium on ensuring the sustainability of traditional institutions.

8 Economy and Livelihoods

The near-complete destruction of coconut plantations and severe damage caused to the areca nut plantations has forced the people to consider other forms of livelihood.[4] The government's plans to encourage vegetable cultivation for supplementing incomes and mixed cropping received support and produced good results. But activities like civil construction and other cash-for-work programmes have yielded inadequate results due to the influence of factors like continued supply of free rations and availability of compensation cash and weakening of traditional authority structures.

The losses of plantations and the compensation paid have created economic disparities. Huge areas under plantations had been submerged and a number of people may now have to work on the plantation land of other families as labourers. For instance, many people in Katchal are now living and working on lands which traditionally belong to the family of one senior tribal chief.

The animal husbandry initiatives of the government with buy-back arrangements have not succeeded fully because, traditionally, poultry, piggery, and fishing have been subsistence activities for the Nicobarese, and the variety of birds and animals

[4]Coconut plantations take about seven years for revival, if the land is not permanently submerged or completely degraded.

being provided were not of their choice.[5] Skill-building training was mostly supply-driven and most of the surveys undertaken for assessing needs tend to be superficial and misleading. The Nicobarese are totally unfamiliar with banking procedures and the loan schemes, which carry unfavourable conditionalities. They have, therefore, been unable to access commercial credit of any significance. The presence of a strong vested interest of non-tribal traders also comes in the way of diversification of skills of the Nicobarese youth.[6]

9 Political Reorganization

The concept of central authority among the Nicobarese is of recent origin. It originated in the seventeenth century in the form of the institutions of 'Captaincy', in response to the exigencies of trade with foreign ships. This system stabilized rapidly, and developed as an instrument of social control cutting across political, economic, and judicial spheres. Captains are elected democratically through a ballot (Singh 2003). Every village has a hierarchy of captains (first, second, and third captains). The first captains of all villages together form the tribal council of the island, which is the pre-eminent political institution of the Nicobarese.

The chairman and all members of the tribal council, except the first captain of Jhula, died in the Tsunami, and so did many of the second and third captains (see Table 5). As a result, there was a complete political vacuum in Katchal. This made it very difficult for the administration to engage with the community. In the early stages of relief work, the SRO had no choice but to pick up active youths from the

[5]The Nicobarese prefer the local fowl known as '*takniet hyum*' which is famed to be the highest producer of eggs per annum among Indian fowls and is easy to maintain. Similarly the indigenous pigs are also a unique breed which are not susceptible to common diseases. However, due to shortage of poultry and livestock after the disaster, mainland breeds like Vanaraja were initially distributed.

[6]Historically, traders from passing vessels visited the Nicobars for barter trade of various kinds. The British gave permission to traders to reside and trade in different islands in the Nicobars. The Nicobars became out of bounds for Burmese and Malay traders after 1947 and for Indian non-tribal private traders after promulgation of ANPATR in 1956. ANPATR prohibits private trade in the Nicobars except by the Nicobarese themselves. To overcome this restriction, the Guajarati firm of Akoojee Jadwet started the practice of partnering with the local Nicobarese cooperative societies which enabled it to trade in the Nicobars. Such partnerships later became widespread and involved Nicobarese only for name's sake. Moreover, government contractors and government employees started illicit shops and trading outlets in the Nicobars, the latter in their spouse's names. Also some non-tribals married Nicobarese women and ran businesses in their names, especially in Katchal. The average Nicobarese's exposure to monetary transactions is fairly recent and this, coupled with their fondness for fancy items and liquor, have often led them to being exploited by the non tribal traders. The acquisition of business skills by the Nicobarese will naturally break the trade monopoly of these handfuls of non-tribal traders. In Katchal, two non-tribal traders dominate all trade and exchange.

community and involve them in the management of relief camps, so as to ensure a degree of political balance between the tribals and the non-tribals.

Authoritative styles of leadership are characteristic of indigenous societies, and factors like age and experience, apart from exposure to the outside world and the ability to negotiate with the outside world, are considered important prerequisites for leadership. The acceptability of the young camp representatives was limited to their role in management of relief work in the camps. The community took its own time to reorganize the political order. The surviving second and third captains became active during the process of site selection for new houses. This process in itself appears to have led to a lot of political churning and realignments, associated as it was with the contentious issue of 'land rights', 'displacement', and 'the presence of the outsiders'.

Once the people settled at various locations in their intermediate shelters, discussions began regarding the selection of future leaders. Consensus was developed first around the name of Michael Yok who was nominated by Albert (an elder from a much respected family of West Bay) for the post of 'Chief Captain' and later elected by the General Body, which assembled at Japan Tikry. Then the villagers also elected village captains for Ponda, West Bay, and other villages. Church pastors and youth leaders appeared to have played an important role in this process. A new tribal council was thus elected in early 2006 (see Table 8).

The composition of the new council is characterized by the following:

- Trends of continuity and change—it has a good mix of new entrants and old second or third captains.
- Old villages have retained their political identity and are represented by their own captains. However, the population of erstwhile Kapanga village has now split into two groups, each inhabiting a different permanent settlement site, and electing its own captain.
- Bifurcation of the offices of Chief Captain and Chairman, Tribal Council (CTC) has been done, with two different persons now holding these posts. The Chief Captain is now responsible for management of internal issues within the island, whereas the CTC's role is to represent and articulate the interests of the people and negotiate with the 'outside world'.

Table 8 Reorganized tribal council of Katchal Island in 2006–7

S. No.	Village/settlements	First captain
1	Minaksh Ram Nagar	Nancy Samson
2	E. Wall	James Kampon
3	Ponda	James Fred
4	West Bay Katchal	Michael Setra
5	Jhansin	Gethsie
6	Jhoola	Nicolas Patel
7	East Bay Katchal	Onismus

Source SRO Katchal

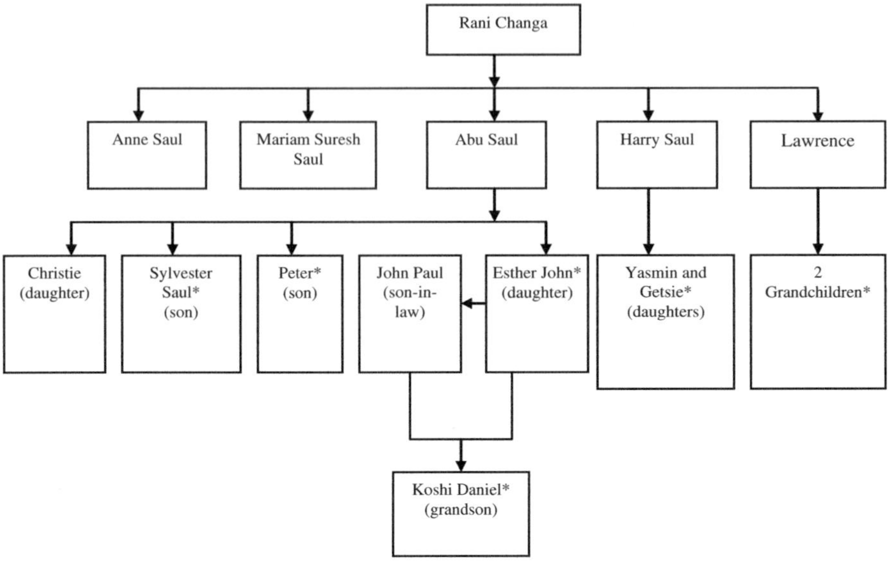

Fig. 1 Family details of the Saul family: survivors out of a family of 95 members

Three factors have been crucial in successful political reorganization of the community and the emergence of a strong and cohesive leadership in Katchal:

- the absence of external interference in the political process by the government and other agencies;
- the complete disappearance of the Saul family which represented the old tribal elite and dominated Katchal's polity (see Fig. 1); and
- the presence of external threats namely, the Sri Lankan 'outsiders', and the rubber plantation.

10 The Issue of Land Rights

In Katchal, as elsewhere in the Nicobars, all natural resources have been traditionally owned by some household or a joint family. These include land resources, creek resources (for example, *rui' la* grass used for thatched roofs), ocean resources, and animal resources. The concepts of individual ownership and community ownership do not exist, nor was there any concept of state ownership as the State itself was a distant, if not an entirely unknown, entity till recently. In the pre-1947 period, the colonial powers, including the British, exercised 'suzerainty' over the islands but left the people in control of all the resources.

The Andaman and Nicobar Protection of Aboriginal Tribes Regulation (ANPATR) was promulgated for the 'protection of the interests of socially and economically backward aboriginal tribes'. In 1957, invoking his powers under the ANPATR, the chief commissioner declared several areas, including the entire island of Katchal as '(tribal) reserved area'. As per Regulation 6(1) of ANPATR, 'no person other than a member of an aboriginal tribe can acquire any interest in any land situated in a (tribal) reserved area'. In effect, while the rights or interests of the non-tribals have been restricted, the members of the aboriginal tribes would continue to enjoy unrestricted rights over the lands and related resources in the (tribal) reserved areas as they did earlier.

Since 1957, the procedure adopted by the government to acquire land in these areas for development projects is as follows. "For obtaining land the local administration makes a request to the indigenous leadership, who, if in favour, in turn pass the request to the actual owner. On compliance, the transfer is made via a 'transfer certificate' issued to the owner in exchange for an agreed compensation" (Singh 2003). In accordance with this procedure, the administration approached Rani Changa,[7] in the 1960s with a request for providing land for setting up rubber plantations on the island. The idea was to rehabilitate Tamil repatriates of Indian origin by providing them employment in the proposed rubber plantations. Acceding to the request, Rani Changa transferred about 800 ha of land in Katchal to the government, which was notified as a 'protected' forest in 1967.

The decision of Changa to part with the land, which the Katchalese contend was not her own, was a hugely unpopular one. The arrival of 48 families of Sri Lankan 'settlers' (who have since multiplied to a strength of about 600), their uncertain entitlements in the (tribal) reserve area, and their consequent agitation for land allotment created apprehensions among the Nicobarese.

According to some scholars, (see Singh 2003), in 1971, the Forests Department of the Government of India notified the entire Nicobar group of islands including Katchal as 'reserve forests' where all activities are prohibited unless permitted. However, Andaman and Nicobar Administration (Forest Statistics 2013) has classified 120 km^2 of the 174 km^2 of land in Katchal as a 'protected forest'. It is noteworthy that the concept of 'protected' forest and 'reserve forests' outlined in the forest laws is aimed at protection of forests, whereas the concept of '(tribal) reserve area' envisaged in the ANPATR seeks to protect the rights and interests of the indigenous inhabitants of these territories. Also, the Nicobarese contend that the island was notified as a 'forest' reportedly without carrying out the mandatory survey to examine and settle the land rights of traditional stakeholders. However, these legalities did not bother the Nicobarese all these years as they did not affect their rights in practice.

The scenario has, however, changed after the tsunami. The Nicobarese have permanently lost their coastal habitations and the associated patterns of livelihoods also cannot be revived in the near future. They are, therefore, in the process of

[7]Rani Changa was the head of one of the five families which are considered to be the original landowners and inhabitants of Katchal Island since the period prior to the Japanese invasion. With the emergence of centralized political leadership in the Central Nicobars under Rani Ishlon, Changa was appointed as her deputy in Katchal, and this elevated her status considerably.

relocating to new habitats and reimagining their futures. In this context, the Nicobarese feel the land surrendered by Changa, which was rightfully theirs, should be returned to them. The traditional land rights of the five pre-eminent families of Katchal, and memories of the alleged tyranny and arrogance of Changa and her family, have been evoked, and, thus, the issue of forcible 'dispossession' of tribal land has rapidly acquired centrality in the political consciousness of the Nicobarese. The deliberate selection of 'E Wall', which is part of the active rubber plantation land, as a permanent settlement site by the Nicobarese is part of a strategy to reclaim the surrendered land (observation of the author as Special Relief Officer (SRO) and discussions with officials of the Rubber Plantation and Tribal leaders). Even the other two sites, namely Survey nos. 68 and 69 (called Minakshi Ram Nagar and Japan Tikry) are also located on 'protected' forest lands in possession of the Forest (Rubber) Corporation.

The emotive political agenda of closing down the rubber plantations (which will automatically lead to the departure of the Sri Lankan settlers) has the potential of bringing the Nicobarese together and strengthening their position on the island. The contentious issue of land rights is a unifying factor, and the Nicobarese leadership is conscious of this fact. There have been proposals to relocate the Sri Lankan settlers to another location (possibly Shaitan Khadi in Middle Andaman) but these have not materialized yet and the presence of the settlers remains a vexatious issue. Sensitivity to these concerns of the Nicobarese is necessary to foster a process of constructive dialogue with the local community, which alone can help resolve these issues.

The integrated development of the new habitation sites demands that adequate land be made available to the community, and failure to do so (as indicated in the preceding section on Habitat and Housing) can adversely affect the development initiatives spearheaded by various agencies, and, indeed, undermine the legitimacy of public institutions in the Islands.

11 Public Infrastructure

Many amenities of life, which were hitherto unknown, are now available to the people of Katchal. Public infrastructure destroyed by the tsunami has been redeveloped and new facilities have been put in place across the islands. Although road and sea connectivity with the isolated habitation of Upper Katchal remains difficult, the construction of a new jetty at Katchal has made access to the island easier. This has improved the supplies of essential items to Katchal. The creation of adequate housing facilities for government functionaries has ensured their availability, though technical staff is still in short supply. At the same time, the Nicobarese struggle to come to terms with increasing governmentality, running from pillar to post for producing affidavits and certificates to obtain or make corrections in documents like Aadhar cards, and the like.

12 Conclusion

In the face of unprecedented challenges unleashed by the tsunami and the post-disaster relief and rehabilitation interventions, the Nicobarese of Katchal have displayed fortitude and resilience in reorganizing their social and political life. While their life has been made easier in some ways, it has also been transformed fundamentally. The issues highlighted above invite us to re-examine our approaches to management of large-scale disasters, especially with reference to indigenous populations.

Large-scale natural disasters are instrumental in the transition of pre-agricultural indigenous communities from the pre-agricultural to modern capitalist modes of economic and social organization. Such transitions can be abrupt and destructive of the social fabric, if the post-disaster aid and rehabilitation interventions entail a sudden and excessive exposure to the outside world and overwhelm the traditional institutions. This can cause a great deal of trauma in cohesive and inclusive communities like the Nicobarese, which otherwise tend to display greater resilience in the face of large-scale natural disasters. As the key player in their external environment, the state can ensure a smoother transition by careful planning and implementation of its interventions.

Uniform and clearly enunciated policies and approaches are generally effective in managing the exigencies of immediate relief in disaster situations. However, strategies, schemes, and packages, which address issues of recovery and rehabilitation need to be nuanced, culturally sensitive, and community-owned if they are not to become counterproductive. Involvement of the community in mere implementation of these interventions is not enough. In particular, care has to be taken to ensure that:

- dependencies are not created in the process of providing aid and assistance;
- the integrity and sustainability of traditional social and economic systems is retained to the extent possible; for instance, the possibility of having composite bank accounts of tribal households, with the head of the joint family operating it, can be explored.
- the interests of the most vulnerable like children, the elderly, and landless labourers, are adequately protected by incentivizing the traditional inclusive practices, and revisiting CRF norms to compensate for loss of livelihoods and productive capabilities.
- the land rights of the indigenous people and their control over natural resources are protected, as they are critical for the survival of the indigenous community and for sustainable development of the islands. Since the ANPATR serves to protect the interests of the indigenous Nicobarese as outlined in the Scheduled Tribes and Other Traditional Forest Dwellers (Recognition of Forest Rights) Act, 2006, it would be useful to examine if there is a need to harmonise the provisions of the forest laws in force with those of ANPATR.

References

Abid, Mohammed A. 2006. *Relief and Rehabilitation: Role of Civil Society Organizations*. Report submitted to the Andaman and Nicobar Administration.

Brahmam, G.N.V., K. Venkaiah, A. Laxmaiah, Ch. Gal Reddy, Sharad Kumar, and B. Siva Kumar. 2005. *Health and Nutrition Status of Tsunami Affected Population Living in the Relief Camps in Andaman and Nicobar Islands: A Rapid Survey*. National Institute of Nutrition, Hyderabad.

Department of Environment and Forests, Andaman and Nicobar Islands. 2013. *Forests Statistics*. Port Blair: Department of Environment and Forests, 2005 and 2013.

Dhingra, Kiran. 2005. *The Andaman and Nicobar Islands in the 20th Century: A Gazetteer*. New York: Oxford University Press.

Digal, Janak. n.d. *Andaman and Nicobar Islands: Post-tsunami livelihood measures*. Port Blair: Andaman and Nicobar Administration.

Directorate of Census Operations, Andaman and Nicobar Islands, District Census Handbook 2011.

DW-World.DE Deutsche Welle: http://www2.dw-world.de/southasia/india/1.168272.1.html.

Justin, Anstice. 1990. *The Nicobarese*. Calcutta: Anthropological Survey of India.

M. S. Swaminathan Research Foundation. 2005. *Action Plan for Development of Post-Tsunami New Andaman*. Port Blair: M. S. Swaminathan Research Foundation.

Nova Trend. *George Weber's Lonely Islands*. US Geological Survey. http://www.andaman.org/BOOK/news_tsunami/2005_03Mar/05_mar.htm#tsunamiaidnic.

Porwal, Vivek. n.d. *Living with Hope: Life in Nicobar Post-Tsunami*. Port Blair: Andaman and Nicobar Administration.

Singh, K.S. (ed.). 1989. *Jawaharlal Nehru, Tribes and Tribal Policy*. Calcutta: Anthropological Survey of India.

Singh, Simron Jit. 2003. *In the Sea of Influence: A World of System Perspective of the Nicobar Islands*. Lund: Lund University.

Singh, Simron Jit. 2006. *The Nicobar Islands: Cultural Choices in the Aftermath of the Tsunami*. Vienna: Oliver Lehmann.

Tata Institute of Social Sciences (TISS). 2005. *Post-Tsunami Situation of Katchal Island*. Mumbai: TISS.

Tyagi, Madhumala Chattopadhyay, and B.V. Ravi Prasad. 2002. *Nutritional Status of Indian Population: Andaman and Nicobar Islands*. Kolkata: Anthropological Survey of India, Government of India.

New Andamans: The Aftermath of Tsunami in the Andaman and Nicobar Islands

Sunita Reddy

1 Introduction

The rehabilitation process in the Andaman and Nicobar Islands (A&N), has to be seen from a historical and political process of strategic 'settlement' of people, to safeguard the territories of these islands under the centrally administered Union Territories of the Indian Republic. Historically too, top-down development plans have been imposed upon these islands, first by British and later by the Central government. The seeds of 'complex disaster'[1] were sown much before the disaster itself. This chapter argues that the developmental plans for these islands have been influenced by the Central government, and the local population's choices and needs have been always compromised, whether it be the indigenous tribes of the islands, or the 'settlers' settled by the government, or the 'illegal' settlers coming to these islands in search of livelihoods. Tracing the history of the island settlements post-independence, the chapter brings to the fore the fragile ecological nature of the islands and the consequences of the 'development' pre- and post-tsunami.

A longitudinal study was carried out between the years 2005 and 2010, where the effect of the tsunami on the people's experiences were observed and recorded at various stages; from moving to relief camps, to temporary shelters, and finally to the permanent shelters. This chapter is based on the documentary evidence of media, government records, and the empirical data collected through in-depth interviews

Aarti Dhar, 'Sustainable Livelihood A Challenge in Andamans: Lt. Governor,' *The Hindu*, Thursday, 28 April 2005.

[1]The definition of 'complex disasters is explained in Simron Jit Singh's paper in this volume.

S. Reddy (✉)
Centre of Social Medicine and Community Health and Adjunct Faculty at the Special Centre for Disaster Research, Jawaharlal Nehru University, New Delhi, India
e-mail: sunitareddyjnu@gmail.com

S. Reddy (ed.), *The Asian Tsunami and Post-Disaster Aid*,
https://doi.org/10.1007/978-981-13-0182-7_10

with the affected population, government officials, and NGOs. The chapter concentrates on the experiences of three sites, where 'settler' population and so called 'illegal' settlers settled in the temporary shelters of Chouldhari, and later settled in the permanent shelters of Badmash Pahad and Bambooflat. The experiences of Nicobarese, who were badly affected in the tsunami in the islands, have already been discussed in Mohammed Abid's paper, "The Tsunami and its Aftermath: Resilience and Rupture of the Social Fabric among the Nicobarese" and Simron Jit Singh's paper, "The Sustainability of Humanitarian aid: The Nicobar Islands as a case of 'complex disaster'" in this volume and thus not covered in this chapter.[2]

2 Aid, Interventions, and the Gap Between the Bureaucratic and Emergent Norm

Disasters are often seen as events, with its processes being ignored. The focus is on the event and the emergency period. The long-term reconstruction and rehabilitation is often neglected and is out of the public spotlight. The focus is shifted to other issues or to ensuing disasters. Understanding disaster as a process helps us to get clues from the historical and political roots, not just on the disaster per se, but the rehabilitation carried on thereafter.

A continuing tradition in disaster research in general, and in anthropology specifically, has been to view hazards and disasters as challenges to the structure and organization of society. It has focused on the behaviour of individuals and groups during various stages of disaster impact and aftermath. However, the current research goes beyond the behavioural approach and searches broader socio-political and historical-structural factors that influence the decisions underlying adjustment to natural hazards (Smith 1996). Through its tradition of lengthy field research and emphasis on social and cultural process from a developmental perspective, anthropology has perhaps devoted more attention to the implications of disasters for long-term social change than other social sciences.

It has been often seen that the event of disaster attracts the attention of the media, and national and international agencies immediately rush in with humanitarian aid. Studies have observed that much of the aid, particularly involving resettlement, has created or compounded dependency as well as exacerbated the disruption among the local people. Serious criticism has been levelled at post-disaster aid that is so culturally inappropriate and sometimes both useless and patronizing (Smith 1996; Singh 2007; Gunewardena and Schuller 2008). It is seen that humanitarian aid and human interventions post-disaster, especially after the 2004 tsunami, have led to a 'dual disaster' (Hyndman and Hyndman 2011), or a 'complex disaster' (Singh 2007) the consequences of the rehabilitation are much more serious in nature than the actual disaster.

[2]Refer to book 'clash of waves' by the author for the details on the Nicobarese.

Studies have shown that after disasters, the disaster-affected are pushed to the periphery. Under globalized and neo-liberal strategies, it is the business class and those in power who capitalize on the catastrophe (Gunewardena and Schuller 2008), and push developmental agendas, which would have been resisted till then, in the state of shock (Klein 2007). Similar findings post Asian Tsunami across the countries have been observed and documented in different chapters in this volume.

Under reconstruction and rehabilitation, providing 'shelter' or 'housing' is the most contentious issue. Construction of permanent shelters is not only about brick and mortar but increasing the capacities of communities. Voluntary organizations and local government have a key role to play in the housing projects. The ideal approach in the housing and the reconstruction projects is to encourage and utilize strong community ties, social cohesion, and communal values. The government should strive to decentralize shelter policies and encourage the local-level participation from planning to implementation within the national framework. It is reported that top-down approaches and schemes are merely symbolic, producing expensive and alien housing units, resulting in either abandonment or alterations in the dwelling units (Davidson et al. 2007). The experiences of 'settlers' and Nicobarese in Andaman and Nicobar Islands too shows similar patterns.

Keeping in view the national vision to build a safe and disaster-resilient India, the Indian national policy framework has been prepared to develop a holistic, proactive, multi-disaster, and technology-driven strategy for Disaster Management. With a paradigm shift, it will now focus on the culture of prevention, mitigation, and preparedness to generate a prompt and efficient response at the time of disasters, involving the community, all government agencies, and non-governmental organizations.[3] Section 23 of the National Disaster Management Act 2005 requires the departments of the state governments to draw up their own plans for disaster management plan NDMA (2007). For the Union Territories, the central government provides relief and rehabilitation assistance.

It is ironic that the paradigm shift is to focus on prevention and preparedness, whereas in this case the planned and the top down interventions led to increased vulnerability. Why some governmental disaster relief efforts are perceived to be successful, while others are widely believed to be failures, are answered by Schneider (1992), who developed an explanation that focuses on the 'gap' between what governments are prepared to do in emergency management situations (bureaucratic norms) and what emerges as the expectations of those 'victimized' by the disaster (emergent norms). The answer to the above questions depends upon the extent to which post-disaster human behaviour corresponds to prior governmental expectations and planning. In the aftermath of virtually every disaster, a 'gap' exists between the emergent norms that guide social interactions and the bureaucratic norms that dominate governmental activity. When this gap is large, the relief effort is likely to be viewed as a failure. When the gap is small, the relief efforts run smoothly and the governmental operations are perceived to be successful.

[3]http://ndma.gov.in/ndma/guidelines.htm. See NDMA website accessed on 20th Oct. 2011.

Further, it is important to reiterate that success and failure in disaster recovery is almost entirely a matter of public perception rather than objective reality. Private citizens cannot be expected to comprehend fully the difficulties and complexities involved in any recovery effort. At the same time, people are naturally absorbed with their own personal problems caused by the disaster. Consequently, individual citizens are likely to view anything short of immediate, direct, and comprehensive help as failure. When one takes a broader perspective, virtually all of the government relief efforts are successful (Schneider 1992). However, given this perspective, it is easy to overlook peoples' experience as biased and subjective and give a clean chit to the states for the interventions carried out and see all of them as successful. But, in some cases, the lived experiences and irreversible changes brought to the lives of people, such as the economic, social and cultural life after the tsunami in the Andaman and Nicobar Islands, shows the unintended and adverse consequences as a result of long term rehabilitation.

Contextualizing post-disaster rehabilitation in the social science literature, and observing the implications of interventions on the affected population, this chapter first looks at the physical, geographical, demographical, historical, and ecological features of the Andaman and Nicobar islands. Taking stock of the history of the islands, it is observed that the seeds of vulnerability were sown much before the tsunami occurred, and bore fruit after the tsunami. The chapter then examines the process of rehabilitation carried out and the experiences of people who were 'given' 'rehabilitation packages' and its implications on their lives.

3 Andaman and Nicobar Islands

3.1 Geographical Location

The Andaman and Nicobar Islands stretch across a total geographical area of 8,249 km^2 in the Bay of Bengal, spread across 572 islands, islets, and rocks; however, only 32 islands are habitable. The islands lie parallel to a geological fault line to the east, crossing the Andaman Sea from north to south and earthquakes[4] are common features. India's only active volcanoes, on Barren and Narcondam islands, are sitting directly on the fault line. Geologically, the Andaman and Nicobar islands represent the highest peaks of an under-water mountain range, which is an extension of the Arakan range in Burma and the Sumatran Basin ranges to the south (Fig. 1) (Curray 2005).

---

[4]The line marks where two tectonic plates are rubbing against each other; the eastern plate, an extension of the huge Eurasian plate, is stationary while the Indian plate to the west is moving north to northeast at the rate of a few centimeters a year, taking the Andaman Islands with it. These movements cause earthquakes and volcanic activity in and around the islands.

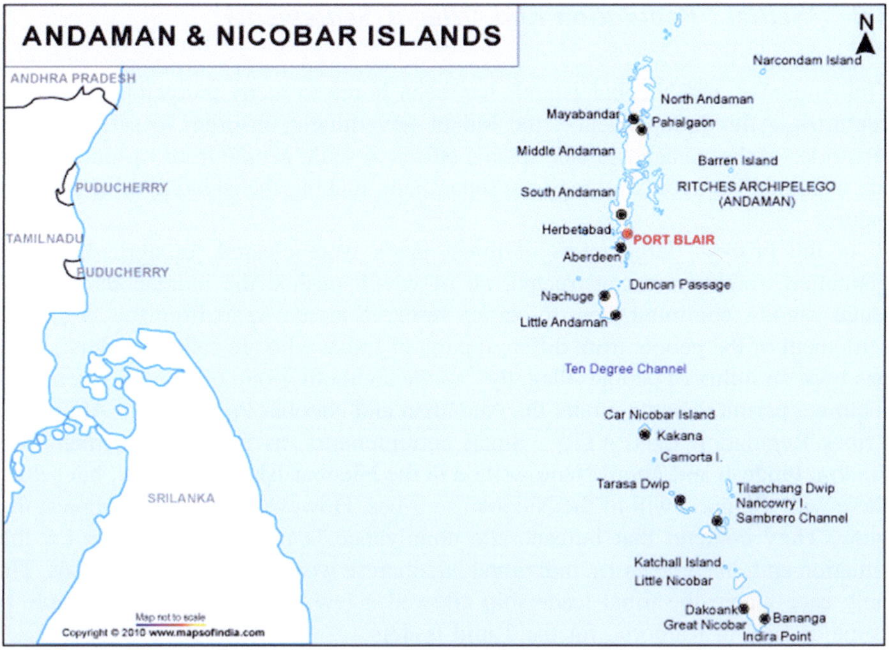

Fig. 1 Map of Andaman and Nicobar Islands. *Source* www.mapsofindia.com

3.2 Historical Significance of the Islands

The strategic position of the Andaman and Nicobar Islands, makes it the centre of a flourishing sea trade between India, East, and South East Asia. The Nicobar Islands were directly on the trade route between the eastern and western hemispheres. Andaman Islands became central to British strategic interests in the region. The first British settlement on the islands was in 1789. A permanent settlement, in the form of a penal colony, was established in 1858 on the Andaman Island following the mutiny on mainland India against the British in 1857. Thousands of mutineers sentenced to life imprisonment were transported to a place where they would pose no political threat to the British. The Cellular Jail, also called *kalapani*,[5] was made for the prisoners. In 1947, the British transferred control of the Andaman Islands to the newly established Indian government, in mainland India, designating it as 'Union Territory' administered by the Central Government from New Delhi.

[5]Kalapani meaning 'black water', as it was understood that once they are sent to *kalapani*, they never returned.

3.3 'Settler' Population and 'Illegal Settlements'

The Andaman and Nicobar Islands has been home to many indigenous tribes for centuries. After Independence, the Indian government, in order to safeguard the territories of these islands, made special efforts to settle people from various parts of the mainland[6] to avoid any regional separatism, making the island a 'mini-India' in nature.

In the process, large tracts of forest lands were cleared for agriculture and plantation, roads were constructed, all of which pushed the indigenous tribes to small islands, confining them to certain reserved areas. Apart from this organized settlement of the people from different parts of India, who are called 'settlers', there has been an influx of people 'illegally'[7] to the southern group of islands, where they require a permit to settle, under the Andaman and Nicobar Protection of Aboriginal Tribes Regulation (ANPATR).[8] Small entrepreneurs, as well as fishermen from Andhra Pradesh and Tamil Nadu, settled in the Nicobar Islands illegally, but settled there with the good will of the Nicobarese tribes. However, Nicobarese contest this view. They contend that bureaucratic connivance is mainly responsible for this situation and, in some cases, individual Nicobarese women married non-tribals. The only case where the tribal leadership allowed a few shops and a small temple to come up was in Kamorta, for the Tamil traders.

3.4 Indigenous Tribes of the Andaman and Nicobar Islands

The indigenous populations of the Andaman and Nicobar Islands have lived and flourished for more than two thousand years (Gautam 2005). However, their numbers have dwindled from thousands to double digits and some of them are on the verge of extinction. Though originally a land of indigenous population—Great Andamanese, Jarawas, Onges, Shompens, Sentinelese and Nicobarese—the 'settler' population now outnumbers them. The current ratio of the 'Particularly

---

[6]For details about the number of settlers from different parts of India at various intervals see, Sircar 2004.

[7]These families settled in the Nicobar Islands and, with some understanding with the Nicobarese, managed permits and settled permanently. They established petty business and were the ones who profited most after tsunami aid came and Nicobarese bought all assets. However, those families who were evacuated were not sent back due to the resistance of Nicobarese and also to avoid further 'illegal' settlement.

[8]In 1956, the ANPATR was passed to protect the rights of the tribals over their lands, preventing any alienation thereof and banning non-tribal access and residence except with a permit issued by the Deputy Commissioner (Dhingra 2005, 89).

Vulnerable Tribal Groups' (PVTGs)[9] to that of non-tribals is approximately 500:500,000, a stark testimony to the scale at which the dominant majority has outnumbered them (Venkateswar 2004). The growth of non-tribal population in the Nancowry group of islands has been 47 per cent in the decade of the 1980s (Dhingra 2005, 25). Those who lived comfortably and in harmony with Nature for centuries have been almost wiped out in the past few decades due to the contacts with the so-called 'civilized' people. Now there are only 92 Onges, 43 Great Andamanese, 350 Jarawa, 100 Sentinelese and over 250 Shompens, who constitute PVTGs (Sircar 2004). The other major indigenous tribe is the Nicobarese, now over 25,000 in number (Dhingra 2005, 25; Ramachandran 2000). Nicobarese do not come under PVTGs, because of their large numbers and better economic conditions. They have been residing in coastal areas and were worst affected by the tsunami.

The history of the islands, discussed above, shows the influx of people from time to time, who abused and manoeuvred the native people and the environment for their own advantage. The habitat of the indigenous people of the Andaman and Nicobar Islands, was invaded by British and Japanese colonizers and, after Independence, was taken over by the Indian Government as a Union Territory with similar policies, more or less colonial in nature[10] (Venkateswar 2004).

3.5 Depletion of Ecological Treasures and a Disaster in the Making

The Andaman and Nicobar Islands are an environmental treasure, with varied flora and fauna and rich biodiversity. Out of the total 8,249 km^2 area of Andaman and Nicobar Islands, 2,929 km^2 is reserve forest and 4,242 km^2 is protected forest (Dhingra 2005). According to the Planning Commission report 2008, only 50,000 ha of land is available for agriculture.

The Islands have unfathomed ecological treasures. They are endowed with some of the finest tropical evergreen forests in the world and are home to a large number of rare and endangered species.[11] The islands also have some of the finest mangroves,

[9]Particularly Vulnerable Tribal Groups (PVTG), earlier called 'Primitive Tribal Groups (PTGs) are the Scheduled Tribe groups who were identified as more backward communities among the tribal population groups by the Government at the Centre in 1975. So far seventy-five tribal communities have been identified as 'PVTG' in different States of India. These hunting, food-gathering, and some agricultural communities, who have been identified, as more backward communities among the tribal population groups need special programmes for their sustainable development.

[10]Sita Venkateshwar uses the term 'Indian Colonizers' with similar policies as that of British colonizers.

[11]The Islands are said to be a repository of an immense variety of species; with over 6,540 species of Fauna, out of which 834 species are endemic, 2,500 species of Flora, of which 261 species are endemic, 61 species of mangroves, 364 species of Corals, 300 variety of species of fishes, more than 1,000 molluscs, 1,500 insects and a number of medicinal herbs and plants.

which accounts for 18% of the country's total mangrove area (Ramakrishna et al. 2010). The pristine islands, rich biodiversity, and forests have been eroded due to developmental activities, whether directly or indirectly (Nair 1989). The exploitation of timber began with the establishment of a penal settlement on the islands, and the forest department was started in 1883 (Majumdar 1975). Although 70% of the forests of the Andaman Islands are still forested, much of the forest is degraded with secondary growth. Over the past thirty years, soil, meteorological, and forestry experts have stressed the need to keep the islands under forest cover. Despite all pleas, large-scale clearing of the forests for cultivation and encroachment of land with the legal and illegal settlements are underway (Dhingra 2005). Accelerated immigration from mainland India and the growth of a number of forest-based industries have led to extensive settlements and the conversion of forest areas into revenue and agricultural lands. Over time, there has been a growth in population from 14,628 in 1881 to about 400,000 by 2001, largely due to the encroachment and illegal settlements in demarcated 'tribal reserves'. Pristine forests have been lost to the timber and plywood industry, while expanding settlements, and commercial logging for over a century have destroyed large areas of the island forests (Dhingra 2005).

Apart from clearing forests for settlement, over the years the destructive consequences of sand mining for construction is readily visible in the disappearance of several beaches, with its effects on wildlife on that region, as well as steadily increasing coastal erosion (Ali and Andrews n.d.). Evidence shows the sharp decrease in mangrove cover, destruction of coral reefs as a consequence of the run-off from land–based activities such as logging, agriculture, and unchecked use of pesticides (Sekhsaria 2002). All these, combined with complete disregard for the indigenous population had already paved way for a disaster in making.

3.6 Overcrowding of the Islands and Encroachment of Land

The island population as per 2011 census was 379,944 and has a density of 43 per square kilometre, having a high literacy of 82%. Some estimate that the carrying capacity of the islands has already been exceeded. There is a continuing influx of approximately five per cent per annum. The first lot of settlements had been made on available vacant land; the later settlements required the clearing of virgin forests. An assessment of land available for allotment in Great Andaman for agriculture, grazing, and plantation was 142,920 acres. Between the years 1949 and 1955, 41,800 acres have been cleared for settlements (Dhingra 2005, 77).

The influx of large number of immigrants, largely in search of livelihood as manual labour has led to encroachment[12] (Dhingra 2005, 103). The issue of

[12]Official records show 3,737 encroachments covering 5,885 acres of revenue land and 1,794 encroachments on 7,117 acres of forest lands, and 18,000 applications for regularization, wherein 7,212 encroachments on revenue land. Deputy Commissioner's (DC) office has given away 27,728 houses sites and the DC office has not yet closed all the files.

encroachment of land and illegal settlements became even more obvious after tsunami. It is estimated that as much as 80% of land is encroached. Anyone who wants to live on the island without a permit had to pay 500 Indian rupees per month. To start a petty business, it goes up to 1,500 rupees per month. A construction business has similar rates.[13]

Over the years, the islands have exceeded the carrying capacity and placed burden on its natural resources like fresh water, as they are limited on the islands. Electricity is generated through diesel and there are problems of waste disposal. Most of the food items and all consumables are brought from the mainland. The economy of the islands is heavily subsidized by the centre. The main economic activities of the people are agriculture, animal husbandry, fishery, tourism-related activities, and employment in government services. Public facilities like schools, hospitals are free of cost for all islanders.

Given this ecological and historical account, it can be said that the Andaman situation is a classic example of unplanned natural resource exploitation and disregard of basic ecological principles. It is primarily an economy determined by the prerogatives of mainland development (Venkateswar 2004; Guha and Gadgil 1989). The seeds of vulnerability were sown much before the 2004 tsunami happened. The development agenda from the mainland continued after the tsunami in the form of rehabilitation packages.

4 Rehabilitation Process

4.1 Impact of the Tsunami

The tsunami killed more than 12,400 in India: three quarters were women and children.[14] Andaman and Nicobar Islands administration brought out a list of dead and missing persons: 3,513 were considered to be dead, out of which 2,955 were Nicobarese, and 558 were non-tribals, living in the Nicobar Islands. However, in terms of infrastructural losses, and loss of livelihoods, both were affected in the Andaman and Nicobar Islands. Large tracts of agricultural land and plantations were destroyed and the land rendered useless, submerged for years. Table 1 shows around 2177.70 ha of paddy land and 5891.01 ha of plantation land have been affected by the tsunami. Livelihoods of people were severely affected, as nearly 6,000 ha (14,826 acres) of land in the Nicobars and about 1,800 ha (4,447.8 acres) in the Andamans was permanently submerged. 10,000 houses were completely destroyed, with a population of 50,000 directly affected by the tsunami. Apart from this, 85 schools and 34 Health Centres were permanently destroyed. Further,

[13]Simha Vijay, 'The Great Andaman Cover Up,' *Tehelka The People's Paper*. 12 February 2005.

[14]Pallava Bagla, 'Building Back Better—After the Tsunami.' UNICEF India. http://www.unicef.org/india/education_1085.htm.

Table 1 Loss of property: private and public

Loss of private property	
Paddy land affected	2177.70 ha
Plantation land affected	5891.01 ha
Permanently submerged land in Nicobar	6000 ha
Permanently submerged land in Andaman	1800 ha
Boats fully damaged	938
Boats partially damaged	765
Loss of livestock	157,577
Loss of cattle	3786
Loss of pigs	38,446
Loss of goats	16,623
Houses fully damaged (approx.)	10,000
Loss of public property	
Schools destroyed	85
Health centre destroyed	34
Power capacity destroyed/shut down	37 MW
Power houses inundated/washed away	20
Jetties damaged	24
Fully/partially damaged	15
Minor damages	9
Roads	
Villages affected	257
Roads fully damaged	210.80 km
Roads partially damaged	143.45 km
Water supply	–

Source Fact Sheet Andaman and Nicobar Administration, Page i

20 power houses, 24 jetties, and 210 km of roads were fully damaged in the Andaman and Nicobar Islands. All major sources of livelihood, including poultry, animal husbandry, farming, plantation, and agriculture have been affected by the tsunami.

In the South Andaman region, an area of 1331.81 ha was submerged, severely affecting farmers who depended on their land for livelihoods.

4.2 *Challenges in Rehabilitation*

Infrastructure loss, like buildings and roads, can be 'built back better' but the real challenge is to revive sustainable livelihoods and resettle people. In the case of the Andaman and Nicobar Islands, the livelihood losses were in the form of submergence of land, loss of crops and plantations, and loss of livestock. Reviving livelihoods was major challenge in the islands due to the inherent problems; limited markets, distance from the mainland, less scope for alternative livelihoods due the islands' nature, and the logistics of transportation of materials by sea routes. Reclaiming the land was near impossible and copra cultivation would take years to bear crop. People were scared of any new innovations as the market had to be created first. Diversification was thought of with regard to cultivating hybrid variety seeds and in livestock business.

4.3 Relief Measures

Assistance of INR 38.3656 billion was requested from Government of India (GOI) by the Union Territory (UT) Administration of Andaman and Nicobar Islands towards relief, rehabilitation, and reconstruction activities necessitated by the earthquake and tsunami disaster.[15] Repair of jetties alone needed INR 523 crores.[16] As relief packages, the Andaman and Nicobar administration distributed different kinds of ex-gratia payments. These include INR 100,000 for identified dead and missing persons to next of kin, INR 5,000 for seriously injured, INR 2,000 from LG's relief fund, INR 10,000 for purchase of household articles, INR 3,000 for temporary relief, INR 10,000 relief for small business, INR 200,000 for orphans below 18 years, and INR 100,000 for widows and unmarried girls. A total of INR 575.762 million was disbursed, out of the sanctioned amount of INR 603.478 million.[17]

In addition to ex-gratia payments mentioned above, the administration also provided assistance to orphans, children, and lactating/pregnant women in the form of supply of nutrients. 2,754 children and 526 women were sought to be benefited from the scheme, which was coordinated and implemented by the Department of Social Welfare.

The immediate task of the administration was to build temporary shelters, due to heavy monsoon on the islands. By the end of April 2005, temporary shelters had been constructed and the survivors were shifted. The affected population was reluctant to shift to the temporary shelters due to the large distance from the main town in Port Blair. They were given free relief materials and free ration.[18] Along

[15]Andaman and Nicobar Administration Records. Annexure 1.

[16]Ramakrishnan T. 'Centre committed to rebuilding Andamans: Kapse,' *The Hindu*, 18 February 2005.

[17]Andaman and Nicobar Administration Records. Copy of master report on ex-gratia payment as on 10.08.2005. A&N Administration records.

[18]Free ration, at the rate of 500 g of rice, 100 g of pulses, 300 g of vegetables, 50 g of milk powder, 40 g of edible oil and 30 g of sugar, per person per day, was distributed every week. Later it was given once in 15 days or once in a month. People also complained that the ration amount was reduced subsequently, whereas the official view was following the guidelines of central calamity relief fund, the amount of free ration was reduced and was given once in 15 days or once in a month. However, the revised guidelines say, in case of natural calamities other than drought,

with these, the households were also provided with the other essential items like tarpaulin tents, clothes, blankets, utensils, mosquito nets/repellents, soaps, and detergents.

By 25 January 2006, there were a total of 207 relief camps for 46,023 people. Many had also been evacuated to Port Blair (13,928) and to the mainland (5,618). The relief camps, which were established in Port Blair for the evacuated people were managed with the assistance of NGOs. Relief camps in the form of tents were set up in various islands. In the relief camps, community kitchen was set up; food was provided by the local administration and also NGOs.

4.4 Aid Agencies

As was the case in most tsunami-affected countries, aid agencies came forward to provide relief and reconstruction in the islands. The Andaman and Nicobar Islands administration initially did not allow international organizations and other NGOs to provide humanitarian aid, due to the protected nature of the islands under various rules and laws,[19] and also due to the strategic defence establishments in the Islands. But, by 2 February 2005, 50 NGOs worked with the Administration for relief operations. Of the NGOs which have been working, 31 were national, 11 were local, and 8 were international. Since the Nicobar District is a restricted area under the PAT regulation 1956, foreigners, and Indians too, were required to acquire a permit to enter the southern group of islands. Later on, the established organizations were issued special permission.

In the islands, like other affected places, there was (un) due attention and immediate focus by national and other agencies for the first few weeks. For first six months, there was lot of chaos in the islands, with media coverage galore including day-to-day footage of the disaster. The media was sometimes mediating, and sometimes intrusive, in the process of relief distribution. The presence of multiple agencies, government, Non-Governmental Organizations, International Non-Governmental Organizations and Community Based Organizations had created pandemonium, and a lack of coordination. Most of the NGOs had come with their own relief materials and distributed to the people directly. On top of this, the

provisions like food for a maximum period up to 15 days and in case of natural calamities other than drought of a severe nature for a maximum period up to 30 days should be given. In case of the Islands, it continued for almost 5 years. The most ironic aspect of this is that it was linked to shifting to permanent shelter rather than livelihoods. The free food aid was given till 2009 until people shifted to permanent shelters.

[19]There are a number of regulatory controls applicable to the islands including: The Indian Forests Act, 1927; Andaman and Nicobar (Protection of Aboriginal Tribals) Act, 1956; Wildlife Protection Act, 1972; Forest (Conservation) Act, 1972; Prevention and Control of Pollution Act, 1974, 1981; Environmental Protection Act, 1986; and Coastal Regulation Zones, 1991.

frequent visit of VVIPs added to the chaos, while all junior functionaries were busy making security arrangements for visits to different islands.

The NGOs provided lot of relief material and helped the administration in managing the relief camps and construction of temporary shelters. However, only a few remained for long-term rehabilitation. The immediate and initial enthusiasm, and the humanitarian fervour, fizzled out as the years passed, without taking note of the proper rehabilitation of people. By the end of three years, the agencies started winding up and closed their account books and exited. An administrative officer reported that the contribution of NGOs was much less significant, and led to chaos and duplication (Abid 2006).

4.5 Debates Over Development in the Islands After the Tsunami

The Andaman and Nicobar administration saw opportunities in two sectors for island development; fisheries and tourism. It suggested high-value, low-volume tourism and suggested the construction of an international convention centre. Another suggestion was to develop fishing for commercial use using trawlers, to earn foreign exchange. It was anticipated that developing tourism on a massive scale would, by default, generate employment, and also rebuild the economy, especially by matching its scope with some of the well-known international island tourist spots. The Andaman and Nicobar administration put up a proposal for INR 50 billion for developing tourism in these Islands.

Though this development agenda was proposed much earlier in 2002 (Andaman and Nicobar Development report), the tsunami brought in the opportunity to push through this development agenda, in the post-tsunami state of shock. Klein (2007) defines 'The Shock Doctrine': the use of public disorientation following massive collective shock—disaster in this case—to push through highly unpopular high intensity tourism.[20] Such was the case in the Andamans too where the people near to the coast were moved to temporary and permanent shelters, resorts were constructed near the beach, as on Havelock Island, and some hotels and guest houses in Port Blair.

Aggressive tourism promotion was planned; the opening of 15 islands and 50 new sites was proposed to attract eco-tourists, and even filmmakers; bids were invited from private investors and overseas bidders. However, these ideas of aggressive tourism and fisheries are laced with a lot of hurdles, due to the fragile nature of the islands, scarcity of water and electricity, insufficient hotels and guest houses, and also concern for the welfare of the indigenous tribes who are protected under PAT regulation. The long-standing idea of an agreement between the pristine

[20]Klein examines how, in post-tsunami Sri Lanka fishermen were pushed back to make space for the tourism industry, violating the law of coastal zone regulation.

Andamans and Phuket to have a twinning pact, opening two islands to tourism was suggested. The twinning baffled many because it has completely ignored the fact that the Andamans are among the world's most fragile and ecologically valuable islands (Bharati 2005; Reddy 2007). However, due to resistance from the local media and activists, the proposed opening of more islands to tourism did not take off.

Efforts to revive tourism began by the way of conducting festivals, and inviting artists and travel writers from the mainland. The local newspaper reports; "*the Andaman administration has been working directionless since the day of tsunami. The six day tourism festival titled 'Island on March' in early 2005, had wasted INR 5 million of public money, in various shows. Expenses of 65 artists including their boarding, lodging, were borne by the public money, whereas only 150 people watched the show*".[21] This festival is organized almost every year. How much revenue it is generating is a question mark.

At Marina beach in Port Blair, water sports, speedboats, and water scooters were procured and run by the tourism department. Boats to Ross and Viper islands from Port Blair were privatized. However, very few new players have come in; it is mostly those who were already in this business expanded their buisness.

In order to expand tourism on a large scale, various projects and proposals have been launched. As per the records of the tourism department, by end of 2010, 50 units Yacht Marina (INR 526.4 million), 35 room luxury boats (INR 455 million) and 15 seated mechanized luxury boats were proposed to be launched. At Havelock, one camp with 20 luxury cottages worth around INR 225 million was approved and the work was in progress at the time of research in 2010. A canopy walkway project worth INR 58 million was finalized. Approval for development of tourist circuit between different islands worth INR 80 million was being sought. Sea planes, an international golf course, and equipment for scuba diving and water sports, have been procured.[22]

It is astonishing to see how millions of rupees are spent on tourism development. However, the revenue made out of tourism may not be commensurate to the investment. The losses in terms of environment or ecological degradation have not been calculated to estimate the overall impact of tourism on the islands. Since these islands are centre-sponsored and highly subsidized, millions of rupees have been poured in, without making much business sense. There seems to not be much scope for people to come, from the far-off mainland, spending more than thirty thousand rupees on airfare, except for those who are provided Leave Travel Compensation by their employers, to visit to these islands. Instead, some would prefer to visit other South Asian destinations for tourism. However, the tourist numbers have been gradually increasing.[23] During the peak season all the rooms are full and tourists find

[21]'Several lakhs of public money dumped in dustbin, tsunami victims crying for bread and milk,' *The Andaman Express*, 5 April 2005.

[22]As per interview with an official of tourism department, 17 October 2010.

[23]98,180 tourists visited in 2003; the number dipped to 32,381 in 2005, and steadily increased each year to 155,726 in 2009.

it difficult to get accommodation. In Port Blair, there are 107 hotels small and medium having 2,504 rooms. In Havelock there are 23 hotels with 375 rooms, Baratang has 6 hotels with 31 rooms and Neil Island has 8 hotels with 92 rooms.[24]

It was observed that newly procured buses, for ferrying tourists on the Andaman Islands, were lying unused at the government guesthouses for months, gathering dust and with fading paint. It was observed, in October 2010, that the general infrastructure has developed, and the availability of accommodation has gone up. A lot of prosperity around Port Blair was seen, with many luxury cars and car showrooms. The narratives of the tsunami-affected people show that they are not part of this development in any way and continue to live in isolated 'tsunami houses,' struggling to make ends meet. The following section looks at the livelihood rehabilitation process.

4.6 Livelihood Rehabilitation

It was reported by the administration that the planning of rehabilitation process was quite encouraging and inclusive. However, it really had not made any difference to the lives of people at the end of six years. Industries' department carried out a detailed survey in the intermediate shelters. A micro-level assessment was done to capture the current socio-economic dimensions of the affected communities for rehabilitation planning. The objective of the survey was to assure at least one daily-wage job per family for 300 days over one year starting from November 1, 2005. Further, responsibility was fixed with different government departments like fishing, industry and agriculture to provide livelihood as per the option exercised by the people. The Andaman and Nicobar administration identified 6,313 family units, who were still without livelihood assistance as on May 2006, to be assisted and identified for intervention.[25] Initially 'man-days' were generated where the manual labour to create ponds, drainage, clearing of debris, was to be carried out. It was reported by the administration that 2,166 family units did not show much interest in any type of assistance, because they had been paid ex-gratia and were getting free rations. The survey also found out that most people were expecting support from the government in the form of a grant, a major reason for them to decline any livelihood assistance.

The agriculture department recommended various new initiatives like organic farming, and shrimp cultivation on the islands. It started procuring tons of organic manure, neem cakes, pump sets, power tillers, and farm implements, and distributed these to all farmers free of cost. Further, coconut and areca-nut seedlings, and

---

[24]According to an interview with an official of tourism department, 17 October 2010.

[25]For livelihood needs of the farm sector, 2,196 families were identified, for non-farm sector plantations 2,627 families were identifies. For off-farm sectors like fisheries and animal husbandry, 490 families were identified.

planting material were distributed to the farmers. Multiple cropping, high-value agriculture like floriculture, aromatic plants, and spices was proposed. A package of 25 crores was utilized over the period of 2005–2006. However, the shrimp cultivation did not materialize, as it required people to organize and get training.

The official records of the Department of Industries showed 3,532 people (1,382 male and 2,150 females) were given skill-building training programme in association with NGOs. Of the 228 training programs, 100 were undertaken by the organization CARE, 70 by the Department, and rest were provided by various NGOs by the end of May 2006. Training in more than a dozen trades[26] were imparted as per the records of the Department. Given the rehabilitation packages and efforts by the Andaman and Nicobar administration and NGOs, the following section discusses the experiences of the people about the livelihoods and shelters in which they were accommodated.

4.7 Linkages of Shelters with Livelihoods

The livelihoods of the people and their settlement in temporary/permanent shelters cannot be delinked. Problems multiplied with the relocation of these affected families in temporary shelters, which were far away from the main town and away from their fields and the seashore. Living in these shelters have brought about changes in the patterns of livelihoods. Bimola Mondal from Chouldhari shared:

> In the first three months of tsunami, our land was filled with seawater. After one year we started the paddy cultivation again, but could not continue due to the distance of agricultural land from this temporary shelter. There is no meaning in doing agriculture by sitting here. Now we are planning to do the paddy cultivation and vegetables in the coming years. Because we are not cultivating and no livestock, there is hardly any income. Only a pan shop is there and it is very difficult to manage the expenditure with such less income.[27]

According to Sita Mondal from Chouldhari:

> If we were getting permanent shelter at Badmash Pahar, it would create lot of problems of transportation to our agricultural lands especially during the period of rainy season. We have to take another route to reach our field as Badmash Pahar area is surrounded by agricultural fields. It will take lot of time to reach to our land for cultivation from the permanent shelters.[28]

Sumita Das from Chouldhari, a widow, has also shared similar views, saying:

[26]These were: Carpentry, Fashion and Apparel Designing, General Engineering, Computer Application, Fruits & Vegetable processing, Plumbing, Household Wiring, Tailoring and Readymade Garment, Bakery Products, Repair and Servicing of Gas Stove, Repair and Servicing of TV and other Electronics, Equipments, Jute Bags and Jute Handicrafts, Areca nut sheath cup/plate making, Repair and Servicing of Hero Honda Motorcycle.

[27]Interviewed on 18 December 2006.

[28]Interviewed on 9 January 2007.

> Our cultivable land is almost 13 kilometers away from this shelter, due to this it is very difficult to look after the crops. So we wanted to get a shelter there in our own land.[29]

Thus, a very important challenge faced by the farmers has been the distance between the temporary shelter/permanent shelters and their agricultural lands. They preferred to go back to their damaged houses, repair them, and stay there. Thus, it was also observed that a few temporary shelters were unoccupied.

The fishermen also faced many challenges. Fishermen had lost boats and nets worth thousands in the tsunami and the so-called ‘illegal’ settlers who were not allowed to go back to the Nicobar Islands faced different challenges. It was difficult to continue fishing activities, due to restrictions imposed by the authorities in the area close to the shelters. As told by Haran from Chouldhari:

> It is very difficult to do fishing in this area because in most of the areas, the government has prohibited fishing, so we have to go far away. In Nicobar region, there were no such prohibitions and they get lot of fishes without going far away from the house. Because of the distance I cannot work more than four days; earlier I use to go for fishing every day.[30]

The present temporary shelters have created hindrances to practice livestock and poultry farming among the settlers, due to lack of space and the sharing of limited space. They found it difficult to keep animals and birds in restricted places. With no livestock to rear, women in the households were left with no income, as keeping poultry was one of the important sources of income in Port Blair. Another reason for not being able to practice poultry had been the death of the birds provided by the veterinary department just after the tsunami. As told by Bijoy Majumdar from Chouldhari:

> After Tsunami the government gave us some seeds and fertilizers for cultivation of vegetables along with some tools. Veterinary department gave 20 chicks but all died out of some diseases in a few days time.[31]

Gokul Basu from Chouldhari was a boat engine mechanic, but, after shifting to the new shelter far away from the coast, he faces problems in resuming his occupation.

> Even though I bought some tools for boat repairing, I am not getting much work here in this place. There is not even a single workshop for the boat repairing nearby where I can work’.[32]

Deepa Joydhar, a widow with five daughters, has been earning for 12 days per month, which has been insufficient; she had to sell all her property to get her daughters married. The government-sanctioned the widow a pension of INR 500 was not sufficient to support a family.[33]

---

[29]Interviewed on 13 December 2006.

[30]Interviewed on 9 December 2006.

[31]Interviewed on 7 December 2006.

[32]Interviewed on 12 December 2006.

[33]Interviewed on 9 January 2007.

The tsunami-affected people had their own expectations from the state. They were expecting a government job for one member in each family, as expressed by Sathi Mondal from Chouldhari, which never materialized.

> We do not have a permanent income from outside; the amount we are getting from the daily wage is not even sufficient for the children's expenses, at least one person from each family should have been given a permanent job.[34]

Single women like Piyali Samundar from Chouldhari expressed a need to get vocational training within the shelters due to her immobility because of small children.

> If NGOs could have offered some kind of trainings like, sewing, pickle making, packing, papad[35] making, etc., in the intermediate shelters, that would be more beneficial for us, because we have small children, due to which we cannot go far away for training.[36]

Some who had received training, however, reported that they did not have enough capital to start their own business or set up a shop. The families in which there was no one with a permanent government job, demanded that one member be given a job to have some kind of security for the future.

4.8 'Tsunami Houses'

There were varied opinions about the permanent houses[37]; the designs, quality and on the twin-sharing issue. People were against twin-sharing where there was hardly any privacy, and no space to rear livestock, and more chances of quarrels. Shelters were rife with problems and led to inter- and intra-community conflicts, especially among the 'settlers'. There was a clash of interests, heartburn, and jealousy of the house owners against those who were tenants: the tenants, too, were getting permanent houses, though they had not lost anything. The same kinds of packages were given to those who were settled by the administration decades ago and to the ones who had settled 'illegally' few months back. This became a bone of contention among the people; however, for the administration all those affected by tsunami, irrespective of their past records, were to be treated equally.

There was also a grievance that the recent migrants from the mainland, who were engaged in business activities, had been provided with ex-gratia, while the 'settlers' did not get their due share, recounted Mondal from Chouldhari,

[34]Interviewed on 19 December 2006.

[35]Papad is a thin, crisp Indian preparation sometimes described as a cracker or flatbread. It is typically served as an accompaniment to a meal in India.

[36]Interviewed on 19 December 2006.

[37]The cost of each permanent shelter came to 700,000 to 800,000 in Port Blair and nearly 1.5 million in the Nicobar islands.

> We got just 24,000 rupees for the loss of our 57 years hard work in this field. At the same time, those who came here from main land for business purposes and stayed in rented houses got more than what we have got as compensation money. They are staying in temporary shelters and will get permanent shelter in future. In Chouldhari, there was a Lohar (blacksmith) who came to this area just one month before the tsunami, from mainland, and started a shop in the market. Even though the tsunami did not affect his shop, he got compensation from the government, which was more than what we got.[38]

The outsiders called the permanent houses constructed by the administration 'tsunami houses'. The affected people were seen as the chief beneficiaries of these houses. However, the housing did not solve the problems; in fact, it aggravated the situation in most cases. The narratives below show the implications of getting 'tsunami houses', and the change in economic status from BPL to APL because of the houses allotted.

The respondents from Badmash Pahar said that the rich and those who have money have the BPL (Below Poverty Line) card. All the members in the shelters have APL (Above Poverty Line) card. When they asked for BPL card, they were told: '*You have got Rs. 10–15 lakhs (1–1.5 million) worth permanent shelter, how will you get a BPL card*'. Another respondent from Badmash Pahar expressed: '*What is the use of having a permanent house when you don't have anything to eat?*'; '*Every department has vacancies but they are not giving any jobs'*. They further said, '*khane ko dana nahi, toh is ghar me rehene ka kya maja?*' ('If there is no morsel in our mouths, what is the fun in living in this house?'),[39] questioning the whole intervention.

Many permanent structures, in the vicinity of the permanent houses, were constructed and were not yet functional. It was observed that, by 2010, a few permanent shelters were locked up and were not used. A multi-storied school was opened in Badmash Pahar and there were many other schools in the vicinity. For health care, a sub-centre in Bambooflat too was not yet functional. The 'settlers' who could reclaim their lands, repaired the houses near their agricultural fields and continued to stay in their older houses, retaining and locking the permanent houses[40] given to them. These settlers however, benefited from various schemes of the Agriculture and other Departments, but those whose land could not be reclaimed, and whose houses completely damaged, had to move to the permanent shelters, and were not able to get back to sustainable livelihoods.

---

[38]Interviewed on 13 December 2006.

[39]Interviewed 18 October 2010.

[40]A few houses were permanently locked, as they were not allowed to rent it out. Since their names figured in the list of beneficiaries, most of them got their temporary shelters and also permanent houses.

4.9 *Loss of Neighbourhood and Networks*

The so-called 'illegal settlers'—mostly Tamil and Bengali families—who were evacuated from the Nicobar Islands, were not sent back to these islands. Under the PAT regulation, and resistance from the Nicobarese, they had to be settled in Port Blair. The families had adjustment problems with the neighbours due to various regional backgrounds. It was reported that for the widows, it was a nightmare to leave their children at home in new surroundings to go for work. Loss of neighbourhood and social networks led to psychological problems, especially among women. Women found the shelters and toilets very unfriendly, and feared lack of privacy. Petty quarrels with the neighbours and commotions due to drunken men were common features.

Life in the temporary shelters was very difficult; in the one-room accommodation, large families had been staying together, with complete lack of privacy, for almost five years. The shelters were very hot and inhabitable. The premises of the shelters were full of disturbances for young children, who found it difficult to concentrate on school work, and spent most of their time chatting in groups, because of the clustering of the houses. Most parents worried about their studies and their future.

Those who did not have any land and had no other choice were shifted to permanent shelters. To make the shelters earthquake resistant, the walls and the roof were made of bamboo sheets and only the kitchen was concrete. It was observed in Oct. 2010, after just one monsoon, that all the roofs were leaking and the water splashed inside through the windows and doors. There was no awning over the windows to cover, and after shifting, most people tried to cover their windows and doors temporarily to stop the rain coming in. The leakage of the roofs of the permanent houses was due to the last-minute hurry to complete construction work, hence contractors did not do a good job: the quality of the permanent shelters was sub-standard.

People wanted to build their own houses according to their needs and choices. The 'settlers', especially the agriculturists, were of the view that if given cash in instalments, they could have built their own houses near their fields. The official concern was that the cash might be misused and also feared that people would start cutting trees to build their houses. Due to various legal restrictions, they felt it was best to locate patches of land, built temporary, and later, permanent shelters, and give them to the affected families.

5 Discussion

Due to the fragile ecosystem of the islands, there were a few restrictions and reservations against leaving it to the affected to build houses as per their wish. In the process of rehabilitation, millions of rupees were spent by the administration and various other agencies. Roads, and other infrastructure like jetties and public buildings were constructed better than before. But some of the structures, like school buildings and health centres lay unused. Some of the permanent shelters were locked up, as they were far away from their fields.

Regarding revival of livelihoods, though the official records showed that numerous training programs had been conducted and various new initiatives had been taken, the experiences of the people showed that the benefits did not reach them. It was observed that the tourism industry was growing, but the beneficiaries were not the tsunami-affected individuals, but already existing entrepreneurs. The tsunami-affected people were far away from development; in the permanent shelters, isolated and left behind. Instead, the intermediaries, contractors, petty businessmen benefited and prospered in this 'New Andaman'. The tsunami came as an opportunity for many others while the life of the tsunami-affected population became precarious, as they struggled to survive in the so called 'tsunami shelters'.

The tsunami and its aftermath opened a Pandora's Box of problems. The covert issue of illegal settlement and agricultural-land conversion, to which the local administration had been turning a blind eye all these years, came to the fore. However, the administration decided to support everybody who got affected by tsunami, irrespective of their 'legal' or 'illegal' status of settlement. They were all included in the list of beneficiaries and were treated at par. This led to bitterness among the long-time settlers, who had lost everything, against recent migrants who were illegal settlers. For construction of temporary and permanent shelters, large tracts of land had to be cleared and these shelters were stigmatized as 'tsunami houses'. The people who resided in these were seen as people who did not require any further assistance, after getting a permanent house worth hundreds of thousands of rupees. However, the decision to continue to provide free ration till they shifted to a permanent house, led to dependency and did not alleviate the food security problem. Thus, even 5 years after the tsunami, though people had houses, their livelihood issues were left unresolved. Most of them were struggling in petty jobs and labour work. Their demands to have one person per family in permanent government employment were not fulfilled, adding to the woes of widows, and single women with children.

There was a disjunction in the planning process for housing and livelihoods. The decision to construct houses away from the sites of livelihood added to the problems, both for the agriculturists and the fishermen. The distance of the shelters from the source of livelihoods added to the cost of transportation. The security and safety of their fields and livestock was also a major concern.

Though relief was available in plenty and was often repetitive, some policies had serious flaws, like the decision to provide free ration till the affected were shifted to permanent shelters had serious implications. It took almost five years to shift to permanent shelters, and availability of free ration discouraged people from working, and led to dependency. The food aid was not linked to revival of livelihoods or food security, but moving to permanent shelters was taken as a proxy for food security.

The outsiders and the non-affected were of the opinion that they had been provided many things free of cost, even a house worth hundreds of thousands of rupees; what else could they want? However, the tsunami-affected families questioned how a permanent shelter could solve their food-security problems? Most of them were in petty jobs or were employed in daily-wage labour. The new 'tsunami houses' were empty with no assets inside; the linking of free food to permanent shelters made no sense.

The administration should have helped the communities to construct their own houses and monitored them strictly, letting them build according to their needs. This would have filled them with fervour, interest, and complete participation and would have catered to the needs of the people and would have been more economical. The wait wouldn't have been so long and the clash of interests could have been minimized.

Despite the massive cost of reconstruction, elaborate livelihood plans, construction of temporary and permanent shelters, the affected people were ultimately left ghettoized in 'tsunami houses' with no sustainable livelihoods. The long-term development plans of reviving livelihoods and shelter reconstruction were not suitable to the needs and choices of the people, and were thrust upon them. The decision to provide free food till they shifted to permanent houses, led to a disjunction in establishing food security and livelihoods for the people. Coyne's (2013) 'Doing bad, by doing good' provides a perspective on how, in the Andaman and Nicobar Islands, the tendency towards linear technological thinking—as opposed to organic thinking that recognizes the inescapable complexity of socio-economic conditions—led to various negative implications.

Thus, the relief and rehabilitation effort in the islands, though immense, were guided by mainland prerogatives and plans which were not to the satisfaction of the people affected. For the administration, 'numbers' mattered: dead, missing, evacuated, settled in temporary and permanent shelters. However, people's experience showed that each and every individual affected had their own problems and concerns that were not necessarily addressed by the administration. According to bureaucratic norms, the rehabilitation is a near success; with regard to emergent norms, however, a wide gap exists.

References

Abid, M. 2006. *Relief and Rehabilitation. Role of Civil Society Organizations*. Report submitted to the Andaman and Nicobar Administration.

Ali, R., and H. Andrews. n.d. Casefile: Andamans Report. In *Development and Ethnocide: Colonial practices in the Andaman Islands*, ed. S. Venkateswar. Denmark: IWGIA (2004).

Bharati, Chaturvedi. 2005. Twinning may cost the Andamans dear. *HT Nations New Delhi*, 25 July 2005.

Coyne, C. 2013. *Doing Bad by Doing Good: Why Humanitarian Action Fails*. Stanford University Press.

Curray, J. R. 2005. Tectonics and History of the Andaman Sea region. *Journal of Asian Earth Sciences* 25 (1): 187–232.

Davidson, Colin H., Cassidy Johnson, Gonzalo Lizarralde, Nese Dikmen, and Alicia Sliwinski. 2007. Truths and Myths About Community Participation in Post-disaster Housing Projects. *Habitat International* 31 (1): 100–115.
Dhingra, K. 2005. *The Andaman and Nicobar Islands in the 20th Century a Gazetteer*. New Delhi: Oxford University Press.
Gautam, R.K. 2005. The Primitive Tribes of Andaman and Nicobar Islands. *Primitive Tribes in contemporary India: Concept, Ethnography and Demography* 1: 35–73.
Guha, R., and M. Gadgil. 1989. State Forestry and Social Conflicts in British India: A Study in Ecological Bases in Agrarian Protest. *Past and Present* 123: 141–177.
Gunewardena, Nandini, and Mark Schuller (eds.). 2008. *Capitalizing on Catastrophe: Neolibral Strategies in Disaster Reconstruction*. New York: Altamira Press.
Hyndman, J., and Hyndman, J. 2011. *Dual Disasters: Humanitarian Aid After the 2004 Tsunami*. Sterling, VA: Kumarian Press.
Klein. 2007. The Shock Doctrine: The Rise of Disaster Capitalism. *Capital & Class* 33: 172–173.
Majumdar, R.C. 1975. Penal Settlement in Andaman. In *Gazetteers Unit*, Dept. of Culture, Ministry of Education and Social Welfare. New Delhi: Government of India Press.
Nair, C.T.S. 1989. Environmental Issues in Forest Land Use in the Andaman Islands. In *Andaman, Nicobar and Lakhsadweep: An Environmental Impact Assessment*, ed. C.J. Saldanha, 94–111. New Delhi: Oxford Publishing Company.
NDMA. 2007. https://ndma.gov.in/images/ndma-pdf/DM_act2005.pdf. Accessed on 20 May 2018.
Oliver-Smith, A. 1996. Anthropological Research on Hazards and Disasters. *Annual Review of Anthropology* 25 (1): 303–328.
Ramachandran, R. 2000. A People in Peril: High Levels of Hepatitis B Infection Among Sections of the Already Minuscule Tribal Population of the Andaman and Nicobar Islands Raise Medical and Social Concerns. *Frontline* 17 (16).
Ramakrishna, Raghunathan, C. and Sivaperuman, C. 2010. *Recent Trends in Biodiversity of Andaman and Nicobar Islands*: 1–542 (Published by the Director, Zool. Surv. India, Kolkata).
Reddy, Sunita. 2007. Mega Tourism in Andaman and Nicobar Islands: Some Concerns. *Journal of Human Ecology* 21 (3): 231–239.
Schneider, Saundra K. 1992. Governmental Response to Disasters: The Conflict Between Bureaucratic Procedures and Emergent Norms. *Public Administration Review* 52 (2): 135–145.
Sekhsaria, P. 2002. Logging Off, for Now. *Frontline* 19. Available at: http://www.flonnet.com/fl1901/19010650.htm. Accessed on 28 Feb 2011.
Singh, Simron J. 2007. The Political Economy of Aid. A Comment on Post-tsunami Nicobars. *Light of Andamans* 32 (9–10).
Sircar, Pronob K. 2004. *The Primitive Tribes of Andaman and Nicobar Islands*. New Delhi: Akansha Publishing House.

Post-disaster Housing: Lessons Learnt from the 2004 Tsunami of Sri Lanka

Nisha Arunatilake

1 Introduction

The December 2004 tsunami that hit Sri Lanka was unprecedented in terms of the damage it caused to people's lives and property. Over 31,000 people lost their lives and more than 500,000 were displaced. Close to 100,000 houses were partially or completely destroyed (Centreal Bank of Sri Lanka 2004). In addition, schools, hospitals, and other types of buildings, roads, railways, bridges, communication networks, and other infrastructure were extensively damaged. Geographically, most of the coastal belt in the Northern, Eastern, and Southern provinces, and parts of the Western province were severely affected (IPS 2006). The damage was greatest in the Eastern province, which is situated directly facing northern Sumatra where the earthquake that caused the tsunami took place (Weerakoon et al. 2007).

Sri Lanka has never experienced a natural disaster of the magnitude of the 2004 tsunami. This made the challenges of reconstruction and rehabilitation in the aftermath of the tsunami more complex than providing immediate relief (IPS 2006). The preliminary needs assessments, by the Asian Development Bank (ADB), the Japan Bank for International Cooperation (JIBC), and the World Bank (WB), estimated the asset-damages to the country to be US$1 billion (4.5% of the country's GDP) and the medium-term financial needs for relief, rehabilitation, and reconstruction to be US$1.5–1.6 billion (7.5% of the GDP) (Weerakoon et al. 2007). These estimates closely corresponded to the needs assessments of the

The author is grateful to the IPS staff involved in the tsunami surveys and related studies. Special mention should be made of Paul Steele, Dushni Weerakoon, Kanchana Wickramasinghe and Priyanka Jayawardena. The paper draws on earlier IPS studies on post-tsunami rehabilitation and reconstruction.

N. Arunatilake (✉)
Institute of Policy Studies of Sri Lanka, Colombo, Sri Lanka
e-mail: nisha@ips.lk

S. Reddy (ed.), *The Asian Tsunami and Post-Disaster Aid*,
https://doi.org/10.1007/978-981-13-0182-7_11

Government of Sri Lanka (GOSL). The availability of funds was not a problem. The amount of external assistance that was promised—a total of US$2.2 billion over the next 2–3 years, in addition to assistance from within Sri Lanka—seemed more than adequate to cover the costs of reconstruction. The greater challenge was to plan and execute the rehabilitation and reconstruction process in a fair and efficient manner.

The planning and implementation of the rehabilitation process was made difficult due to several factors. Sri Lankan institutions have never before been involved with a rehabilitation project of this magnitude. That the tsunami-affected areas were spread along a vast expanse of the coastal belt of the country made needs assessment more difficult. In the Northern and the Eastern Provinces, the tsunami-affected areas closely overlapped with those affected by the long-drawn-out secessionist conflict in the country. Further, despite an ongoing Cease Fire Agreement (CFA) at the time of the tsunami, the affected areas that were under the control of the Liberation Tigers of Tamil Eelam (LTTE), were difficult to access, due to security reasons and the need to liaise with the LTTE to enter those areas. The distance from the capital, Colombo, and the poor condition of the roads in those areas, were other reasons that impeded the access. From a macroeconomic point, the country was not doing well when it was hit by the tsunami (IPS 2006). The economic growth has begun to slow from the second quarter of 2004, partly due to the policy-uncertainty following the election of a new government in April 2004, but also due to the slow pace of reforms and the weakening performance of the macroeconomic indicators. Inflation in the country has been increasing from mid-2004, due to a larger-than-anticipated budget deficit (of 8.2% of GDP), resulting from increased current expenditures. The balance-of-payment deficit increased from 0.4% in 2003 to 3.3% in 2004, due to oil price increases and declining capital inflows. The currency depreciated by 8.5% despite interventions to improve the exchange rate (Jayasuriya et al. 2006). The progress in the post-tsunami rehabilitation process can only be assessed in the context of the above-described political and economic environment of the country.

Following the immediate relief phase, housing was the country's main concern. The Task Force for Rebuilding the Nation (TAFREN)[1] was given the task of planning, coordinating, and implementing the housing programme for the tsunami-affected. On 15 March 2005, TAFREN came forth with a set of policies and guidelines for housing and township development. One main feature in the guidelines was the definition of a 'buffer zone'—henceforth referred to as the 'no-build zone'—within which the building of new houses and reconstruction of damaged or destroyed houses was not allowed. People whose damaged or destroyed houses were within the no-build zone were to be relocated to nearby areas, while those outside the zone were to be given assistance for rebuilding. Another feature of this two-pronged housing strategy was that the rebuilding of houses outside the no-build zone was homeowner-driven. Under this scheme, funds

[1]One of the three Task Forces set up by The Centre for National Operations (CNO) that was set up by the Ministry of Public Security, Law, and Order, to handle tsunami relief and reconstruction efforts.

and technical assistance were provided to homeowners to reconstruct their destroyed or damaged houses. The relocated houses were donor-driven; that is, the agents that came forward to assist in the relocation scheme were provided land and infrastructure facilities by the government to build houses and communities.

The Institute of Policy Studies (IPS) carried out a survey in April 2005 of 600 households in 14 Grama Niladhari Divisions[2] (GNDs) in six affected districts across the Southern and Eastern Provinces in an attempt to obtain views on the government's post-tsunami housing programme (IPS 2006). These households were revisited in July 2006 to gather information on housing and livelihood recovery progress, aid distribution, community development, and preparedness for disaster management (IPS 2006). The main aim of the survey was to obtain the households' views on the progress and process of post-tsunami reconstruction and rehabilitation. The aim of this paper is to document the details of the post-tsunami housing programme of the GOSL and to highlight the views of the beneficiaries on the process of housing and its progress. The post-tsunami housing programme will be assessed in four specific dimensions: (1) how well the beneficiaries were aware of the post-tsunami housing programme and how well they could access assistance; (2) the speed with which beneficiaries were resettled; (3) the welfare effects of the housing programme; and (4) donor coordination and implementation issues.

The paper will highlight the GOSL's tsunami housing programme in 2005, and its 2006 revision in the second and third sections. The fourth section sets out the data and methodology, the fifth section gives the results, and the last section concludes.

2 Government's 2005 Policy on Housing

2.1 The Two-Pronged Process[3]

The main feature of the document published by TAFREN on 15 March 2005, detailing the policies and implementation guidelines for the post-tsunami housing programme was the definition of a buffer zone within which no house was allowed to be built or reconstructed.[4]

The policy encouraged household-driven housing reconstruction in areas where rebuilding was allowed. The affected households were given cash grants in stages,

[2]Grama Niladhari Division is the smallest administrative unit in Sri Lanka.

[3]This section summarizes the "Assistance Policy and Implementation Guidelines—Housing and Township Development" by the Presidential Secretariat.

[4]In the districts of Kilinochchi, Mannar (in Northern Province), Puttalam, Gampaha, Colombo, Kalutara (In Western Province) and Galle, Matara, Hambantota (in the Southern Province) the buffer zone was 100 m landwards from the mean high water line, while in the districts of Jaffna, Mullaitivu, Trincomalee, Batticaloa and Ampara (in the Eastern Province) the buffer zone was 200 m from the mean high water line. The Survey Department, the Urban Development Authority (UDA) and the Coastal Conservation Department (CCD) were given the responsibility of demarcating the buffer zone.

and the homeowners were given the task of managing their own reconstruction process. In areas where reconstruction was not allowed, the households were be relocated to a house on land in close vicinity of the original house, keeping communities together as much as possible.

All 'households' whose houses were either lost or damaged by the tsunami were entitled to receive assistance. A 'household' was defined to be all individuals who were living in one house prior to the tsunami. In the case of extended families living under one roof, only one house was to be provided (see Tables 1 and 2 for a summary of the housing policy and its process).

Table 1 Housing reconstruction assistance policy

Within buffer zone	Outside buffer zone
No construction is permitted	Fully damaged: if repair cost more than 40% of replacement cost
New house on land allocated by government	Fully damaged: grant of SLR 250,000 Partially damaged: grant of SLR 100,000
Beneficiaries are not required to demonstrate ownership to land	All beneficiaries must show ownership of land
House for house: extended families living in the same house not entitled to more than one house	

Table 2 Housing reconstruction: process flow and managing agency

Donor-driven relocation (new houses/new land)		Owner-driven rebuilding (Old houses/existing land)	
Process	Managing agency	Process	Managing agency
Land identification	DS/UDA	Damage assessment	DAT
Donor screening	TAFREN	Beneficiary selection	GOSL
Land allocation	High-level committee	Grant approval and stage 1 release	Divisional secretary/ state banks
Building and planning guidelines	UDA (district level)	Approval of building plans	Local authority
MOU	THRU (national level)	Grant stage 2 release	DAT; state banks
Agreement on technical details	THRU (district level)	Implementation support	THRU (DSD level)
Project monitoring and facilitation	THRU (district level)	Progress review and certification	THRU (DSD level)
Certificates of conformity	UDA (DSD level)	Grant stage 3 and 4 release	DAT; state banks
Allocation of houses	GOSL	Certification of conformity	UDA

Notes: *GOSL* Government of Sri Lanka; *UDA* Urban Development Authority; *THRU* Tsunami Housing and Resettlement Unit; *DS* District Secretary; *DSD* Divisional Secretary Division; *DAT* Damage Assessment Team; *Div Sec* Divisional Secretary
Source Presidential Secretariat (2005)

2.2 Donor-Driven Relocation

No (re)construction of houses that were partially or fully damaged was allowed within the no-build buffer zone. All the households located inside the zone were given a house which was built with donor assistance on land allocated by the state (IPS 2006). Households were not required to show land-ownership of their pre-tsunami house. The new houses were to be built according to guidelines issued by the Urban Development Authority (UDA). Most importantly, they were to have a floor area of 500 ft^2, with access to electricity, running water, sanitation, and drainage facilities (IPS 2006). In addition, the proposed houses in urban and rural settlements were to have facilities such as road systems, recreation, and the like.

The Process
The District Secretary (DS) and UDA were to identify lands suitable for community relocation within the same Divisional Secretariat Division (DSD). A list of approved lands was to be made available for the eligible donors (selected by TAFREN) from which the donors could choose a location for their assistance. Once a donor selected a location, he was given guidelines for planning the rebuilding. A memorandum of understanding was signed by the donor with TAFREN. Once the houses had been completed, they were allocated amongst the eligible tsunami-affected households by the GOSL.

2.3 Owner-Driven Rebuilding

All households that were outside the no-build buffer zone were eligible for a grant by the state to rebuild their houses. To be eligible for the grant, the households were required to show ownership of the land. The grant was given in stages according to the extent of damage to the house as follows:

(1). A household with a fully-damaged house, where the cost of repair to the house was deemed to be more than 40% of the replacement cost of the house, was given a grant of SLR (Sri Lankan Rupees) 250,000, to be disbursed in four stages (IPS 2006).
(2). A household with a partially-damaged house, where the repair cost was less than 40% of the replacement cost of the house, was given a grant of SLR 100,000 disbursed in two stages (IPS 2006).

Households that were within the no-build zone were also given the option of using this grant to build on land outside the no-build zone, as long as the land was owned by them and was located within the same DSD.

The households which successfully utilized the housing grant were eligible to receive a concessionary loan of SLR 500,000.

Beneficiary Selection Process
A three-stage process was adopted for selecting beneficiaries for the housing programmes. A Damage Assessment Team (DAT) was set up by the Divisional Secretary in each Grama Sevaka Wasama (GSW, the smallest administrative unit of the government) for damage assessment. The team consisted of a divisional technical officer, the Grama Niladhari (GN, the officer-in-charge of the GSW), a donor representative appointed by the DS, and a member of the Village Rehabilitation Committee (VRC). The VRCs were especially established in each GND to assist in the post-tsunami reconstruction efforts and consisted of community leaders.

In the first stage, the DAT assessed the eligibility of a household for the housing programme. Those eligible[5] were assessed to determine under which housing programme they were eligible to receive a house. In the second stage, the GN and the DS published the preliminary beneficiary lists. Grievances relating to the assessments by the DAT were recorded and addressed at a meeting of the VRC. Grievances that could not be resolved at the VRC level could be referred first to the Divisional Grievance Committee, and those still unresolved at that level to the District Grievance Committee, whose decision was taken as final. In the third and final stage, the beneficiaries were issued beneficiary certificates and a unique identification number. A panel of auditors nominated by the GOSL/TAFREN was appointed for the purpose of assessing that the procedures were duly followed and that funds were disbursed adhering to proper accounting standards.

Donor Coordination
All donor agencies, which were engaged in housing construction in a given district, were required to register with the DS of that district to avoid duplication of assistance. Government agencies at various levels, bilateral and multilateral agencies, individuals and other agencies were considered as donors for this purpose. Any donor that wished to provide houses without adhering to the government-stipulated procedures were still required to register with the District Manager of the tsunami Housing and Resettlement Unit of the DS.[6]

3 The Government's Revised Policy on Housing

Towards the end of 2005, it was apparent that declaration of the initial no-build buffer zone was holding back the progress of housing development. In March 2006, the Reconstruction and Development Agency (RADA) was established to take over

[5]Homeowners with tsunami-damaged houses who were citizens of Sri Lanka, who have permanently settled in the GSW affected by the tsunami, and were able to demonstrate land ownership, were eligible for the grant.

[6]No donor was allowed to assist to rebuild a house in the no-built zone, and a beneficiary receiving assistance from a donor outside the government housing programme was still eligible for assistance under the government housing programme.

(from the earlier TAFREN) the functions of coordinating the post-tsunami recovery activities amongst the district secretariats and donor agencies. In May 2006, RADA announced a 'Revised Tsunami Housing Policy' whose main aim was to relax the earlier enforced no-build buffer zone and to speed up the housing programme in order to provide a house for all affected people by the end of 2006. The new policy document promised 'a house for all affected people regardless of land ownership'. It defined two zones: land belonging to state reservations within tsunami-affected areas, and land outside state reservations. The beneficiaries were given the option of choosing amongst four housing options as specified below. The costs were to be shared by the government and the donors. Legal inhabitants or encroachers on state land were allowed options 1, 2, and 3, in that sequence. That is, only when option 1 was not possible was option 2 allowed, and only when both options 1 and 2 were not possible, was option 3 allowed. Encroachers outside state land, in zone 2, were given options 2 and 3; and lastly, those on their own land were given option 4. Interestingly, even encroachers in partially damaged houses were given options 2 and 3 for relocation (RADA 2006):

(1) Government land and donor-built house.
(2) Government land and a government cash grant of SLR 250,000 to construct a new house and donor assistance for completing the house.
(3) Government cash grant to purchase land (SLR 150,000 for Ampara and SLR 250,000 for Colombo) and a cash grant (SLR 250,000) to construct a house, and donor assistance to rebuild the house.
(4) Housing construction grant (SLR 250,000 for fully-damaged homes and SLR 100,000 for partially-damaged homes) and donor assistance for rebuilding the house (only for fully damaged houses).

The new housing policy increased the housing needs by 120,000 units, mainly due to the decision to provide houses for those even without legal ownership of land outside the former buffer zone.

4 Methodology

With the objective of informing policy makers on the survivor perceptions of the GOSL post-tsunami housing programme, IPS undertook a survey funded by the World Bank to interview 622 households in 14 GNDs randomly selected from six tsunami-affected districts in two provinces: Galle, Hambantota, and Matara districts in the Southern Province; and Ampara, Trincomalee, and Batticaloa in the Eastern Province.[7] The houses of all interviewed households were destroyed or extensively damaged due to the tsunami. Households both inside and outside the no-build zone

[7]The study initially attempted to survey all tsunamiiaffected districts. But, due to access problems, the survey could not be carried out in the LTTE-controlled Jaffna and Mullaitivu districts.

were interviewed to obtain their views on the housing programme. The interviews were initially conducted in April 2005. In June 2006, the same households were revisited to obtain information on the progress of housing and the householders' views on the housing process (henceforth, the two surveys are referred to as IPS-TS 2005 and IPS-TS 2006).

First, using IPS-TS 2005, the study will analyse the views of the households on the initial housing programme. Specifically, it will examine the awareness of the beneficiaries about the government's policy on housing, issues relating to eligibility requirements, implementation bottlenecks encountered, welfare effects of the housing programme, and disputes and social tensions relating to post-tsunami housing. In addition, using both surveys, the views of the households on the no-build buffer zone and the housing policy and issues relating to donor coordination will also be examined.

5 Results

5.1 Characteristics of the Households

The IPS-TS 2005 survey had 341 households outside the no-build zone and 281 households within the zone. This section describes the household characteristics of those surveyed.

Household Characteristics outside the Buffer Zone

The main pre-tsunami livelihood of the head of the household of those surveyed outside the no-build zone was 'fishing-related activities' (39%), followed by 'not working' (23%), 'trade' (8%), and 'manufacturing' (6%). About 19% were engaged in various other livelihoods. The majority were poor; monthly earnings for most (65%) were SLR 5,000 or less, while a further 25% were earning SLR 5,000–10,000 a month. Culturally, the households were a mixed group: by ethnicity, 21% Sinhalese, 42% Sri Lankan Tamil, 4% Indian Tamil, 30% Sri Lankan Moor, and 3.5% Malay; and by religion, 21% were Buddhist, 39% Hindu, 33% Muslim, and 6.5% Christian. About 16% of households were female-headed pre-tsunami (IPS 2006).

Household Characteristics in the Buffer Zone (of those in Unusable Houses)

According to IPS (2006) the majority (65%) of heads of households were engaged in fishing-related activities prior to the tsunami, followed by trade (8%), manufacturing (5%), and about 15% were in various other livelihood categories and the rest were not working. Most of the household heads were still alive,[8] and just over a quarter were able to continue with their usual activities. These were also a mixed group of households (IPS 2006) by ethnicity and religion. By ethnicity, the majority

[8]5% died during the tsunami.

were Sri Lankan Tamils (52%) followed by Sri Lankan Moor (22%), Sinhalese (19%), Indian Tamil (3%), Malay (2.5%). By religion, majority were Hindu (35%) followed by Muslim (25%), Buddhist (18%), Christian (22%). A small percentage (3%) of the households was female-headed before the tsunami. However, there was significant variation in this number across GNDs.

5.2 *Issues Relating to the Tsunami Housing Programme: Owner-Driven Rebuilding*

Eligibility

One requirement for the owner-driven rebuilding programme was that the household should own the land where the original house was located. The evidence of the 2005 survey showed that, although about 70% of the households owned the land of their original house, about 16% of the households were living on land owned by the state. In addition, about 4% of the households had built on undivided inherited land, while about 7% were tenants and 1% were encroachers.

The eligibility also included proof of ownership of land and proof of Sri Lankan citizenship. Less than half the households were in possession of their deeds, while about 17% had other types of proof of ownership. About 23% had lost their title deeds during the tsunami. Another requirement was proof of Sri Lankan citizenship. A national ID card, passport, driving license or birth certificate was accepted as proof of citizenship. The main identifying document owned by individuals was the national identity card, as most household heads did not have driving licenses or passports. About 41% of those surveyed had lost their ID cards during the tsunami.

Loss of identifying documents and proof of land ownership delayed the progress of the housing programme. But the government took prompt action to rectify the situation by reissuing these documents. The requirement of showing ownership of land also left a fair share of households homeless, as some of the lands the houses were built on were owned by the government. The 2006 housing policy revised this requirement and provided opportunities for those occupying government land previous to the tsunami to rebuild on alternate land provided by the government or land purchased by the household.

Awareness of the Housing Programme

One objective of IPS-TS 2005 was to find out whether the tsunami-affected were aware of the government's housing programme. It was encouraging to see that almost all households were aware of the entitlement to rebuilding grants and loans. However, accessing loans from banks for rebuilding was a problem, as most households had lost their documents during the tsunami. About 23% of those surveyed claimed that they lost their deeds and 41% claimed that they lost their national identity cards during the tsunami.

Ability to Build Own Houses

The owner-driven rebuilding programme put emphasis on houses built by the owners with the help of community-based organizations. The IPS-TS 2005 revealed that most households (62%) were unable/reluctant to manage the rebuilding of their houses. Lack of skills[9] and equipment for house building, concerns regarding the accessibility to funds and their sufficiency, concerns about the availability of material and labour[10] were the main concerns of the households. Close to half the surveyed households preferred pre-built houses, while only 25% preferred credit, and 10% preferred building material. About 10% did not care either way.

Given the escalation in prices following the post-tsunami construction boom, the funds allocated for rebuilding were seen as insufficient to build strong houses. Further, as only 23% of the household heads were members of any community-based organization, they were unable to make use of their social connections for the house-building purposes. However, in some locations the community-based organizations—like religious groups, rural development society, fishery society, local welfare society—were strong and were able to assist the households, particularly in the east.

In most areas, there were skilled and unskilled workers, except in the Eastern Province. However, many skilled workers had lost their equipment during the tsunami.

Welfare Effects

A significant proportion (averaging 37%) of households outside the buffer zone revealed that they would like to rebuild in an alternate location, primarily because they were afraid of living too near the sea. However, the housing programme did not make room for rebuilding in alternate locations.

Despite complaints about the size of the grant, the survey suggests that over 40% of the houses that were destroyed or made unusable were worth less than SLR 250,000. A large percentage of the households, particularly in the Eastern Province, reported having houses worth less than SLR 250,000. This shows that, assuming that the cost of replacement is similar to the value of the house, 40% of the households would be eventually better off in terms of housing. The actual percentage improvement in housing stock is likely to be higher than the observed value given the likelihood of over-reporting house values.

However, some households claimed to own houses that were worth more than SLR 750,000, particularly in the Southern Province. These households would be worse off, as the house-construction grant was limited to SLR 250,000 with the option of applying for a concessionary loan of up to SLR 500,000.

---

[9]Only 9% of households had family members involved in construction, and only 7% said family members were willing to be trained in construction, while only 5% of the households had equipment needed for house building.

[10]Sand was particularly difficult to obtain, and many building materials were rising in price. About one-third of the households were sure of getting help from the community to rebuild, while a slightly higher number were willing to help others to rebuild.

Disputes
The 2004 tsunami, in the worst-affected areas, swept away all buildings that lay in its way, including most boundary demarcations. However, IPS-TS 2005 revealed that boundary disputes were rare and those that existed were resolved through local mediation or resurveying of land. Disputes over post-tsunami ownership were also limited.

The government official, the Grama Niladhari (GN), in charge of a GND, had a large role to play in the rebuilding process. He/she played a central role in making beneficiary lists and needs assessments. In 2 of the 14 GNDs that were studied, there were allegations that the GN was incompetent or corrupt. However, in the majority of cases, the people were satisfied with the work done by the GN.

5.3 Donor-Driven Relocation Programme

Awareness about the Housing Programme
As with the owner-driven rebuilding programme, households within the no-build zone, which were to be relocated, were well-aware of the no-build zone and their housing entitlements. The only confusion was regard to ownership of land in the no-build zone. Many were concerned that once they moved, their land would be taken away by the government. In Hambantota District, some of the affected land has already been identified for a harbour development project, which the inhabitants had been resisting. There was some concern that moving away from this land would facilitate the government's move to acquire this land for the harbour project.

Land for Relocation
One major constraint in the donor-driven relocation programme was the lack of land availability. In some areas, land was identified quickly; these included state land as well as land owned privately. However, in other areas, especially urban, land was a problem.

Most households (90%) preferred single-storey detached houses, as their livelihoods and living practices depended on the availability of a garden. However, contrary to people's preferences, many sites identified for relocation were only sufficient for flats or apartments. For example, more than three-quarters of the households used biomass (fuel wood and paddy husks) for cooking. Living in an apartment or a flat would not allow them to continue using biomass. Further, access to biomass is also limited in crowded living spaces. Alternate sources of fuel are too expensive for most households.

In most communities, (10 out of the 14 GNDs surveyed), the community leaders were aware of the relocation sites. In half of these locations, households were willing to move to the new locations. This was partly due to the fact that some individuals had been displaced prior to the tsunami due to the ongoing secessionist conflict. These households welcomed the opportunity to move to a house of their own. However, in about half the communities, households were not happy with the identified lands for relocation. They wished either to remain in the no-build zone or requested for better land. The main reasons for dissatisfaction with the identified land were greater distance from the sea,[11] and break-up of communities.[12]

At the time of the survey (six months after the tsunami) close to three-quarters of the households were unwilling to rebuild on the same site because of the fear of another tsunami. However, households wished to move to locations which allowed them to continue with their pre-tsunami livelihoods, and where they could move as a community. In addition, they required that the new locations have access to infrastructure such as water, electricity, transport, health, education and places of worship, and not be prone to floods and wild elephants (especially in the east). In GNDs where the people did not know of the relocation sites, especially, the people were anxious about where they would be located.

The beneficiaries have had little voice in selecting suitable lands for relocation. Many people were unsure whether they would be able to keep their old land after relocation.

Willingness to Relocate

A multivariate analysis of willingness to relocate revealed that many factors affected the willingness to relocate.[13] First, the individual characteristics of the head of the household influenced willingness to relocate: male heads of households, heads of households in agriculture, and heads of households who were unable to continue in their usual livelihood were less willing to move. Moors and Tamils, and less-educated heads of households, were more willing to move. Second, psychological factors seem to affect the willingness to move of individuals. Households that experienced flood water levels of more than 12 ft were more willing to relocate. Households with elderly individuals were less willing to relocate—perhaps due to greater attachment to the place of residence. Third, the economic status of the individuals seems to affect the willingness to move. Poorer households (with income less than SLR 5,000) were more willing to relocate. Finally, pre-tsunami housing conditions seem to influence willingness to move of individuals: individuals with either no toilet facilities or only outside toilet facilities were more willing to relocate.

Welfare Effects

In terms of owning property, most individuals would be better off after relocating. About 78% of the households that claimed to own their house also had land tenure. These people would not be worse-off in terms of owning property and land, assuming that the government provides title to the new houses. A little more than one-eighth (13.2%) of the individuals owned houses on government land. If the

---

[11]About 68% were very concerned about the distance to the sea, while 11% were unconcerned about the sea.

[12]Although all relocated houses were moved to one location, they had to leave behind neighbours whose houses were outside the no-build zone.

[13]The highlighted points are based on results that were statistically significant at more than 10% level.

government provides these individuals with houses and land ownership in the new location, they would be better off in terms of owning property. The remaining 9% of the households that owned houses built on land owned by other persons would also benefit from relocating.

In terms of house size, most households would be better off after relocation. The government's housing policy aimed to provide houses with a floor area of at least 500 ft^2. According to the IPS-TS 2005 survey, 53% of the households lived in houses with floor areas less than 450 ft^2 prior to the tsunami. In addition, about 12% of households would benefit from the government's policy of providing drinking water from a protected source.

Some households would benefit from the government policy of providing houses with roofs and walls made of permanent material. Of the damaged houses in all surveyed GNDs, only a little more than half had walls made from permanent material while more than half had roofs made from permanent materials.[14]

Other Issues

The government's policy was to provide a house for a house. However, in cases where more than one family was living under one roof, each family was hopeful of obtaining a separate house.

Relocation has brought forward some social conflicts. In some areas, households were unwilling to break up their small minority communities; in other areas, households in the neighbouring areas were unwilling to accept newcomers belonging to different castes and religious groups. In areas where government-identified land has been rejected, people have identified new land to be relocated. This was the case particularly in Ampara District, where ethnic issues are complex and communities are organized in strong religious groups.

5.4 Progress and Perceptions of Process One and a Half Years on

As described earlier, IPS resurveyed the tsunami-affected households 18 months after the onset of the tsunami to obtain their views on the tsunami housing programme. The following highlight the perceptions of the households on both the progress and the process of the housing programme.

Progress

Eighteen months after the tsunami, close to half of the tsunami-affected households were still in temporary housing (IPS 2006). However, the progress varied across different locations. With less than 6% in temporary housing, progress was best for those living outside the zone in the Southern Province. The tsunami-affected in the

[14]Some houses had both the roof and the wall of permanent material while others had either the roof or the walls of permanent materials.

Eastern Province were worse off with regards to housing, irrespective of whether they were inside or outside the zone (IPS 2006). Amongst those in the Eastern Province, those affected by the conflict were also the slowest to receive a solution to the housing problem (IPS 2006). This was due to a variety of factors. Being located closer to the origin of the tsunami, the Northern and Eastern Provinces were hardest hit by the tsunami (around two-thirds of the deaths and 60% of the displacements were reported from this region). The need to liaise with the LTTE further compounded and delayed the rehabilitation process in the conflict-affected areas, particularly the areas that were under the control of the LTTE. The remoteness of the affected areas, especially their distance from Colombo, was another impediment that slowed the progress.

One main reason for the slow progress in providing housing to the tsunami-affected was the declaration of the 'no-build' zone. Issues with identifying land for settlements and delays in donor assistance contributed greatly to slow the progress of donor-driven housing (IPS 2006). Changes to the buffer zone rules, and issues with obtaining housing material, slowed the progress in the owner-driven programme.

According to RADA, lack of commitment by some NGOs, corruption amongst NGOs, and unilateral decisions to terminate their MOUs due to misunderstandings between NGOs and the government, were some reasons for the delays in the donor-driven housing programme (Weerakoon et al. 2007). For example, according to the *Sunday Observer* of 27 August 2006, international NGOs, who had pledged to build 70,000 houses and signed agreements for 19,600, had only completed 2900 houses. These include: CARITAS that pledged 26,000 houses and completed 72 houses; World Vision Lanka that pledged 10,000 houses, but completed 198 houses; and Red Cross that pledged 15,000 houses, but completed only 169 houses; and Care International that pledged 6,800 houses and completed 125 houses. NGOs have blamed delays on the part of the government in providing suitable land as the main reason for delays in the housing programme. In addition, issues with internal procurement procedures, which the INGOs had to adhere to, may have also slowed progress. These findings corroborate the IPS-TS survey results, which also show that delays in obtaining land and donor assistance slowed progress of the donor-driven housing programme (IPS 2006).

Increases in the cost of building materials and skilled labour, arising from greater demand, also contributed to slowing the progress in housing (IPS 2006). IPS-TS 2006 results also show that costs of labour and material inputs needed for housing construction has increased.

The study has information on the location with respect to the 2005 buffer zone, and house and land tenure for 559 households. Of these, 268 were eligible for the donor-driven housing; 157 were eligible for owner-driven housing; 134 were not eligible for a new house, either because they were not house owners before the tsunami (70% of 134) or because they were outside the 2005 buffer zone, and were house owners without land tenure (30% of 134). The housing progress was the worst for people who were eligible for donor-driven housing; 65% of these were still living in temporary housing. The completion of houses was fastest for owner-driven housing. At the time of survey, only 17% of those expecting houses under the owner-driven housing programme were in temporary housing. However, about 66% of those not eligible to receive a new house also received housing under this scheme (IPS 2006) (see Table 3). Of the people in temporary housing and eligible for donor-driven housing, 46% were expecting to relocate to a house in a new location. Even some who were eligible for the 2005 owner-driven housing programme were expecting to relocate, and others who were not eligible to relocate, were also waiting to relocate (Table 3).

The results also confirm the significant regional variation in housing progress across the country. The Southern Province had met—and in some districts exceeded—the requisite number of housing units with yet more houses under construction. By contrast, the Eastern Province, with the highest requirement of housing, was lagging well behind. The Western Province had also not fared well relative to the Southern Province, most likely due to greater difficulties in obtaining suitable land.

Main Sources of Funds and Access to Assistance

Government was the main source of funds for rebuilding. However, when the source of funds was private parties, such as foreign individuals and the private sector, the house completion was faster. But, only a small share obtained funding through private parties. Building of houses was slowest when individuals used their own funds for house building. Most recipients of housing observed that obtaining funds from the government was fairly easy (see Table 4).

Perceptions of House Quality

The new houses were largely made of better-quality material compared to pre-tsunami houses, especially when houses were owner-funded, or funded by the government or an NGO. However, the new houses owners receiving funds from NGOs or other private sources were less able to participate in designing their house (see Table 4).

For example, many households observed that they were worse off in terms of access to their own source of water. Further, as their new location was more rural, their access to roads, pre-schools, health-care facilities was also reduced (see Table 5 and Fig. 1). Moreover, a large share (80%) observed that their access to employment has also worsened due to relocation. About half of those in the donor-driven housing programme were also not happy with the quality of their new houses. Despite these grievances, no one in the donor-driven housing programme indicated a willingness to sell their new house. The survey results further showed that lifestyles and socioeconomic needs of the house recipients were not taken into account when designing houses. The delays in identifying beneficiaries are partly responsible for this mismatch (IPS 2006).

A large share of the relocated house owners did not have legal ownership of their houses. Less than 6% received legal ownership of their new house.

Table 3 Housing situation as in July 2006 (by eligibility under 2005 housing programme)

Eligibility	Rebuilt		Relocated		Donor-built on old site		Temporary housing		NI[a]		Total	
	No.	%	No.	%	No.	%	No.	%	No.	%	No.	%
Donor-driven	52	19.4	25	9.3	11	4.1	175	65.3	5	1.9	268	100.0
Owner-driven	76	48.4	25	15.9	22	14.0	26	16.6	8	5.1	157	100.0
Not eligible[b]	34	25.4	17	12.7	24	17.9	59	44.0	0	0.0	134	100.0
Total	162		67		57		260		13		559	

[a] *NI* No information

[b] Households not owning a house before the Tsunami (70% of 134) and households owning a house on encroached land (30% of 134) outside the buffer zone were not eligible for a new house under the 2005 housing policy

Source Own calculations based on IPS TS (2005) and IPS TS (2006) data

Table 4 Perceptions on rebuilding of houses by source of funds

	Sample size/%	Completed[1]/%	House bigger[1]/%	Materials better[1]/%	Participated in house design[1]/%
Using own funds or funds from relatives[2]	59/100	10/17	3/5	35/59	44/74
Receiving government funds[2]	134/100	47/35	16/12	63/47	66/49
Receiving funds from NGOs[2]	54/100	20/37	5/9	27/50	15/28
Receiving funds from other sources[3]	7/100	6/86	0/0	2/29	5/71

Notes: 1. The percentages are out of the total number of households given in the first column. 2. Major portion (half or more) of funds from this source. 3. Others include foreigners, local individual donors, priests, leaders and institutions
Source Own calculation using IPS-TS (2006)

Table 5 Quality of houses: relocated households

	Sample size (100%)	Better now (%)	Worse now (%)	No difference (%)
House design	69	42	41	14
Construction materials used for housing	69	27	49	21
Access to services (water, electricity, road)	69	14	63	20
Primary school within 1 km	73	10	62	29
Clinic within 1 km	73	1	59	40

Disputes

The magnitude and type of conflict varied across GDNs (IPS 2006). The information gathered from the survey is insufficient to explain these differences. But anecdotal evidence suggests that the conflicts were higher in areas where aid was distributed in an ad hoc manner, disregarding actual needs. In some cases individuals who were not affected by the tsunami received assistance, while those who were genuinely affected were not given assistance. The disputes were also higher in heterogeneous populations. Some (more affluent) in these areas were resentful that the assistance provided did not fully compensate for their loss, while others were better off after the tsunami. Encouragingly, according to the GNs, most of the disputes could be resolved informally though discussions (IPS 2006). Starting of formal dispute resolution mechanisms from the initiation of aid delivered through the government has reduced the occurrence of long duration conflicts (IPS 2006).

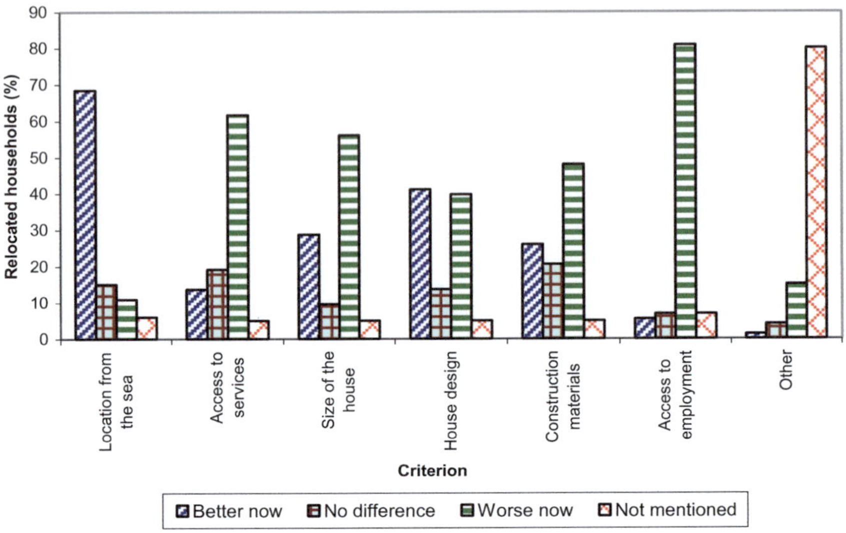

Fig. 1 Perception on location, house quality, and access to services, relocated. *Source* Own calculations using IPS-TS 2006

5.5 *Donor Coordination*

IPS-TS 2006 results show inconsistencies in the government's housing policy and what has happened in practice, which has complicated the monitoring of housing progress even more. Many households (19%) within the zone have rebuilt (although they were asked to relocate), while many outside the zone have relocated (16%) (IPS 2006). There were also households who have received houses outside both these programmes, and some other households which have received houses despite not being eligible for a house (IPS 2006). This is partly due to the ad hoc nature in which aid was distributed by various donors, initially, and the inability (due to lack of capacity as well as the lack of authority) of the local-level government agents to effectively coordinate aid distribution at the local level. According to government officials at the district- and GND-level, the reluctance of some non-government organizations to share information, on beneficiaries and the distribution of aid, has made coordination of aid delivery and monitoring of progress difficult (IPS 2006).

The survey results reveal many inconsistencies in the provision of houses by donors (IPS 2006). Some have not provided houses according to government guidelines. The results also shows that macro-level numbers on the progress of the housing programmes, based on the number of houses needed and the number provided, is misleading. This is because some beneficiaries of completed houses were not among those identified to receive a house (IPS 2006).

Management of aid delivery was centralized and most of the additional resources procured for aid delivery were enjoyed by these central organizations. However, the local level government agencies were an essential part of aid delivery as they had the information on the needs of the people. Not enhancing the capacity of local level institutions to deal with the additional work load led to delays and inefficiencies in the system.

5.6 *Perceptions of the Change in Housing Programme*

The revision of the housing policy has created some confusion, causing beneficiary lists to be revised. The results of the IPS community-level survey show that the eligibility for relocation has decreased from 410 to 239 with the change in the housing policy (IPS 2006). This change has affected households within the 2005 buffet zone more. However, only a small share of households (15%) indicated that the change in the policy affected them a lot; the majority were not affected by the policy (IPS 2006). The change in policy created a lot of confusion; many surveyed (about 25%) were not clear about their housing entitlement (IPS 2006). A large proportion (60%) indicated that they would like to have legal advice on their rights as a homeowner (IPS 2006).

A large proportion (60%) of people surveyed in IPS-TS 2006 were happy about the 2005 buffer zone (IPS 2006). This perception was verified by the GNs at the community level. Paradoxically, GNs who commended the original buffer zone policy were also happy with the change of the buffer zone in 2006. The main reason why people were initially happy with the imposing of the buffer zone was that a majority (close to two-thirds) of them were afraid of living too close to the sea, immediately following the tsunami. Also, poorer households, and households with poor-quality houses, were happy about the prospect of relocating to a well-built house. However, not all were happy with the lands identified for relocation. In a few places, difficulties in identifying suitable lands for relocation, delayed the housing process. In some instances, these lands were a few kilometres away from the sea and had fewer facilities, compared to premises by the sea. In urban areas, the new houses were built in close proximity to each other, with little or no garden space. In such instances, people were happy to return to their original houses. Also, even with the relaxation of the buffer zone, individuals living on encroached land were relocated and given a house or funds for building a house. As such, their situation was not affected by the relaxation of the buffer zone in 2006. Further, with the relaxation of the buffer zone, even individuals living on encroached land outside the buffer zone were given the option of relocating, a benefit that was not available for them earlier.

Results of the IPS-TS 2006 show that despite delays in getting houses, most people have better houses now (IPS 2006). The results show that pre-tsunami houses were generally smaller than the specified 500-square-foot minimum floor area for new houses.[15] Further, new houses were built with permanent material, while a large share of destroyed houses had been built with temporary material (IPS 2006).[16] Also, households that did not legally own houses were able to get legal ownership of houses under the donor-driven program (IPS 2006).[17] The above findings show that, despite delays, tsunami-affected people benefitted from the housing program (IPS 2006).

The main reasons for being affected by the housing policy were due to its coverage: many who were not eligible for a house under the revised policy hoped for a house. People were less concerned about the specificities of the housing policy (IPS 2006).

6 Conclusions

Post-tsunami rehabilitation was a learning exercise for both the government (at its various levels) and its development partners. The quantum of aid received and the lack of capacity of the government to efficiently distribute it to the people and places most needing aid, and the lack of coordination amongst the donors, increased wastage and reduced the effectiveness of aid. The experience of the two-pronged post-tsunami housing programme in Sri Lanka provides clear comparisons of good and bad practices to adopt when distributing aid following a disaster. In the case of owner-driven rebuilding, the government's approach was to involve the beneficiaries in the housing programme from the outset, and provide assistance to rebuild with minimum disruptions to communities. The opposite was true in the case of the donor-driven relocation programme. Whole communities were shifted to different locations, assistance was provided by a multitude of donors, and the involvement of the beneficiaries in the housing programme was minimal. Unlike the owner-driven housing programme where the owners managed the rebuilding process, in the relocation programme, the beneficiaries were not identified until houses were built. This resulted in building houses that were not suitable for the lifestyles of the beneficiaries. Although the owner-driven rebuilding programme had its own problems, such as price escalations and shortages of skilled labour and building material, the programme was largely a success. In contrast, the donor-driven housing programme encountered many problems, and resulted in large delays and disruptions to people's lives. Moreover, the survey findings show that it has also created some unexpected social tensions amongst the beneficiaries and their neighbours. A summary of the research findings and the contrast between the two

[15]About, 53% of the surveyed houses made unusable by the tsunami were less than 450 ft^2, while only 10% were bigger than 600 ft^2.

[16]About 32% of roofs of the surveyed houses were made of *cadjan* or metal sheets, while close to half the surveyed houses had walls made of temporary material.

[17]About 13% of surveyed households inside the buffer zone owned houses on government land, while a further 9% owned houses built on other people's private land.

approaches to housing are highlighted below. Overall, the housing progress was worst for people who were eligible for donor-driven housing; 65% of these were still living in temporary housing at the time of the second survey. Housing progress was best for households eligible for owner-driven housing.

6.1 *Coverage and Access to Assistance*

Encouragingly, almost all those interviewed were well-aware of the post-tsunami housing programme. However, due to the requirement that the households had to show land ownership, to be eligible for assistance outside the no-build zone, deprived housing assistance to close to one-fifth of the households. However, the revised 2006 housing policy rectified this situation. Initially, there were delays in obtaining assistance as many people had lost identification documents and other proof of housing and land ownership. However, these problems were rectified by the speedy reissue of these documents. Those eligible for assistance were of the view that there were no major problems in obtaining assistance. However, many who were not eligible for assistance, such as extended family members expecting two houses, were hopeful of getting assistance. This created problems in clearing temporary housing arrangements.

6.2 *Speed of Receiving Assistance*

The main reasons for the slow progress in housing were the declaration of the no-build buffer zone, and coordination problems in the donor-driven relocation programme. Finding land for relocating large numbers of households was difficult, especially in urban areas. Also, disagreements regarding the procedural requirements of the donor-driven housing programme led to the abandonment of the agreement by several donors, which resulted in long delays, and the subsequent need to revise the housing programme. For example, the donor-driven housing needed to adhere to the housing and environmental regulations of the government. The long processes involved in getting these clearances were not appealing to the donors. The rebuilding programme in conflict-affected regions was slowed due to problems of access to these areas, and problems of coordinating with the LTTE.

6.3 *Welfare Effects*

Despite delays, most people were better off in terms of house sizes as well as materials used for house construction. However, many relocated people were worse off in terms of access to facilities and infrastructure and places of employment. This could lead to new encroachments into state land.

6.4 Coordination and Grievance Resolution

Overly-centralized management (especially in the case of the donor-driven housing programme), and not paying sufficient attention to building capacity of existing institutions, led to duplication of work, unnecessary delays, and wastage of limited resources. Issues with sharing information in a clear and transparent manner between government and non-government agents also created social tensions and problems of coordination and proper targeting. In contrast, establishment of village councils and a clear dispute-resolution mechanism minimized disputes in the owner-driven rebuilding programme.

References

Centreal Bank of Sri Lanka. 2004. Annual Report, Central Bank of Sri Lanka, Colombo.

Institute of Policy Studies of Sri Lanka. 2005. *Listening to Those Who Lost: Survey and Analysis of Rebuilding and Relocation of Tsunami-Affected Households in Sri Lanka*. Colombo: Mimeo.

Institute of Policy Studies of Sri Lanka. 2006. *Post-Tsunami Reconstruction and Rehabilitation: Household Views on Progress and Process*. Colombo: mimeo.

Jayasuriya, S., P. Steele, and D. Weerakoon. 2006. *Post-Tsunami Recovery: Issues and Challenges in Sri Lanka*. Tokyo: ABDI.

Reconstruction and Development Agency (RADA). 2006. *Revised Tsunami Housing Policy*. Colombo: Presidential Secretariat.

Task Force for Rebuilding the Nation (TAFREN). 2005. *Assistance Policy and Implementation Guidelines: Housing and Township Development*. Colombo: Presidential Secretariat.

Weerakoon, D., S. Jayasuriya, N. Arunatilake, and P. Steele. 2007. *Economic Challenges of Post-Tsunami Reconstruction in Sri Lanka*. Tokyo: ADBI.

Part III
Coping Strategies, Systems and Aid Effectiveness

Aid, Religion and Recovery in Post-tsunami Thailand

Monica Lindberg Falk

1 Introduction

This paper addresses aid and recovery after the 2004 Indian Ocean tsunami in Thailand. It is well known that a disaster is always local and the initial help always comes from those who are close by like family, neighbours, friends, and those who happen to be in the locality (see Fischer 1998, 21–22). That was also the case in Thailand, and the Thai people's swift response to the tsunami was widely acclaimed. Local faith-based organizations are commonly among the first contributors of immediate help, although the help they provide is occasionally disapproved of, especially when it is offered only to co-religionists (Wisner 2010, 129). The Thai government was recognized for its effective emergency response directly after the catastrophe, with efforts focused on immediate rescue and offering medical help providing basic needs. However, it was impossible for the government to deliver quick and effective relief alone. The united assistance of the skills and resources of various groups, individual people, relief organizations, and local networks were necessary (UNDP 2005, 35). After the tsunami, military employees and specialized search-and-rescue teams from various nations worked together with their Thai colleagues. We can conclude that local and international NGOs were extremely important in Thailand's post-tsunami relief work (Falk 2013).

Initially, the government concentrated its efforts on restoring Phuket province, which was the most important of the six affected provinces financially, but not the worst-hit province when it came to lost lives and devastation. Thailand declined bilateral and multilateral support from other nations, but the influx of international

Some information in this chapter can also be found in Falk (2010, 2013, 2014, 2015).

M. L. Falk (✉)
Centre for East and South East Asian Studies, Lund University, Lund, Sweden
e-mail: monica.lindberg_falk@ace.lu.se

S. Reddy (ed.), *The Asian Tsunami and Post-Disaster Aid*,
https://doi.org/10.1007/978-981-13-0182-7_12

non-governmental organizations, faith organizations and other aid providers became overpowering. The size of the catastrophe, the lack of coordination of aid distribution, and mismanagement in tackling the catastrophe led to inequities in the distribution of aid, leaving many survivors without support. While some communities were excluded from government relief efforts, others received excessive amounts of donated materials and a lot of money.

Religion can be important in post-disaster situations. Villages in the tsunami-hit localities in Thailand were predominantly Muslim, Buddhist, or Moken/Moklen.[1] Numerous faith-based associations were active in the tsunami-struck regions and many of them were successful in distributing relief, but there were also problems with disorganization, and mismanagement of aid distribution; there were also international faith-based NGOs that only provided aid if survivors converted to Christianity. The majority of the population were Buddhists and, for many Thai survivors, Buddhism became important for sustaining their lives after the tsunami. Buddhism helped them to handle their extremely stressful conditions. Handling questions about life and death are the special expertise of Buddhist monks and nuns. The survivors went to the temples for explanations about why people had died and also about why they had survived. Buddhism gives comprehensive and profound clarifications about life, how to deal with individual as well as collective problems, how to live in harmony, and how to die calmly. The Buddhist temples were crucial in handling corpses, conducting funeral ceremonies, and transferring religious merits from the relatives to those who had been killed. These kinds of activities supported inner relief that many survivors, at least initially, considered more important than material aid. The role of the Buddhist temples was not to distribute material aid; but, for various reasons, some temples became involved in aid distribution (Falk 2015).

The research on which this paper is based was conducted in tsunami-stricken areas in southern Thailand. The fieldwork extended over a period of five years and began in 2005, right after the tsunami.[2] The focus of this paper is mainly on the survivors' recovery in Phang Nga, the worst affected of the six tsunami-hit provinces in Thailand.[3] The principal techniques involved in the research were anthropological methods, in-depth interviews, life stories, and narratives. The research is also based on informal conversations with survivors, their relatives, and relief workers, members of self-help groups and of national and international NGOs. Interviews were conducted with psychiatrists, aid workers, Buddhist monks and nuns, Christian priests, Christian missionaries, and aid workers. Statistical material about the region was collected from the Provincial Administrative Office.

---

[1]Moken/Moklen are a semi-nomadic ethnic group.

[2]This research project was funded by the Swedish Research Council, VR. The book *Post-tsunami Recovery in Thailand: Socio-Cultural Responses* (Falk 2015), is based on findings from the project.

[3]Since the study is primarily based on information collected from personal observation and interviews, consideration of ethical questions is important. I have concealed the identity of the informants and all personal names in this paper are fictitious.

Reports, various articles dealing with the tsunami have also been used to access information.[4]

This paper aims to provide a descriptive and analytical account based on the findings from the anthropological study mentioned above. It commences by giving a summary of how the 2004 Indian Ocean tsunami affected Thailand, and provides an overview of how Thailand dealt with the relief situation and gives examples of how humanitarian aid was distributed. The paper deals with both national and international aid and with how male and female survivors experienced the aid situation. Special attention is given to the link between religion and relief work. The paper discusses why and how religion became significant in post-tsunami Thailand and examines faith-based aid and ethical concerns in post-disaster situations.

2 The Impact of the Tsunami on Thailand

Six provinces in the south of Thailand were hit by the 2004 Indian Ocean tsunami: Ranong, Phang Nga, Phuket, Krabi, Trang, and Satun. After the tsunami, the coastal regions in the six affected provinces were declared emergency disaster areas. The tsunami affected 25 districts and 95 sub-districts, and damaged 407 villages, affecting a total population of 1.9 million people.[5] The waves caused losses, damage, and tremendous suffering.

The population in the coastal areas that were hit by the tsunami is multi-ethnic, and poor fishing families and migrant workers were among those who were hit hard by the tsunami. The figures mentioned in the news and reports about the deaths and destruction after the tsunami varied widely. Local Thai people that I was in contact with held the opinion that the numbers of deaths presented by authorities seemed to them to be too low. One estimate, carried out about one year after the tsunami, stated that approximately 100,000 people in Thailand had lost a family member or home (UNDP 2005, 13).

Of the 407 damaged coastal villages, 47 of them were almost completely destroyed. The people who had survived had not only lost their homes; they had also lost those who were close to them: families, relatives, friends, and neighbours. In Thailand, the government provided financial support to about 1,500 children who had become orphans or lost one parent; however, there were more children lost one or both parents in the tsunami but were not registered by the government. The catastrophe affected the lives of more than 50,000 people. Thai government's official statistics tell that 8,327 people had died or were missing, and that about 8,500 survivors who had been injured, and several hundred individuals were never

[4]I would like to acknowledge my informants for sharing their knowledge and experiences with me. I wish to thank Ms. Nantana Pidtong for her research assistance. I would like to thank Dr. Sunita Reddy and Dr. Harish Nanda for organizing the "International Conference on Re-examining Disaster, Recovery and Reconstruction: Social Science Perspectives on the Tsunami" in New Delhi.

[5]Department of Disaster Prevention and Mitigation website.

found. However, migrant workers who died in the tsunami were excluded since many of them were without proper documents. According to Oxfam more women than men were killed in the tsunami in Thailand (Pimolwan 2006).

In the Phang Nga province 29 sub-districts and 69 villages were affected by the tsunami. The official number of dead in the province was 4,224, with 5,597 people wounded, and in September 2005, 1,696 people were still missing. Out of 2,508 damaged houses, 1,904 were totally destroyed and 604 were partially damaged. Five school buildings were destroyed and 332 large fishing trawlers and 1,544 small trawlers were damaged. Fishponds and fishing nets nearly 785 million baht (about USD 26.2 million) were destroyed. Cattle, sheep, goats, pigs, and poultry worth 13.66 million baht were lost. Rice fields and gardens worth 2.45 million baht, and restaurants and shops worth 20.36 million baht were damaged. Ninety-three hotels, and many piers, bridges, drainage systems, as well as the electricity, water supply and telephone systems, were damaged. Moreover, mangrove forests, regular forests, and coral reefs were damaged (official statistics from Thai authorities).

3 Humanitarian Aid

The disaster activated a relief operation that became the largest in Thailand's history. The Thai government, led by Prime Minister Thaksin Shinawatra, was praised for its rapid response and active handling of the disaster situation. The Thai government quickly declared that they would not request international aid, since Thailand had enough resources and wanted the funds to be given to countries that needed them more. That statement was also a message to the global community that Thailand was financially stable. However, the government did appreciate special types of support from international associates.

3.1 The Thai Government's Relief Efforts

Thailand dealt with the disaster in three steps. In the first stage, the priority was to search for, rescue, and evacuate those who had survived, to bring vital supplies, and identify human remains. In the next stage, the focus was on preventing infectious diseases among those who had survived. In the third stage, the government concentrated on reconstruction and the longer-term recovery of those affected (UNDP 2005, 23).

Initially, the government focused its efforts on restoring Phuket province, which was the most financially important of the six affected provinces. Immediately after the catastrophe, a meeting was held with cabinet ministers and concerned government officials at the southern Tsunami Disaster Victim Logistics Centre. The government decided that the Thai Army would provide satellite communication in the Phuket area. It also decided that warning calls to residents in Phuket and other disaster-prone provinces to safeguard against after-shock damage were urgently needed. A team would be formed to work together with volunteers to search for survivors and corpses in the province of Phang Nga (see Rojanasunan 2012). Transportation would be provided by the Royal Thai Air Force and a Royal Thai Navy cruiser would be deployed to search for and rescue disaster victims, and helicopters would be used for the delivery of food and drinking water, as well as transfer injured survivors (Report: December 26, 2004 The Tsunami on Phuket 2005).

The Department of Disaster Prevention and Mitigation ordered inflatable boats, engines and rescue vehicles as well as other equipment to be sent from Prachuap Khiri Khan province, Surat Thani province and from Songkhla province. The Department also requested active involvement of personnel from other government agencies, as well as soldiers, police officers, members of charity foundations, and staff from provincial sections. The Thai Army, the Thai Navy, the Thai Air Force, the Thai police, Thai Royal foundations, the Thai Red Cross Society, and private foundations were alerted, and were asked to send people and materials to the disaster area to provide immediate help.

Provincial police helped to coordinate searches for disaster victims together with other government agencies and private sector organizations. Two days after the disaster, the police set up a Disaster Victim Relief Centre at the Phuket Provincial Police Office. The police operations were focused on crime prevention, especially theft and looting of disaster-affected properties. Police also supported forces from the Security Volunteer Civil Section. Other responsibilities included safeguarding people against infections caused by decaying corpses, and supporting the identification procedure by recording forensically relevant points (Report: December 26, 2004 The Tsunami on Phuket 2005, 66–67).

The Fishery Research and Development Centre, established in Phuket province by the Department of Fisheries, set up a support unit to coordinate rescue, relief, and rehabilitation in the Andaman Sea coastal region. This centre gave high priority to helping the fishing communities in Phang Nga province. In addition, Phang Nga province created two support centres, one to assist fishermen and another to assist aquaculture entrepreneurs. The officers of units at central and provincial levels were mobilized, and 138 experts manned 12 vessels. They searched the seas around the archipelagos for survivors and retrieved the corpses of victims (Phang Nga Tsunami 27).

3.2 *National Foundations and Relief Work*

The Buddhist Supreme Patriarch of Thailand donated two million baht to support the survivors.[6] Their Majesties and the Thai royal family gave 30 million baht through the Rajaprajanugroh Foundation, under royal patronage, to help disaster

[6]Converted to US Dollars, 1,000,000 baht is about 30,000 US Dollars.

victims in the six disaster-hit provinces. The royal foundation responded to the need for basic necessities and distributed, for example, 2,250 care packages of milled rice, dry food, and clothing for the relief of the tsunami victims. Of these care packages, 1,000 were distributed in the worst-hit district in Phang Nga province, 750 packages at the Krabi Provincial Hall, and 500 packages were given to survivors in Phuket province. Royal assistance delivered by the Thai Air Force consisted of 1,000 care packages, 54 rolls of white fabric required for wrapping corpses, and 275 coffins (Weber 2005, 59). Through the disaster relief sections of the Chaipattana Foundation, under royal patronage, funds totalling almost 21 million baht were granted to various foundations, also under royal patronage, to assist the disaster victims. The Thai Red Cross Society, under royal patronage, sent out requests by email for the donation of blood, medicine, white fabric, paper tissues, canned food, underwear, sanitary pads, shoes, drinking water, blankets, formalin for the handling of corpses, and coffins (Weber 2005, 59). The Chaipattana Foundation opened a special account named the 'Southern Region Restoration Fund' to administer the numerous contributions from benefactors, including both Thai and foreign organizations, donated to support the tsunami relief projects initiated by their majesties and, upon royal command, supervised by the royal family. The disaster relief staff at the Chaipattana Foundation was divided into three groups and each group took care of two provinces. Their first assignment was to evaluate the damage and to appraise the aid provided by other organizations, particularly on remote islands with precarious transportation links, so as to make assessments of requirements and to propose ways and means to provide help (Woothisen 2005).

Several private sector foundations and organizations, as well as government agencies, collaborated to help schools and students affected by the disaster. The contributions comprised both aid to meet urgent needs and aid to provide support in the long term. The Thai king personally provided funding for the reconstruction of school buildings (Weber 2005, 61).

Immediately after the tsunami had struck, the Phang Nga Chapter of the Thai Red Cross Society mobilized teams and made appeals for help from the general public, business people, and government officials to cooperate in cooking food for the victims. The distribution of donations was organized and implemented to deliver consumer goods, medical supplies, milled rice, dry food, clothes, kitchenware, electrical appliances, bedding, utensils, and beverages. Later, the emphasis shifted to restoring occupations, mental rehabilitation, and housing. These tasks were carried out collectively by the Chaipattana Foundation, the Thai Red Cross Society, the Red Cross Society Federation, the International Red Crescent Society, and other domestic as well as foreign organizations. The Phang Nga Red Cross Chapter actively supported such projects as the Phru Tiao Red Cross Village and the Ban Nam Khem Infant Nursing Centre (Nga 2005, 5).

3.3 International Cooperation

International cooperation was common in the aftermath of the tsunami. One example of this was the Japanese National Defence Force, which provided three vessels and supported the Thai Navy in search and rescue operations to retrieve corpses floating in the sea. Other examples are the United States, and Asian and European countries that also sent rescue teams of more than 400 experts, who brought rescue apparatus, tracker dogs, and search equipment (Weber 2005, 39).

A centre was set up to register persons reported missing and to update lists of missing persons found alive or dead. Volunteers helped in this task as interpreters and translators. The centre's registers of missing people—compiled at the Provincial Halls of Phuket, Krabi, Ranong, Phang Nga, Trang, and Satun, as well as in each and every district office in the disaster areas—were shared to exchange information and coordinate follow-up actions on missing people with the help of hospitals. Relatives of missing persons who were in Bangkok could contact the Residential Registration Office of the Department of Provincial Administration, Ministry of Interior. The search for missing persons was coupled with a DNA identification service (Weber 2005, 39).

3.4 International Relief Work

Despite enormous support from, and mobilization of, the Thai people, the scale of destruction forced Thailand to accept international assistance. Thailand especially appreciated technical help from international associates. The international help for tsunami recovery in Thailand commenced early in the process of relief work. The concern that foreign governments, international NGOs, private companies and private individuals showed were tremendous. Assistance came from different sources for reconstruction after the tsunami in Thailand. One motive for the strong international empathy was the high number of dead and missing foreign tourists in Thailand. The tourists came from about 40 countries and made up one-third of the total who had lost their lives or were missing. This loss prompted foreign governments and people from other countries to offer their help (Falk 2013). Those who had lost people in the catastrophe were united in helping with material aid as well as with other kind of assistance. International and local NGOs carried out significant relief work, and did an exceptional job in assisting those who had survived. In Thailand, about 45 international associates provided technical support and that aid and assistance was made up of almost 200 distinct projects (UNDP/World Bank 2005).

Immediately after the tsunami, organizations and volunteers concentrated their work on providing instant assistance and solving urgent problems, such as handing out essential things and daily supplies. Many volunteers from Thailand, and abroad, provided support to the survivors in the disaster-affected villages. The different aid providers had their own policies, based on their organization's aim and beliefs, that determined who should obtain help and what principles that would be used for each kind of assistance. The aims of each organization were diverse and they guided the different principles they used. Some organizations coordinated the assistance by themselves and operated directly in the affected villages. Other organizations delivered aid to the local authorities for distribution, while several others collaborated with government organizations. Some allowed the survivors and communities to have their own opinion, so that they could explain what they needed. For some organizations, it was significant that aid was distributed equally including avoiding duplication of relief. The majority of the survivors I communicated with had no faith in the local authorities' way of distributing aid because they saw nepotism and various forms of dishonesty (Falk 2013).

3.5 Relocation and Housing Issues

At the end of February 2005, more than two months after the tsunami struck, 60,000 people were displaced by the disaster (Nga 2005, 33). Six months after the disaster, about 7,000 survivors were living in provisional shelters, and the figures for nine months after the tsunami showed that 2,900 persons were still living in shelters (UNDP 2005, 15). The Thai government had received about 3,350 applications for housing assistance and by November 2005 it had built 2,700 houses (UNDP 2005, 32).

In the wake of the disaster, there were arguments over land, and numerous households received threats and were prevented from returning to rebuild their houses. Many of the houses were rebuilt on the same place as before the disaster and, for a long time, these people lived with the fear that a tsunami would hit again. They watched the news on television and listened to the radio and were prepared to run immediately if there was an alert. Some of the survivors could not cope with the anxiety and stress and, therefore, deserted their houses and moved to live further from the sea, which they found safer and more relaxed. That was the reason why many of the rebuilt houses were left vacant or hired by newcomers (Falk 2014). There are also examples of whole villages that were rebuilt on new land, at a safe distance from the sea. When they felt more secure, they moved back to their old place because that was more convenient for fishing.

3.6 Livelihood and Compensation

Disasters typically hit the poor hardest. Phang Nga and Phuket were two provinces that had a prosperous tourist industry and people from other provinces in Thailand were recruited to work there. Migrant workers from neighbouring Myanmar also worked at unskilled jobs in these economically booming provinces (Falk 2014).

Many of the survivors were originally from south, central, northern, and north-eastern Thailand; they worked mainly with construction work or were employed in the service sector (Pimolwan 2006).

The Thai government treated men and women differently and awarded higher compensation for a dead man than for a dead woman, referring to men as the breadwinners. In reality, Thai women are equally important as breadwinners in society and both men and women work outside the home. The government discriminated against women who lost their husbands by refusing to compensate them for their missing fishing boats. The compensation went only to men despite the fact that it was not uncommon for husbands and wives to fish together (Falk 2015).

In the long-term aftermath phase, a pressing problem for many of the survivors was the loss of their livelihoods. It was hoped that unemployment and the lack of jobs would be short-term, but for many it was difficult to find an occupation.

Furthermore, the compensation that they received did not match the damages. In Phang Nga, fisheries suffered 50% of the entire damage but received only 30% of the total value of aid. In contrast, Satun province had only 6.6% of the total damage but received 39% of the total compensation. Fisheries in Krabi received 65% of the total compensation, and that was much more than they should have since their damage was much less. One suggested reason for the uneven compensation could be that Phang Nga was not given the highest priority because of the many ethnic minority groups in the area. Another reason for unequal disbursement might have been the result of the distribution of budgets by central and local governments (Jayasuriya and McCawley 2010, 192–193).

4 Religion in Post-disaster Recovery

Religion has been proved to play an important role in times of crises. It is generally related to the search for meaning behind disasters and it stimulates social interconnectedness (see Falk 2015). Religious beliefs can give the survivors a sense of control. Rituals and ceremonies have been proved to have healing capacities both individually and collectively. Religious institutions are often also central in providing material help. After the tsunami, Thailand was open to all sorts of aid organizations and, soon after the catastrophe, the number of faith-based aid agencies in the tsunami-hit region mushroomed.

In post-tsunami Thailand, religion had both positive and negative impacts. Religion became a refuge for many survivors and a vital factor in the healing process. The Buddhist temples became a sanctuary for the survivors, a place to keep the bodies of the deceased, and also a place with temporary shelters (Falk 2015). However, religion did also prove to be destructive and an obstacle in the recovery phase, very often in relation to the conversion activities of Christian missionaries in the area (Falk 2013). Many, but not all, adhered to the established ethical codes of conduct, which forbids promotion of religious or political viewpoints while carrying out disaster work. Most aid providers followed the rules but there were also those who did not care about ethical concerns or the code of conduct. Those who carried out unethical activities of providing aid in return for conversion often increased the suffering among the devastated and destitute survivors.

After the tsunami, the influx of Christian missionaries to the south of Thailand was noticeable. Some of them took advantage of the vulnerable situation of survivors, who had not only lost family members, relatives and friends but also their homes and their material necessities. The missionaries approached the survivors and offered them material aid and in return, the survivors would have to convert to Christianity. The new religious affiliation of the converts impacted negatively on their relations with people they were close to and they had to exclude themselves from activities that included Buddhist monks. Those who converted in return for material aid were often met with suspicion by the people of the neighbourhood, who thought it was unfair to convert in order to get aid (Falk 2013). Some of the local people I talked to questioned the conversion to Christianity and did not believe that the converts had given up their Buddhist faith. The converts were considered, by others, to have changed their religious belonging in exchange for material goods and that caused rifts in the communities (Falk 2013). The new regulations the converts had to follow restricted them from attending Buddhist ceremonies and, as long as the Christian priests were around, the converts adhered to the rules. When the priests left, they gradually returned to their old routines and visited the temples.

5 Buddhist Temples in Post-disaster Phang Nga

Buddhist monastic life at the temples in Thailand is firmly regulated and establishes a boundary that separated the community of monks and nuns from the lay people. Buddhist temples are a place of shelter in difficult times and lay people visit monks and nuns for counselling. Buddhism in Thailand is gendered throughout and it is only men (monks and novices) that constitute the monastic community *(sangha)*.[7] At some temples, the Thai Buddhist nuns (*mae chi*s) are welcome to live in a separate section but they are not considered members of the *sangha*. Some *mae chi*s live in 'nunneries' that are governed by the *mae chi*s themselves (Falk 2007). Very few temples in the tsunami-hit areas had resident *mae chi*s. The gendered Buddhism in Thailand had implications for how monks and nuns could assist the survivors (Falk 2015, 81–88).

The local communities supported the national relief effort and local Buddhist temples provided shelters and played an important role for individual survivors in many ways. Some Buddhist monks and nuns were wounded in the disaster and

[7]In Thailand the *sangha* refers to the assembly of monks.

there were some monks who were killed by the tsunami. However, most temples were not hit by the tsunami and the temples became one of the main places for people to take refuge. People stayed in tents and temporary shelters, both on the temple grounds and in close proximity to the temples (Falk 2010). Temples were also the site for keeping the numerous dead bodies that were brought to the temples and cremation was carried out at there. Temples became a base for the forensic teams that worked with identification and also for volunteers who came to help out with the crises. As soon as the Thai people learned of the tsunami catastrophe, both lay and ordained people, who felt that they had something to contribute, travelled to the south of Thailand to assist the affected people and the small Buddhist village temples.

The majority of Buddhist monks, who went to help after the tsunami, came in groups and they stayed at the village temples. But there were also monks who went individually to help. Most of them came to help the survivors to cope mentally with their suffering and to go on with their lives. Some of the monks who went individually, stayed outdoors in tents or in temporary *kutis* (huts). Most of the monks who travelled to the affected regions came through the *sangha*'s organization. The *sangha* acted quickly, and monks from the Buddhist universities in Bangkok, and from many temples all over Thailand, were sent the tsunami-stricken areas to help. This mobilization was made possible through the *sangha*'s nationwide network of temples and monks. The monks were organized in groups before they went to the villages hit by the tsunami, and planned how long they would stay in the south before they needed to be replaced by a new group of monks. The *sangha* exemplifies how important it is to be prepared and organized in order to be able to function effectively in a post-disaster situation. The *sangha* proved to be an efficient institution with functioning leadership and teamwork. The number of monks who came and assisted at the local temples was significant, and it would have been difficult for the village temples to handle the initial severe situation after the disaster without their help (Falk 2015). The affected villagers turned to the Buddhist monks for explanations about the tsunami and monks provided help with understanding and accepting what had happened and also with finding consolation. Ceremonies and rituals were of vital importance, especially for those who had lost family members, relatives, and friends. The monks became mediators in rituals transmitting religious merits from the living to those who had perished or were missing in the tsunami (Falk 2015, 91–96). The Buddhist monks stressed the significance of being helpful to other people as a way of helping oneself. After the tsunami, self-help groups emerged, some of them set up by Buddhist monks. More women than men participated in these groups. The Maya Gotami Foundation led by a Buddhist monk brought the Saori-weaving project to the tsunami-hit area.[8] In 2004, the monk had introduced the Saori-weaving technique in Chiang Mai and Kanchanburi provinces as a kind of meditation technique for those who, for various

[8]The Maya Gotami Foundation was founded by Ajahn Gavesako in 1990 and the foundation runs a variety of educational and other social projects.

reasons, could not use traditional meditation techniques. The free-weaving technique had positive effects on the survivors who had difficulties in concentrating as a consequence of the tsunami. This self-help group developed from being a mental health project into an enterprise where survivors earned their living by selling their woven products.

5.1 Temples and Aid Distribution

Thai Buddhist temples are open for everyone and, after the tsunami, they provided shelter for all irrespective of ethnic belonging and religious faith, regardless of having official documents or not, something that people who belong to minority groups in Thailand do not always have. Some Buddhist temples cooperated with other faith organizations and were trusted to distribute their aid. But a temple is not an obvious place to seek material aid. Some Thai temples are more engaged in social matters than others and temples that already had experiences of working with social projects were more prepared than other temples to run relief projects or distribute aid that different organizations and individual donors offered them. There were several reasons why NGOs and other donors approached the Buddhist temples and asked for help with their relief activities. Some NGO representatives that I talked to said they wanted to minimize the risk of corruption and therefore turned to the temples. The village authorities would have been the obvious collaborators but the NGO representatives did not trust their management. In their post-tsunami study on Thailand, Jayasuriya and McCawley found that NGOs had difficulties motivating villagers to participate in their activities (Jayasuriya and McCawley 2010, 213). I had similar information from the survivors who found many of the NGOs' projects were short-term activities and therefore not meaningful for them to engage in. Earning a livelihood was the main concern for many survivors and they were therefore looking for projects with long-term commitment. Members of NGOs mentioned that lack of coordination between NGOs created problems that led to inequitable and inefficient delivery of aid (Falk 2013). NGOs reported that they also faced another kind of problem with aid recipients who made fake claims of damages and injury. Some went so far as to use corpses that belonged to other families and claimed compensation and relief aid (Falk 2013). The difficulties NGOs and other donors encountered made some of them search for reliable cooperation in the Buddhist temples.

In the immediate post-tsunami period, many of the temples took care of both those who had died and those who had survived. Initially, those who had survived the waves needed the most basic things such as food, lodging, medical care, and clothes. Donations came from local people from villages in the neighbourhood and also from many well-known international NGOs as well as from Thai NGOs. They all donated aid to be administered by Buddhist temples. The temples ran the relief projects for the NGOs, which demanded skills and required a lot of organizational work. The temples ran projects that covered an extensive range of activities, such as building houses and boats, handing out fishing gears, managing funds for schooling and school equipment, organizing libraries for children and distribution of baby food (Falk 2013). Since the monks follow certain monastic rules, they are not allowed to handle money; therefore, the temples need lay people to take care of monetary transactions.

At one temple in Phang Nga province, many survivors stayed at the temple's temporary shelters and each person was given a special notebook to sign in order to keep track of food and other supplies that they had received. Approximately 180 survivors stayed at that specific temple's provisional housing for many months, and people belonging to an ethnic minority group stayed there for more than a year. The temple received donations from individual Thai people and foreigners, and the temple was meticulous about registering all donations and lay people kept records of the transactions that the relief aid generated. The temple's abbot was aware that the activities at the temple would cause problems for the temple and the monks, because it involved a lot of money and also lay people, both men and women. The abbot was particular about keeping records and being transparent about the donations and how they were spent. However, temples that cooperated with international and national aid organizations could not avoid accusations of corruption. It was common that lay people questioned how temples handled the aid that was received.

The minority Moken/Moklen had lived in the Phang Nga province for a long time and, like members of other minority groups in Thailand, some but not all of them were Thai citizens. The Moken/Moklen live close to the sea and the tsunami had hit them badly. Aid organizations arrived in Phang Nga with a different kind of assistance assigned only for the Moken/Moklen. Some Christian missionaries took a special interest in the Moken. The missionaries had recognized the Mokens' extremely vulnerable situation and decided to assist them and simultaneously sought to convert them to Christianity. Minority groups in Thailand have long been a target for missionaries' conversion and the tsunami become a golden opportunity for the Christian missionaries. The missionaries provided all kind of things that the Moken needed or aspired to have, and several Moken, and also Moklen, groups became Christians (Falk 2013). One consequence of that is the Christian Moken/Moklen have stopped participating in the traditional ancestor ceremonies that used to be central to their identity.

6 Survivors' Experiences of Post-tsunami Aid

The Thai survivors were not alarmed by the Thai government's announcement that Thailand had declined international aid. In the first phase following the disaster, the survivors did not think of where the assistance came from, they were just concerned with getting help regardless of the source. The Thai government acted fast and was praised for its efficient work. Since the government prioritized Phuket, there was a delay in the distribution of aid to Phang Nga. The poor coordination of aid hindered effective distribution and the aid did not always correspond with what the survivors needed.

The Thai government's aid delivery was deeply dependent on the elected local authorities, who operate at sub-district levels. Many survivors suffered and were dissatisfied with how aid had been distributed by the local government. The villagers had no insight into how the aid were allocated and had no faith in how the aid was handled. In interviews, survivors in Phang Nga complained about not receiving what was promised. The unequal distribution of aid, implied that specific small neighbourhoods were not part of government aid programmes, in contrast to other villages that received assistance that they did not need, something that upset the villagers from the ignored areas.

The survivors expressed concerns that the Thai government treated men and women differently. The government paid higher compensation for a dead man than for a dead woman, labelling men as the 'breadwinners' (Falk 2015, 28). In actual fact, Thai women are just as important in society as breadwinners and often bear the entire economic responsibility for the children. However, it is not unusual that men and women share the burden of many domestic tasks and both men and women work outside the home. While men, as husbands and fathers, are the recognized 'heads' of households and occupy most positions of community leadership, women nonetheless exercise considerable economic and personal autonomy in their roles as respected mothers and the managers of household finances (Falk 2015, 144).

After the tsunami, there was a flood of willingness to assist those who were hit. In the roadsides in Phang Nga were a large number of signs that revealed the numerous organizations that had been active in the region with relief distribution. Many of the signs had a Christian cross and that disclosed that they were a faith-based Christian association. The majority of the survivors that I met over the years in the tsunami-stricken areas had been contacted by Christian missionaries. They came to allocate aid and among them there were many evangelical Christian groups, and the survivors said that the Christians tried to exploit the disaster for their own gains, through proselytizing; this was a perception that aggravated existing religious conflicts or created new ones. As they arrived soon after the tsunami, the overwhelming presence of Christian groups was noticeable in a region where other beliefs dominate. The influx of missionaries and faith-based aid workers put interdenominational interactions to a test and emphasized the many different approaches faith-based associations implement with regards to humanitarian work. Some groups followed the Red Cross Code of Conduct, which forbids promotion of religious beliefs or political perspectives while working with disaster assistance. However, the priority of many Christian groups was to win new converts and they used aid as a way to coerce survivors into converting to Christianity. It was said among the villagers that international aid associations donated for example money, houses, and boats to villagers that converted to Christianity. However, villagers were very sceptical of the religious motives of the converts and considered them to be false conversions motivated by the assistance that the missionaries offered. Because of that, some international aid organizations faced negative attitudes that ended in clashes with the local people (Falk 2013).

Monks and survivors at a Buddhist temple in the tsunami area experienced how destructive conditional aid could be. The temple had been careful about not permitting families to receive excessive resources, which had been the case elsewhere, and the temple had the principle that the families should not receive more than they had lost and therefore could a family not get more than one house and one boat. That caused problems for a group of people who were promised houses from a European foundation if they could raise money and purchase land. After almost one year, they had managed to buy land but the foundation then stated the families must convert to Christianity in order to get the houses, something that they refused to do. That resulted in a lot of suffering and difficulties for the families and also for the temple, which had been involved in helping the families and had given them a place to stay. Finally, the Buddhist temple succeeded to get an agreement with three Christians NGOs and together they managed to finance houses for the group of 43 families (Falk 2013). If the Red Cross Code of Conduct or a similar code of conduct had been followed, survivors would have been less vulnerable, and they would not have been exploited by organizations that only wanted to provide aid on the condition of religious conversion.

7 Conclusion

This paper has focused on aid and recovery in post-tsunami Thailand with special attention being paid to the significance of religion in disaster relief work, a topic that is often overlooked. The Thai government reacted rapidly and managed to efficiently respond to the disaster. Thailand had enough resources to handle the disaster and declined bilateral aid. However, numerous international donors were active in Thailand and many cooperated with local partners and allocated assistance without any problems. Among the difficulties in relation to international aid assistance were problems with poor organization, discordant aid associations, badly managed allocation of aid, and the lack of necessary long-term plans. These difficulties resulted in problems for the survivors' self-sufficiency. The affected people were critical about the mismanagement caused by the fact that the many aid agencies that operated in the area did not cooperate with each other. The survivors claimed that if one aid association preferred to assist only one group of villagers, that situation could later trigger conflicts in the villages (Falk 2013). Issues concerning housing and livelihood were the main problems for the survivors. They would have preferred to be more involved in the house construction process and wanted long-term livelihood projects. It is of great significance that aid distribution was coordinated and planned effectively. The assessment of the people in need after a catastrophe is also of great importance to avoid increasing the suffering. The Thai Red Cross Society's task was to organize and coordinate some of those international relief organizations that had connections with the Thai Red Cross. However, other associations' relief actions did not have any coordination. Fisher concludes that the early post-disaster situation is characterized by altruism. But, in later phases, selfish behaviour, such as looting, seems to be usual (Fischer 1998, 21). The size of the catastrophe, lack of coordination of aid distribution, and mismanagement in tackling the catastrophe, led to inequities in the distribution of aid, leaving many survivors without support. While some communities were not included in the government's relief efforts, others received disproportionate amounts of donated materials and money. Some of the relief schemes were set up in ways that supported individuals rather than the whole community, which sometimes resulted in conflict within the community. Villagers who used to cooperate and work together did not continue to do so; instead, they started to compete for donations. Some villagers' decision to become Christians were said to not be based on belief; other villagers believed that they had only converted because it was a requirement for receiving aid from the Christian organizations, and this caused antagonism within the community.

Religion had both positive and negative impacts on the survivors' lives after the tsunami. Initially, the Buddhist temples became central to the relief work and the monks and nuns were needed in capacities that were outside their roles as monastics. However, the Thai Buddhist *sangha*'s gender inequality became a weak point; they could have been effective in assisting the lay community if nuns were acknowledged and fully accepted in their religious role. As a result, the numbers of nuns in the tsunami-struck area was low. The followers recognized Buddhist monks and nuns for their individual skills. Most monks and nuns in the Thai countryside live in relatively small village temples and follow the daily temple routine. They have daily contact with the villagers and some of them help the lay society by being engaged in social issues. The main focus of some is on meditation practice, and some monks have a reputation of possessing healing powers.

The small village temples became crowded after the disaster and the temples hosted not only the monastics but also lay survivors and the deceased. In the month following the tsunami, the temples were the centre for identification of the deceased, and were the place forensic specialists carried out their work. It became the place to gather information about missing persons, and photographs with contact addresses were displayed on notice boards for those who had information to provide (Falk 2015). The temples also became the places where many of the volunteers stayed when they helped out after the disaster. Temples also became involved in managing the distribution of material aid, in providing temporary housing for survivors, and dealing with a wide range of relief projects. That can be seen as a consequence of society's shortcomings in being able to handle the material post-disaster relief work.

The Buddhist temples became a refuge for the survivors and, with the help from monks and nuns, the survivors gained control over the terrifying situation and learnt ways to cope with the situation. At the temples, they had a sense of belonging and they were not alone and shared their suffering with other survivors. Monks and nuns dealt with life, death and suffering, and helped the survivors recover mentally.

This support was at the core of Buddhist monks and nuns expertise and was invaluable for many survivors, despite the fact the importance of immaterial aid risks was being overlooked in relief work.

References

"December 26, 2004 The Tsunami on Phuket". Phuket: Phuket Province 2005.

Falk, M.L. 2007. *Making Fields of Merit: Buddhist Female Ascetics and Gendered Orders in Thailand.* Copenhagen: NIAS Press/ Washington: University of Washington Press.

Falk, M.L. 2010. Recovery and Buddhist Practices in the Aftermath of the Tsunami in Southern Thailand. *Religion* 40: 96–103.

Falk, M.L. 2013. 'Thai Buddhists' Encounters with International Relief Work in Post-tsunami Thailand. In *Buddhism, International Relief Work, and Civil Society*, ed. Hiroko Kawanami, and Geoffrey Samuel. New York: Palgrave Macmillan.

Falk, M.L. 2014. Gender and Buddhism in the Wake of the Tsunami. In *Contemporary Socio-Cultural and Political Perspectives in Thailand*, ed. Pranee Liamputtong. Dordrecht, Heidelberg, New York and London: Springer.

Falk, M.L. 2015. *Post-tsunami Recovery in Thailand: Socio-Cultural Responses.* London and New York: Routledge.

Fischer, H. 1998. *Response to Disaster: Fact Versus Fiction and its Perpetuation—The Sociology of Disaster*, 2nd ed. Lanham, MD: University Press of America.

Jayasuriya, S., and P. McCawley (eds.). 2010. *The Asian Tsunami: Aid and Reconstruction after a Disaster.* Caltenham, UK, Northampton, MA, USA: The Asian Development Bank Institute and Edward Elgar Publishing.

Nga, P. 2005. *1st Year Tsunami Memorial.* Phang Nga/Bangkok: T&D Studio Phang Nga/EPAC Co Ltd.

Pimolwan, I. 2006. Unpublished presentation at the 2nd Annual Conference of Population and Social Research. Asia Hotel, Bangkok, Research Institute of Population and Social Research, Mahidol University, Thailand (in Thai).

Rojanasunan, P. 2012. *Dr. Porntip Rojanasunan: The Dead Do Talk.* Singapore: Marshall Cavendish Editions.

UNDP. 2005. *Tsunami Thailand One Year Later—National Response and the Contribution of International Partners.* Bangkok: UN.

Weber, K.E. 2005. *Tsunami 2004: Nam Chai Thai.* Bangkok: Thai National Identity Board, Royal Thai Government.

Wisner, B. 2010. Untapped Potential of the World's Religious Communities for Disaster Reduction in an Age of Accelerated Climate Change: An Epilogue & Prologue". *Religion* (special issue: Religions, Natural Hazards, and Disasters) 40 (2): 128–131.

Woothisen, O. 2005. The Royal Benevolence Wiped Away Tears. *Tsunami Chai Pattana Andaman* 22–35.

Memory Interventions: Gujarat and Sri Lanka

Edward Simpson and Malathi de Alwis

1 Introduction

Memory projects have become almost routine in the aftermath of disasters. They may aim to effect catharsis or remembering; and, as such, are often also seen as a method of encouraging future disaster-preparedness or mindfulness. Memory has become part of the healing work associated with psycho-social responses to natural disasters: the 'soft' work of reconstruction. Memory is often casually associated with 'remembering'. We wish to take a rather different approach by suggesting that some kinds of memory projects are actually attempts at homogenizing and making realities, rather more than they are about cultivating or preserving something vulnerable or precious.

Our particular focus is on memorials and we approach the subject as social anthropologists. Scholarship on memorials has illuminated how they are social objects, products of particular times and places, and open to constant reinterpretation. For example, the historian Inglis (1992) has described how the bronze statue (originally called *The Homecoming*) at the end of Station Road in Cambridge (UK) was erected as a monument to victory after the First World War. In the following decades, debates and conflicts over the object gradually transformed its public meaning to a memorial for the dead of the two wars. The image of the upright and resolute soldier returning home triumphant from war is clearly quite at odds with the losses and tragedies of war that the memorial is popularly held to represent.

The authors wish to express their grateful thanks to Samanmali Kumari Jayawardana and Dilip Vaidya.

E. Simpson (✉) · M. de Alwis
SOAS University of London, London, UK
e-mail: es7@soas.ac.uk

S. Reddy (ed.), *The Asian Tsunami and Post-Disaster Aid*,
https://doi.org/10.1007/978-981-13-0182-7_13

Obviously, memorials have histories, and we are well aware that one man's memorial is perhaps a symbol of another's oppression, defeat, or loss (see also Simpson and de Alwis 2008). However, we wish to challenge two commonly held assumptions in the substantial literature on memorials and memory. The first is the common-sense assumption that memorials enable us to remember an event. Clearly, objects do not remember; it is individuals and collectivities that do. Objects may comprise a record, and, perhaps, act as an aid to memory from their positioning, materiality, inscriptions, and ethos. For sure, memorials are commonly tied to particular events, but what kinds of memory can memorials be said to actually represent or, conversely, evoke?

People who experience catastrophe may bring their own inchoate and grotesque memories of particular moments and tragedies with them when they visit a memorial. They may also bring with them memories which are simply too terrible to articulate, and remain hidden from view. Memorials bring together a varied range of individual experiences into a single public representation. Inscriptions of the date, place, and time of the disaster become the focus of collective and public attention and, as such, are far easier to deal with than the stench of rotting bodies, or the confused horror on the face of those at the edge of life.

In this way, memorials become sites through which the diverse experiences of individuals can be mediated. Individual memory in extreme form is flattened, disguised, or even pacified with something more publicly acceptable. Arguably, a fundamental function of memorials is to allow people to communicate meaningfully and on common ground. The device becomes a signifier of an event; a shorthand. The unthinkable and the haunting become thinkable and recognizable in the object. The same object is also recognizable to others, and allows bystanders to share in an imagined community of disaster-affected people.

The memorial does not, of course, remove individual memory; but it creates a public platform through which disasters can be discussed and commemorated without individual memory acting as a disruptor. Thus far, we have assumed memorials are erected by those who have been directly affected by a memorialized event. However, as we shall see, those who put it there are often other people altogether. Therefore, a memorial is often more reflective of the concerns and aspirations of those who put it there and who control the public activities around it.

A second theme in the scholarly literature we intend to question is the possibility that memorials speak to the fundamental cultural building blocks of the memorializers. This is often glossed as the *mentalité* approach. It would be extravagant to completely reject the possibility of a relationship between the cultural ideas of people and their memorials; however, we wish to deconstruct the familiar assertion that the form, design, and siting of a memorial, and the practices that grow up around it can be read in a straightforward way as visible indicators of submerged meaning revealed by tragedy. We have seen that the debates and arguments that surround the formation of memorials can be revealing of underlying social processes and pre-existing struggles; but memorials themselves do not represent such critical engagements any more than they stand for the extremes of individual memory.

By looking at memorials as devices that have an uneasy relationship with memory (in any conventional sense of the term), and as products of cultural compromise rather than as epitomes of culture, we wish to turn the emphasis away from the objects themselves and, rather, to the processes that allow the object to come into being and to then stand for something.

2 A Note on Natural Disaster and Social Process

Memorials are created to mark a diverse range of events from the assassination of individuals to the loss of millions in wars. Different types of memorial tend to signify distinct types of events (see Foote 2003). Natural disasters are one such kind of event that is commonly memorialized. Given our comparative frame, it is worth pointing out that the earthquake in Gujarat and the tsunami crashing onto the shores of Sri Lanka were also quite different kinds of events. Both were 'natural' or 'divine', depending on the perspective taken, but the demography of loss and trauma varied quite markedly in each case. However, there are some generalizations to be made from the literature on natural disasters as a way of framing our material.

The most obvious generalization is that post-disaster environments commonly exacerbate social trends that were waxing prior to the disaster. This is as true in Gujarat and Sri Lanka, as it is in the extensive comparative literature. It, thus, follows that the starting points for many of the debates surrounding memorials, their authenticity and legitimacy that have been taking place in India and Sri Lanka over the last few years, at the grass-roots level, have been influenced by social currents and counter currents that predated both disasters. Older squabbles and ideas have not been reproduced exactly, because both disasters encouraged those affected to reappraise the identities of heroes and villains, and to bring only select images and metaphors from the pre-disaster days into the present to accompany the experience of mourning and loss.

The earthquake and the tsunami are moments of rupture, offering the possibility of a break with the past, especially in terms of personal memory. There is, so to speak, more past in the present than previously. Nostalgia has become a cultural trope in both locations and has been imbricated upon the landscape of disaster through a network of memorials. The combination of disaster, nostalgia, and memorials has, in both instances, also led to a resurgence and reassertion of regional identities. In India, we have seen a campaign for regional independence in the worst affected areas of Gujarat; in Sri Lanka, both national and local politics have struggled to realign administrative and political boundaries within the charged atmosphere of reconstruction and loss.

Having said that disasters exacerbate longstanding social processes, and draw cathartic attention to local identity politics, it is also the case that the ruptures of calamitous events can also usher in revolutions and other radical forms of innovation. Disasters produce conditions of uncertainty and opportunity, promoted initially by loss and confusion, and later by the sense of rejuvenation. It is these critical moments into which humanitarian and development aid organizations step.

Global orthodoxies on post-disaster humanitarian activity have increasingly pushed the idea that immediate relief can and should be part of longer-term initiatives aimed at reconstruction and rehabilitation. Thus, there is continuity in care, so to speak, and closer relations between affected populations and humanitarian organizations are formed, thus, in theory, improving communication and mutual understanding. The consequence of such relationships tends to be that local memorials emerge from the interaction between outside organizations with a humanitarian impulse and local populations. When seen thus, memorials fuse discourses belonging to different locations and scales. The emergent object also marks a relationship as well as a disaster and, as has often been noted, the relationship between mobile and resourced humanitarians and local populations affected by disaster is seldom equal.

3 The Gujarat Earthquake

On 26 January 2001, the district of Kachchh in the western Indian State of Gujarat was hit by a massive earthquake (for an extensive discussion of this event and its aftermath, see Simpson 2013). Over 13,000 lives were lost and nearly 1,000 villages were badly damaged. The disaster was sociologically indiscriminate, affecting urban and rural areas, cutting across caste and religious boundaries, and affecting both rich and poor. Tragically, because it was a national holiday to mark Republic Day, crowds of children were out on the street celebrating the nation. Nearly 200 children died in Anjar as they walked in a procession through the street waving the Indian tricolour. The district of Kachchh, located in the far west of the state of Gujarat, accounted for most of the fatalities and a great deal of property loss. Around 1 per cent of the population, in this sparsely-peopled and relatively inaccessible region, lost their lives.

At the centre of the district, stood the town of Bhuj. It had an administrative function, busy bazaars, and was the ancient seat of kingly rule. It was famous for its ornate and extravagant architecture and indigenous craft traditions. The town was well preserved and was popular among visitors, both domestic and foreign, for these reasons. More than 2,000 people died in Bhuj, or about 1.72% of the town's population. In Anjar, in the east of the district, the official death toll stands at 3,862 (around 5% of the population), and for nearby Bhachau it is 7,054 people (around 20% of the population).

In the region the earthquake became popularly and variously known as 'Shiva's dance of destruction', and was variously likened to 'a distant flour mill' or 'an atomic explosion'. Some clearly believed that the end of the world was nigh; or that Pakistan had unleashed a terrible weapon. In the years to come it was quite evident that the disaster made people question why the disaster that happened to them. Why there? Why not elsewhere? Theories emerged quite quickly about who was to blame and who was being punished. Some suggested that the earthquake was divine repayment for the hedonism and consumerism of modern society. The old ways had been forgotten; the old gods neglected. These kinds of conversations meandered for years. Interestingly, however, neither the relative intensity of the earthquake, nor the accounts of blame in the aftermath, played into the ways in which memorials were conceived and constructed.

In the first years after the disaster, USAID and other organizations funded a project to collect photographs and stories of the Bhuj that had been partially lost to the earthquake. The resulting book (Tyabji 2005) contains a great deal of important material about the past, and also documents some of the significant conditions prevailing in the region at the time. Collecting and admiring old photographs was not however a passion shared by the majority of the town folk and the project became oriented around the interests and narratives of those who thought such things were important. Likewise, the concerns of architects and conservationists involved with the preservation of material heritage were sometimes judged to be indulgent or dismissed as metropolitan in comparison to the rather more pressing needs of those who had lost their homes in the disaster.

The collection and preservation of heritage was led by humanitarian concerns and individual actors from outside the affected region. Their efforts were important and have led to publications and the protection and restoration of some historically important structures in Kachchh. They were also backward-looking projects which enlivened a particular vision of the past of a royal kingdom and of a particular form of regional identity. It is unsurprising that the image of the past was predominantly elite; but this excluded many of those who saw themselves as subordinated to, and by, this evocation of history.

In what might be thought of as the 'pre-memorial era' in Gujarat—the space between the disaster and the temporary closure promised by an official memorial—there have been countless lesser memorializing events (see Simpson and Corbridge 2006). Many of these are very personal: shrines in houses for dead family members, and the maintenance of special objects that were once cherished possessions of those now dead. Others have been organized along community lines to coincide with reconstruction projects, such as rituals now memorialized by a permanent structure to mark the very centre of new neighbourhoods (see Simpson 2006).

The various state governments (who served as humanitarians) from elsewhere in India that financed reconstruction projects in Kachchh also built memorials as part of these projects. On the outskirts of the new village of Indraprastha, for example, is a monument to the dead built on top of a small hillock, called 'Tiger Hill' (see Fig. 1). On a plaque set into the foundations of a pole for the national flag, the formal monument lists the names of those villagers who perished in the earthquake. The hillock has been planted with sweet-smelling flowers, and, power and water permitting, a fountain cascades down an artificial watercourse. The significance of this monument is that it establishes a direct connection between those who died because of the earthquake and those who sacrificed their lives for the nation. The bloody, partially televised, battle for Tiger Hill, adjacent to the Srinagar–Leh highway in Kashmir, was a turning point in the border skirmishes between India

Fig. 1 'Tiger Hill', Indraprastha, Gujarat

and Pakistan in 1999. Soon after the Indian Army captured the peak, the press was full of images of the tricolour planted in the scrubby soils of the summit. The Indian soldiers who lost their lives in this battle became the most celebrated of martyrs.

Looking at the region as a whole it is striking that there is no permanent memorial to the earthquake that attempted to speak for everybody. In the first decade after the disaster, a host of smaller memorials were scattered throughout the region. Many of these had no local resonance, carrying messages which were often as idiosyncratic as that built into the Tiger Hill monument already mentioned. Most of these memorials reflected the ideas of various outside agencies that had come to reconstruct the region after the earthquake. There was little excitement amongst local people about the possibility of building a memorial. Every now and again, debate would flare up about the location or appropriateness of particular kinds of memorial. In the early years, a group lobbied for the preservation of some urban ruins as a memory of not only of the earthquake but of how life was in the years before the disaster. Others campaigned for the creation of a traditional house built in the vernacular style which was to be filled with objects associated with all those who had lost their lives. It must be said that both of these ideas had been borrowed from elsewhere—a Holocaust memorial and a monument to another earthquake—and were not spontaneous cultural uprisings. Both these ideas disappeared once the offer of international funding was withdrawn. The absence of a memorial at the epicentre was also striking, but the location for that was contested and remote, and would have brought little political mileage for anyone sponsoring such a structure. Quite quickly, somebody constructed a memorial in the town of Bhachau—at the so-called gateway to the district—but a year later that had been demolished to make way for the statue of another kind of political hero.

A few years after the earthquake, the government sanctioned two memorial projects. A competition was announced, architectural plans drawn up, and fine ideas were developed which linked aspects of the tragedy to particular kinds of space and place. In some ways, these memorial projects can be understood as the state attempting to rein in the confusion of memory and political protest that characterized life in Kachchh in the years immediately following the earthquake. The first was to be a plantation on the outskirts of Bhuj containing a tree to mark the life of each of the 13,000 dead. The park was to be laid out around contemplative spaces and other internationally-circulating clichés dreamed up by a firm of architects from a faraway urban centre, which had been awarded the contract to develop the idea. Fantastically, it was proposed that different species be planted to represent different castes and religious communities. This plantation was originally planned for a large demilitarized hillock just outside Bhuj, until someone pointed out that there was no water there and trees need water. Suitable alternative land could not be found and the project has moved slowly ahead in modified form.

The second project was more ambitious and involved the preservation of the ruins of the village of Adhoi, in the east of the district, as a walk-around memorial to the disaster. At first glance, the ruins and empty silence of the almost-uninhabited settlement evoked the tragedy of death and the destruction of the village way of life (see Fig. 2). However, what the visitor could actually see were not simply the tragic ruins of a village devastated by an earthquake. The first part of the illusion was that Adhoi was atypical of villages in this part of the district, and the haunting ruins were predominantly those of the large second homes of wealthy families residing in Mumbai (earlier Bombay), most of which were, importantly, still standing, and, thus, appeared as dramatic ruins, unlike in other villages where poor-quality buildings simply collapsed leaving singularly unimpressive piles of rubble.

Fig. 2 The preserved 'Ruins' of Adhoi, Gujarat

The second part of the illusion was that most of the damage the visitor could see was caused by the pillaging of building materials after the earthquake, and not by the earthquake itself: earthquakes do not carefully remove wooden window and door frames. Finally, many of the houses of the resident poor that formed less impressive ruins had been cleared and their occupants resettled at some distance from the original site, contributing to the feeling of empty desolation in old Adhoi.

Pillaging aside, on the surface, ten years ago the unusual silences of the ruins might have suggested a desire among local people to create a memorial. One did not however have to ask too many questions to see that the idea had been deeply divisive. For some, it had meant an opportunity to start again somewhere else; for others, perhaps the less needy, it was an opportunity to obtain compensation money from the government. There were court cases and counter cases and, unusually, a war of words which took place in the classified advertisement pages of the local newspaper. In the face of government orders restricting the ruins and the muscular politics of some local men of influence, it was also clear that life was beginning to return the ruined village. This, it seems to us, is all utterly typical of a memorial in gestation. In time, support for this memorial dwindled and today there is little evidence of it being maintained.

4 The Tsunami and Sri Lanka

On 26 December 2004, an earthquake off the west coast of Northern Sumatra triggered a Tsunami of unprecedented proportions that devastated two-thirds of Sri Lanka's coastline (de Alwis and Hedman 2009). It left 36,000 dead, 3,800 missing, and 15,000 injured. It reduced around 75,000 houses to rubble, and damaged even more, displacing over a million people and leaving entire districts without basic services.

As with the earthquake in Gujarat, the tsunami was seen as a sign of divine retribution for the betrayal of cultural traditions and the embrace of immoral, consumerist, 'Westernized' practices. The fact that many places of worship remained unscathed was used to drive this point home, despite engineers arguing that this was due to superior building materials used in their construction; unlike in the primarily lower-income housing along the coast which was unable to withstand the force of the waves.

The areas worst affected by the tsunami—the northern and eastern regions of the island—had, for the past 30 years, been the site of a civil war between the Government of Sri Lanka (GoSL) and the Liberation Tigers of Tamil Eelam (LTTE), while most of the southern regions affected were limping back from a youth insurrection from 1987–90 which had led to the 'disappearance' of around 40,000 persons. These embattled regions were now not only re-inscribed by the ferocity of the ocean, but also by subsequent waves of reconstruction and memorialization, which reiterated socio-political tensions and fissures.

Priority was given to the speedy rehabilitation of the Sinhala-dominated southern regions (with newly-built houses far exceeding those destroyed in Hambantota, President Rajapakse's constituency), rather than the Tamil- and Muslim-dominated northern and eastern regions. Government ministers vied with each other to 'twin' the patronage they extended to new housing settlements (funded by international and local NGOs as well as local business enterprises), with the erection of tsunami memorials along the main coastal road. Many such memorials are now locally known by the names of politicians; the most (in)famous one, near the site where a train carrying over 1,500 people was swept off the track, was commonly known as the 'Fernandopulle Memorial' after the minister who oversaw its construction (see Fig. 3). However, a collective, representing those who perished in this train, would annually erect a counter-memorial next to the Fernandopulle Memorial, declaring that these deaths were not merely due to the tsunami, but also to 'those in authority neglecting their responsibility' (see Fig. 4). This collective was successful in getting the state to appoint a tri-partite Committee comprised of three High Court Judges to inquire into the culpability of the Department of Railways (for a discussion of the hearings and subsequent verdict, see de Alwis 2009).

The badly-battered train, which the Fernandopulle Memorial in effect replaced, had become an impromptu memorial soon after the tsunami. Thousands, both locals and foreigners, flocked to it and it quickly became a money-spinner for some villagers who had begun to sell entrance tickets, concessions, souvenirs, and so forth. Others took to begging from the visitors. When it became a site of contention

Fig. 3 Section of mural depicting the train being swept away, from the 'Fernandopulle Memorial', Peraliya, Southern Province, Sri Lanka

between those who wished it to be made into a permanent memorial/museum in situ, and others who wished it to be taken away as it was leading to further fissures within their village, the state moved the carriages, promising to turn it into a museum. It now lies rusting and forgotten in a disused rail yard at the Hikkaduwa Station.

The villagers who heatedly fought over the train largely ignore the Fernandopulle Memorial (it is rumoured that some villagers smashed all the light fittings on the very day it was opened), as well as a rather ad hoc photo gallery/museum, which has been set up by a Dutch activist, across the road from it. Many have relocated, and others who remain prefer to memorialize their dead as they see fit: a group of 10 families is sponsoring an illuminated pinwheel behind a seated Buddha statue encased in glass; a Tsunami-commemorative construction by a local monk with funds donated by a Taiwanese monastery; a schoolteacher hung mourning flags on the ruins of his ancestral home in order to memorialize his wife and son ('It's a symbol of my grief as well as of my anger at the government which has been so neglectful'; personal communication, May 2007).[1] He was also scathing about what has now become

[1]He had been instructed to remove these now-ragged, black-and-white flags in preparation for the President's visit to inaugurate the Fernandopulle Memorial which is located opposite his property, but had refused.

Fig. 4 Counter-memorial at Peraliya, Southern Province, Sri Lanka

accepted as the national Tsunami memorial —a 54-foot Buddha statue constructed by a politically powerful local monk under the patronage of the Ministry of Buddha Sasana with funding from a monastery in Kyoto—built further down the road from him. The SLR 55 million spent on it should have been used to materially resuscitate the Tsunami-affected (rather than seeking to effect their mental rehabilitation, as mentioned in the Japanese prelate's message on the plaques at the site), was his view.[2]

This dynamic interplay between the state and a critical citizenry was absent in the north and east, particularly in areas populated by the Tamil minority. In Batticaloa District, for example, the populace not only had to contend with the militarized face of the state, but also the LTTE and its competing breakaway factions. Resourcefulness in the face of isolation and vulnerability was exemplified in the Tsunami memorializations; fishermen's cooperatives, sports clubs and village committees took the lead in constructing memorials by collecting funds from among themselves. The series of simple, columnar memorials, poignant reminders

[2]The message from President Rajapaksa, on the plaque beside the statue, proclaims this statue as symbolizing the 'Buddhist temples destroyed by the tsunami and World Heritage Buddha statues destroyed by terrorists'. The latter seems to be a reference to the Bamiyan Buddha statues destroyed by the Taliban and supposedly replicated by this statue.

of decimated coastal communities, extending from Thiruchendur to Navalady, are the result of such collective community efforts (see Fig. 5). Their verticality is striking and not replicated elsewhere along the coast. These memorials are now frequently reduced to ghostly sentinels at night due to community coffers no longer being able to afford the astronomical electricity bills; the national memorial beside the southern coastal road is illuminated at state expense.

Fig. 5 Navalady memorial, Batticaloa, Eastern Province, Sri Lanka

Despite the tsunami affecting people of all religions, ethnicities, classes and regions, the national tsunami memorial only represents one religion and ethnicity. This, of course, is hardly surprising given the state's pro-Sinhala/Buddhist/southern policies, not only in terms of post-tsunami reconstruction and memorialization, but also when it comes to memorializing other extraordinary moments of violence: a national 'Shrine to the Innocents' memorialized deaths and 'disappearances' during the Southern youth uprising (this no longer exists, having been bulldozed two years ago to make way for an entertainment complex, see de Alwis 2015), but no national memorial commemorates civilians killed during various anti-Tamil and anti-Muslim riots and pogroms nor the victims of the Civil War.

5 A Comparative Conclusion

In both contexts, memorials have been constructed and have proved controversial. Although our delineation of the two locations is rather schematic, the comparison raises a number of questions. Why should there have been a greater incidence of memorial construction in Sri Lanka than Gujarat? Memorials in Sri Lanka often depict the wave and the destruction it brought. In contrast, memorials in Gujarat are generally abstract objects. What are the reasons for this difference? In Gujarat, many memorials were erected by governments, while in Sri Lanka memorials were more commonly erected by people. Why? Given the political climate in Gujarat at the time, it is striking that many of the memorials produced were strictly secular in the orientation; in contrast, memorial practices in Sri Lanka took on strong Buddhist colours. In neither location was there evidence of memorials being constructed for the Muslim dead.

Memorial construction in the two countries followed different lines. Can such different patterns be explained by the type or impact of the disaster? Is this simply a game of numbers? As more people died in Sri Lanka at the local level than in India, was there a greater need to build memorials? Rather, might a more efficacious explanation be found in the cultural drivers of different societies? In Gujarat, we might suggest, fewer memorials have been constructed because of the influence of Hindu ideas about death and the erasure of individuality. However, we also described the common trend in Gujarat to maintain shrines to the deceased and to retain some of their cherished possessions. In Sri Lanka, however, Hindus have also been as enterprising as their Buddhist neighbours in constructing memorials for the tsunami.

A towering wall of water is easier to represent than an earthquake, which might explain why the wave appears so often in Sri Lankan memorials. Can this be the only explanation for a lack of visual imagery in Gujarat? Might there not also be an argument to be made using culture as a resource and to say that visual representation is a more common genre of expression in Sri Lanka than in Gujarat? A strong state in Gujarat may leave little space for civil society to construct memorials in the earthquake landscape. Conversely, a buoyant civil society in Sri Lanka had the agency and political capital to place memorials in contested disaster locations whereas the state could not take such a risk in the circumstances. It might also be possible to argue that people in India, in comparison to Sri Lanka, have traditionally had more expectations of the state, especially when it comes to the management of risk. However, in both locations, minority communities have been historically conditioned to not depend upon the state or to make particularly obvious interventions in contested landscapes.

On the first anniversary of the disaster in Gujarat, the government conducted Hindu rites to quieten angry gods and goddesses. With this exception, memorial practices in India have drawn more upon secular traditions of statecraft than they have religion. This is somewhat surprising given the reputation Gujarat has developed for engendering Hindu nationalism within the apparatus of the state. Is this Indian secular democracy in action? Perhaps; but it might also be the case that such secular commemoration is an echo of the practices and aesthetics of an earlier colonial era. The colonial government perhaps remains the greatest of all the memorializers in India, at least in relatively recent times. However, the British built memorials in Sri Lanka and this legacy has not resurfaced in the aftermath of the tsunami. So, a more evocative angle would assume that statecraft in Sri Lanka, given the recent troubled history, is popularly legitimized among its supporters by the symbols of Buddhism, and, therefore, memorials simply reflect this. Similarly, we could also argue that the fact that there are no Muslim monuments to the dead in either country (despite a large percentage of the dead being from this community, in both countries) is a result of religiously-prohibited restrictions on visual and iconic representation.

Earlier we argued that memorials do not simply reflect memory any more than they do culture. Material suggests that, when viewed comparatively, neither culture nor memory can be discounted. However, history, politics, and a whole series of other forces, come to weigh on the process of debating and constructing memorials. In our field sites, there is evidence of borrowing from other traditions and disasters. Ideas about memorials are transmitted through international disaster relief organizations, and governments tend to remember the previous memorialization act it was responsible for, creating genealogies of traditions. In both India and Sri Lanka, it is clear that compromise, negotiation, and chance play influential roles in the memorial business. The process itself generally seems to override actual cultural concerns of culture and memory and something hybrid and specific (and therefore often highly unusual) emerges as a result.

In both India and Sri Lanka, memorials have become part of political protest against the state and the perceived injustices of reconstruction; likewise, in both countries, governments have used memorials in an attempt to impose a controlling narrative on the disaster. On the one hand, therefore, memorials have been used by governments as methods and techniques of control; on the other hand, people out there in the wild, so to speak, have developed memorials as part of a greater process of government critique.

Memorials have become a way in which people and the state communicate. The various memorials we have described are also focal points for those who have experienced loss. We discussed earlier how memorials became a meeting point for individuals stripped of their subjective experiences. Some writers have seen memorials as bringing people together (see Winter 1995, 50–3, for example). We have seen instances of this, but more frequently, memorials have exacerbated old conflicts and encouraged new social cleavages.

In southern Sri Lanka, many tsunami memorials were constructed without public consultation. In such a context—that is, where states seek to be the sole arbiter of meaning—a reworking of Yves Helois' formulation that war memorials are exercises in 'biopolitics' seems particularly apt. Helois (*pace* Foucault) argues that war memorials are sites of symbolic exchange: death is both deconstructed (horror, trauma and individuality is buried) and reinvested with meaning as an abstraction, a collective sacrifice: *a nos morts*, the collective spirit is now embodied in the state (quoted in Winter 1995). The monument on 'Tiger Hill' in Gujarat is particularly instructive in this regard, as it seamlessly folds the memorialization of the earthquake into the memorialization of the battle for Kashmir. In a similar, but different register, the national tsunami memorial constructed by the Sri Lankan state is a Buddha statue which not only seeks to recuperate tsunami-destroyed Buddhist temples within the country (no similar measures are mobilized for the loss of Hindu, Christian, and Muslim places of worship), but makes claims on a wider Buddhist universe while simultaneously providing an indictment of 'terrorism' by invoking Bamiyan (footnote 2).

Other memorials discussed here seem more to be forms of cultural compromise that have emerged haphazardly out of conflict and dispute as well as agreement. As such, memorials can only be thought of as beacons of cultural meaning because their construction reflects processes of negotiation and perhaps prevalent ideas of visual and ritual aesthetics. Memorials also reflect the regulative capacities of national planning regimes. In a sense, the material we have presented demonstrates that memorials, as manifestations of compromise often without precedent, are creative and extraordinary acts that may routinize and synthesize similarly extraordinary events and atypical collective emotions and will. At other times, memorials appear as generic or copycat responses to tragedy which nonetheless incorporate memory, broken landscapes, and individuals within a master narrative.

Earthquakes—perhaps because of the ways they are viscerally experienced and thus literally 'embody' memory—seem to be almost universally met with calls for contemplative spaces. This often happens quite quickly, and often on the first anniversary of the disaster. In Nepal, for example, following the earthquake of 2015, the President used the anniversary to lay a foundation stone for a memorial park before conducting feasibility studies (there is clearly political millage and media opportunity in appearing to be making things better).

Earthquakes are often marked with memorial parks. The 'Wailing Wall' at Tangshan, China, constructed in 2008 to commemorate 242,000 people who died in the 1976 earthquake, is notable for its ambition and beauty. Such parks generally combine open spaces with bold reminders of the weight and appearance of concrete and other building materials. They also often incorporate exhibitions about seismology and pictures or the names of those who died. In this sense, individuals are literally incorporated in a master, and state-led, narrative about the disaster. One of the other striking things about such creations is the amount of time it generally takes to bring them into being. Following an earthquake in Peru in 1970, it took quarter of a century to erect a memorial (again in park form) that did not horrify the local population; in the case of the Tangshan memorial, three decades had passed before the final design competition was announced. The park planned for Bhuj to remember the 2001 earthquake has not yet been completed.

In Sri Lanka, the burgeoning of civilian-initiated tsunami memorials, in the years that immediately followed the tsunami, ironically reflected the absence of civilian-initiated collective memorials to a greater number of civilians who had died in the Civil War, up to that point. The primary reason for this, as de Alwis has argued elsewhere, is that the majority of the Sri Lankan populace perceived 'natural' disasters to be "acts of God", (despite an acknowledgement that they were often exacerbated by human intervention such as de-forestation, sand and coral mining, and illegal settlement in coastal conservation zones) that "lay waste to lands and peoples irrespective of ethnicity, class, caste, religion or gender while those who die in civil wars are rarely perceived as innocent. Thus, to commemorate their death(s) publicly was sure to bring on the wrath of either the Sri Lankan state or Tamil militant groups" (2016, 100).

However, many of these tsunami memorials are now lying neglected as state- and civil society-led initiatives encourage local communities to memorialize those who died in the civil war as a means of enabling reconciliation within the paradigms of the latest buzz phrase, 'transitional justice'. How long, or how productive, such a process would be is still to be seen.

Much of what we have written about here has already turned to dust. The debates and protests in Gujarat and Sri Lanka are distant memories or, more probably, forgotten. In Gujarat, a memorial has been built in Bhuj on a hillside over-looking Bhuj. The strength of the place is in the view which is a beautiful reminder, not so much of the disaster, but of how much the town has grown and prospered since the earthquake. Looking back on those times, debate and conflict over memorials can be seen forms of catharsis which are part of a long process of getting the memorial right.

References

de Alwis, Malathi. 2009. 'A Double Wounding? Aid and Activism in Post-tsunami Sri Lanka' in *Tsunami in a Time of War: Aid, Activism and Reconstruction in Sri Lanka and Aceh*, ed. Malathi de Alwis and Eva-Lotta Hedman. Colombo: ICES.

de Alwis, Malathi. 2015. 'What's Beneath the Paving Stones? Colombo's Surfeit of Secrets'. In *Paper presented at the Burden of the Urban conference*, Tata Institute of Social Sciences/ Warwick University, Mumbai, December 2015.

de Alwis, Malathi. 2016. 'The Tsunami's Wake: Mourning and Masculinity in Eastern Sri Lanka' in *The Gendered Terrain of Disaster*, ed. Bob Pease and Elaine Enarson. London: Routledge.

de Alwis, Malathi and Eva-Lotta Hedman (eds), *Tsunami in a Time of War: Aid, Activism and Reconstruction in Sri Lanka and Aceh*. Colombo: ICES, 2009.
Foote, Kenneth. 2003. *Shadowed Ground: America's Landscapes of Violence and Tragedy*, 2nd edn. Austin: University of Texas Press (1997).
Inglis, K.S. 1992. The Homecoming: The War Memorial Movement in Cambridge, England. *Journal of Contemporary History* 27 (4): 583–605.
Simpson, Edward. 2013. *The Political Biography of an Earthquake: Aftermath and Amnesia in Gujarat, India*. London: Hurst.
Simpson, Edward. 2006. 'The Rituals of Rehabilitation: Rebuilding an Urban Neighbourhood After the Gujarat Earthquake of 2001', in *The Meaning of the Local: Politics of Place in Urban India*, ed. H. Donner and G. De Neve, London: UCL (Cavendish Publishing) Press.
Simpson, Edward, and Malathi de Alwis. 2008. Remembering natural disaster: Politics and culture of memorials in Gujarat and Sri Lanka. *Anthropology Today* 24 (4): 6–12.
Simpson, Edward, and Stuart Corbridge. 2006. The Geography of Things that may Become Memories: The 2001 Earthquake in Kachchh-Gujarat and the Politics of Rehabilitation in the Pre-memorial Era. *Annals of the Association of American Geographers* 96 (2): 566–585.
Tyabji, Azhar. 2005. *Bhuj: Art, Architecture and History*, Ahmedabad: Mapin.
Winter, Jay. 1995. *Sites of Memory, Sites of Mourning: The Great War in European Cultural History*. Cambridge: Cambridge University Press.

Conflicting Paradigms and the Danger Discourse: Re-thinking Indian Disaster Management Framework in the Post-tsunami Era

Pravin Kushwaha

1 Paradigms of Disasters, Development, and Vulnerability

Contemporary discourse around the theme of disasters has evolved from a simple perception of a disaster—as an 'act of god' or a 'natural calamity'—to being perceived as the outcome of a very complex process of production of 'vulnerability', as human societies and their environment engage in complex interplay. Interestingly, despite systematic disaster-research throughout the second half of the twentieth century, a few key terms used in this field, such as 'disaster' and 'vulnerability', still lack common consensus. In the recent past 'vulnerability' has emerged as a popular theory in disaster studies and there is growing consensus among policy makers and planners that improper development increases vulnerability to disasters. The policy process requires a complex, evolving, and dynamic understanding of the assumptions involved in these theories, the ideologies guiding these principles, and their interactions with human as well as natural environments.

It is well documented that early societies have considered natural calamities to be disasters that were 'acts of God' to punish human beings for their 'moral misbehaviour'. Such considerations visualized disasters as 'external' and 'inevitable' threats. In other cases, as a mechanism to deal with them, communities simply avoided sites of frequent disasters, like floods, by shifting to newer locations for instance. Scholars of disaster research argue that overall, twentieth century witnessed the most organized and large-scale attempts to reduce the damaging impacts of disasters, primarily by using engineering methods like constructing dams to avoid floods, building earthquake-resistant buildings, and so on. These large-scale attempts in modern times can be traced back to a disaster-research tradition that primarily developed in the USA between the nineteenth and

P. Kushwaha (✉)
Transdiscilinary Research Cluster on Sustainability Studies,
Jawaharlal Nehru University, New Delhi, India
e-mail: pravin.kushwaha@gmail.com

S. Reddy (ed.), *The Asian Tsunami and Post-Disaster Aid*,
https://doi.org/10.1007/978-981-13-0182-7_14

mid-twentieth century, slowly evolving from pre-modern attempts, to protect communities from natural calamities.

Given these memories of the past, the first paradigm (the 'engineering' paradigm) of modern disaster studies reinforced the ideas of already existing structural engineering, on a much larger scale, to control unwanted physical processes. Subsequently, during the latter part of the twentieth century, as disaster research spread across the world, three more paradigms of disaster studies appeared: the 'behavioural' paradigm, the 'development' or 'vulnerability' paradigm, and the 'complexity' paradigm (see Hewitt 1983; Gilbert 1998; Warner et al. 2003; Smith and Petley 2009). Interestingly, none of these paradigms of disaster studies completely replaced the previous ones. Rather, all of them continue to co-exist in practice. In other words, disaster practitioners are often associated with the two conflicting paradigms: the 'behavioural' paradigm and the 'vulnerability' ('development') paradigm. The behavioural paradigm is primarily based on the experiences of developed (western) societies; and is known for its 'quick-fix', 'techno-centric' solutions. On the contrary, based upon the experiences of developing (non-western) societies, the vulnerability paradigm opposes the behavioural paradigm by stressing the need to address the "root causes" (Wisner et al. 2004) of disaster vulnerability as a result of the failure of political and social systems. However, in practice today, both behavioural and vulnerability paradigms are observed as suggesting application and subsequent transfer of western expertise in non-western societies. More specifically, conventional approaches continue to conceptualize disasters as acts of external forces as well as events limited in time and space, thereby the focus remains on attempting to mitigate the impact of certain select natural hazards and technological accidents. In other words, practitioners often revolve around predicting and controlling hazards with the help of modern science and technology, followed by the demonstration of scientific and technological capabilities through rescue and relief operations. It reflects the dominance of the physical corrective engineering approach of the 'engineering' paradigm as well as behavioural adjustments of the 'behavioural paradigm'.

Critics of both the above mentioned paradigms argue that neither merely spatial considerations (prediction and physical protection) of engineering and behavioural paradigms, nor mere inclusion of temporal dimensions of history of 'root causes' in the structural paradigm (vulnerability paradigm) are sufficient for the conceptualization of vulnerability of people and societies to disasters. Recently, disaster researchers are echoing a more nuanced approach of the complexity paradigm based upon historical inquiry into the dynamic complexity-oriented processes. A process-oriented historical perspective of society, highlighting man–nature interactions along with various technological as well as other deliberate interventions, is suggested to give a comprehensive understanding of disaster-vulnerability for better, context-specific, policy formulations.

Dorothea Hilhorst calls this approach "mutuality" that refers to "mutual construction of society and environment". The concept of mutuality assumes that the real-world-systems consist of a large numbers of individual elements interacting with many other elements in a variety of ways. These interactions create emergent behaviour of the complex system of man-nature interactions. It is difficult to understand the interaction of the components in this system, but the emergent behaviour is easy to recognize. Hilhorst further argues that one way of understanding the complexity is to "devise models to capture and control it". On the other hand, "instead of capturing and controlling complexity, the challenge then becomes to acknowledge multiple realities. In order to devise quick ways to immediately reduce the vulnerability, many students and institutions may land up researching upon the attempt to control the complex system. On the other hand, the ways to device means of acknowledging multiple realties (emergent behaviour) are more promising because they reflect better realities" (Hilhorst 2006).

The paradigms of disaster studies discussed above, appeared one after the other, but could not actually replace old ones completely. Because of the various ways of looking at disasters by the actors involved in such practices, all these paradigms seem to be operating in conflict with each other in practice. In this situation, it becomes important to refine the concepts and develop consensus on certain basic challenges faced by the discipline of disaster studies. More precisely, one of the most important tasks is to enrol improved definitions of a disaster and vulnerability for better policy formulations.

In an attempt to broaden the horizon with better conceptualization of disasters, Quarantelli stresses on the need to further refine the concept of 'disaster'. He urges that it has to be based upon "a new typology of crisis situations" (within which disasters are only one type) in the form of "a new paradigm". The suggestion is to develop a 'genotypical' conception (common non-visible factors) of disasters based on characteristics of origin, rather than 'phenotypical' (surface or manifest characteristics), which is generally based on physical appearance because "the field has long been misled by a 'phenotypical' distinction drawn from the common senses" (which creates a typological distinction between natural and technological disasters). Further, out of the different kinds of typologies—such as empirical, logical, heuristic, ideal type, and so on—Quarantelli prefers the 'ideal type' with four phases of disasters: mitigation, preparedness, response, and recovery. These four phases are 'ideal types' because they have no clear-cut boundaries, and often overlap. The aim is to address sudden and/or chronic situations, agent or response focus, placement of newer types of future crises, complex emergencies or compound disasters (complex emergencies' or 'compound disasters' are "complex mixtures of civil strife, famines, genocidal activities, epidemics, and large-scale displacement and movement of refugees") and implicit or explicit value positions. The 'paradigm shift' also requires "adopting some currently much talked about intellectual trends in the social sciences" such as chaos theory, eco-feminist views, cultural theories (as explained by Oliver-Smith), theoretical approach to 'risk' (Ulrich Beck's 'risk society') and theoretical formulations that stress emotional-affective factors (taking an actor's perspective into account), and so on (Quarantelli 1998). Thus, in defining a disaster, the real challenge is to adopt a definition that avoids compartmentalization and fixity to any one or two perceptions. Also, it is more important to rely on a definition that can encompass the upcoming challenges or threats in a rapidly changing globalized world.

Like disaster, the discourse on theorizing 'vulnerability' is also complex in nature. Hilhorst and Bankoff summarize the nature of vulnerability as follows:

> Vulnerability changes through time in unpredictable ways and in varying directions: increasing, decreasing, accelerating, oscillating, concentrating, or diffusing. It varies with the interplay of three different time frames: long-term, short-term, and cyclical change. (Hilhorst and Bankoff 2006, 6)

The origin and evolution of vulnerabilities in various societies are deeply rooted in the ideologies and the practices involved. Despite lack of consensus on any single model or theory or definition, the two characteristics of vulnerability are very well-recognized: (1) recognizing/defining vulnerability as a social construction, and (2) vulnerability as an internal character of communities. Hilhorst and Bankoff explain the political nature of vulnerability in two ways: first, that "the material production and distribution of vulnerability is the result of political processes", and second, that "the labelling of vulnerable people is also a political act" (Hilhorst and Bankoff 2006, 7).

Reflecting on the 'cultural construction' of nature–society relations, in which "the nature of disasters is rooted in the co-evolutionary relationship of human societies and natural systems", Anthony Oliver-Smith argues that "societies and destructive agents are processual phenomena rather than an isolated event demarcated in an exact time frame". Oliver-Smith further elaborates that the meeting of a human population and a potentially destructive agent does not inevitably produce a disaster. Actually, a disaster is the result of a "historically produced pattern of vulnerability", which depends on multiple factors like "location, infrastructure, socio-political structure, production patterns, and ideology that characterize a society" (Oliver-Smith 1998). Similarly, Michael K. Lindell argues that major challenge in recognizing the impact of disasters is the notion of community as a social unit. He points out that communities are not homogeneous; rather, its subunits such as households and businesses vary in their vulnerability. Ultimately, for defining vulnerability, people "can be differentiated by many different characteristics", and therefore, "a comprehensive theory of hazard vulnerability" would require identification of varieties of variables including its measures, proximal causes, distant causes and variables merely correlated with these causes (Lindell 2011).

In an important observation, Greg Bankoff claims that the depiction of some regions of the globe as unsafe due to increasing numbers of disasters is a "western discourse". He explains that, between the seventeenth and nineteenth centuries, the warm and moist climate of tropical and equatorial regions was depicted as disease-ridden by newly-arrived Europeans. Later, in the late nineteenth century, 'germ theory' explained the cause of diseases as 'bacteria' rather than 'climate', and a 'cure' was sought in Western medicine. But, by the end of the twentieth century, this notion of Western security was diluted with the emergence of antibiotic-resistant strains of known diseases, spread of the AIDS pandemic, and other new viruses (Bankoff 2006, 26). He summarizes the debates of 'developmentalism' and 'vulnerability' as 'cultural constructs' of 'Western discourse' as following:

> Between the seventeenth and early twentieth centuries, this discourse was about 'tropicality' and western intervention was known as 'colonialism'. Post-1945, it was mainly about 'development' and Western intervention was known as 'aid'. In the 1990s, it was about 'vulnerability' and Western intervention is known as 'relief'. (Bankoff 2006, 33)

Bankoff also opines that, similar to the 'tropicality' discourse, the discourse of vulnerability divides the world into two in which the geographical distribution is the same as that of the tropicality and developmentalism discourse (Bankoff 2001).

Thus, disaster studies have essentially been inserted into the larger narrative of development that continues to rely upon techno-managerial approaches. Most often, such perspectives do not recognize the role of everyday social and political processes in creating contexts for disasters or vulnerability. In other words, development discourses have tended to downplay the role of deeper social and political processes in shaping and influencing disaster events and their causes. Invariably, it is this technocratic refrain, as I point out below, that runs through most of the so-called 'paradigm shift' approach of the Indian government.

2 Unpacking the 'Paradigm Shift' in Indian Disaster Management Framework

The renewed disaster management framework of Indian government is lauded as an exemplary initiative. The adoption of the 'Paradigm Shift' approach, the enactment of the Disaster Management Act (DMA), 2005, the formulation of National Policy on Disaster Management, 2009, and the preparation of the National Disaster Management Plan, 2016, are the major initiatives that indicate a new era of disaster management in India. Principally, new initiatives are aimed at developing a comprehensive mechanism of disaster governance in India. The emphasis is on recognizing preparedness, prevention, and mitigation, in a more integrated manner as compared to the previous focus on post-event responses. Despite these seemingly drastic shifts, a lack of clarity both in terms of a certain key definitions, as well as in the overall perspective, continues to prevail. In particular, the notion of 'technology-driven strategy' that often dwells around hazard-centric practices continues to dominate the new approach. The discourse of production of disaster vulnerabilities as an outcome of various socio-technical as well as socio-political processes is yet to take centre stage in policy formulations. It appears that despite adoption of the 'paradigm shift' approach, those involved in policy formulation are yet to emerge from the traditional mind-set that continues to direct them towards framing narrower conceptions of disasters.

The roots of the emergence of a new disaster management framework in India lie in the setting up of a High Power Committee (HPC) on Disaster Management in 1999. The HPC submitted its report in 2001 and proposed the setting up of a separate Department of Disaster Management and a National Emergency Management Authority at national level. HPC identified various disasters and grouped them into five sub-groups: (1) water- and climate-related disasters,[1] (2) geologically-related disasters,[2] (3) chemical-, industrial- and nuclear-related disasters,[3] (4) accident-related disasters,[4] and (5) biologically-related disasters.[5] The HPC report further states that:

> Even after enlistment and deliberations over thirty odd disasters, there were further suggestions for inclusion of more types of disasters such as civil strife, communal violence etc. However, the members were of the view that it was not possible to make the list exhaustive, but the basic condition of it having been designed to cater to all major categories of disasters should suffice for the purpose of the HPC. The HPC felt that the systems developed based on these listed types of disasters, further classified into five groups should suffice in catering to any additional types of disasters that may emerge. (NCDM 2002, 71)

Thus, we can say that HPC conceptualized disasters as an encompassing concept covering all kinds of disasters, and also recommended regular refinement through inclusion of additional upcoming disasters. This conceptualization was translated into a roadmap for a 'National Disaster Framework' covering an institutional mechanism, mitigation and prevention measures, a legal/policy framework, preparedness and response, early warning systems, human resource development and capacity building (GOI 2004, 3–4). Later, in December 2005, the Disaster Management Act (DMA) of India was enacted.

The most important development in all these initiatives taken up by the government is the adoption of the 'paradigm shift' approach in disaster management practices as the guiding principle. In the inaugural speech of the First India Disaster Management Congress (29 November 2006), Prime Minister Manmohan Singh elaborated this approach as the following:

> …the time has come for a paradigm shift in disaster management from a 'relief-centric' and 'post-event' response, to a regime that lays greater emphasis on preparedness, prevention and mitigation…Such an approach should place emphasis on improving early warning systems, ensuring the reach and efficacy of dissemination, creating awareness and building

---

[1]Subdivided into: (1) Floods and Drainage Management, (2) Cyclones, (3) Tornadoes and Hurricanes, (4) Hailstorms, (5) Cloud Bursts, (6) Heat Waves and Cold Waves, (7) Snow Avalanches, (8) Droughts, (9) Sea Erosion, (10) Thunder and Lightning.

[2]Subdivided into: (1) Landslides and Mudflows, (2) Earthquakes, (3) Dam Failures/Dam Bursts, (4) Mine Fires.

[3]Subdivided into: (1) Chemical and Industrial Disasters, (2) Nuclear Disasters.

[4]Subdivided into: (1) Forest Fires, (2) Urban Fires, (3) Mine Flooding, (4) Oil Spills, (5) Major Building Collapses, (6) Serial Bomb Blasts, (7) Festival-Related Disasters, (8) Electrical Disasters and Fires, (9) Air, Road and Rail Accidents, (10) Boat Capsizing, (11) Village Fire.

[5]Subdivided into: (1) Biological Disasters and Epidemics, (2) Pest Attacks, (3) Cattle Epidemics, (4) Food Poisoning.

> capacities at all levels of public administration. I am, therefore, very happy that the draft National Policy[6] on Disaster Management, places great emphasis on efficient management of disasters, rather than only focus on immediate response to disasters.[7]

This new approach also recognized the inclusion of mitigation strategies in the development process and that mitigation itself has to be multidisciplinary. An important observation one can make is that although India's DMA, 2005, defines several terms,[8] certain important key terms like hazard, vulnerability, risk, resilience and community are completely missing from it. The Act further defines a disaster as:

> a catastrophe, mishap, calamity or grave occurrence in any areas, arising from natural or manmade causes, or by accidents or negligence which results in substantial loss of life or human suffering or damage to, and destruction of, property, or damage to, or degradation of, environment, and is of such a nature or magnitude as to be beyond the coping capacity of the community of the affected area (Gazette of India 2005, 2).[9]

Inclusion of a community-based approach and coping capacity is a welcome step in the above definition, but the focus on disaster as a sudden event still reflects a narrower conception. A broader process-oriented conception of disaster indicating historical production or cultural construction of vulnerabilities at various structural levels in the community/society as defined below is yet to take the centre stage.

> A process/event involving the combination of a potentially destructive agent(s) from the natural, modified and/or constructed environment and a population in a socially and economically produced condition of vulnerability, resulting in a perceived disruption of the customary relative satisfactions of individual and social needs for physical survival, social order and meaning. (Oliver-Smith 1998, 186)

---

[6]Broad features of the Draft National Policy on DM included: (1) a holistic and proactive approach, (2) each ministry/department set apart an appropriate fund under the plan for specific schemes/projects for vulnerability reduction and preparedness, (3) each project in hazard-prone areas will have mitigation as an essential term of reference, (4) community involvement and awareness generation, (5) close interaction between the corporate sector, NGOs, the media and the government, (6) institutional structures to be built up and development of inter-state arrangements for sharing of resources during emergencies, (7) culture of planning and preparedness for capacity building, (8) construction designs as per Bureau of Standards codes, (9) all lifeline buildings, e.g., hospitals, railway stations, airport control towers, etc. to be disaster-resistant or retrofitted, (10) revision of Relief Codes in the states and developing them into disaster management codes/manuals (see GOI 2004, 9–12).

[7]'Emphasise Prevention Says PM,' *Times Foundation* 29 November 2006. http://timesfoundation.indiatimes.com/articleshow/729689.cms.

[8]Various definitions mentioned in the Disaster Management Act, 2005, include 'affected area', 'capacity building', 'central government', 'disaster', 'disaster management', 'district authority', 'local authority', 'national authority', 'mitigation', 'national executive committee', 'national plan', 'preparedness', 'prescribed', 'recommended', 'resources', 'state executive committee', 'state government', and 'state plan'.

[9]Gazette of India (2005), 2.

> A disaster is a function of the risk process. It results from the combination of hazards, conditions of vulnerability and insufficient capacity or measures to reduce the potential negative consequences of risk. (International Strategy for Disaster Reduction[10])

Such conceptualizations appear to be closer to the conception of the new approach of the 'paradigm shift', but they are yet to be realized in definitions of disasters adopted in India. The argument is to adapt to a larger context of process-oriented dynamics of complex vulnerability of the coupled human–environment systems. Turner et al. suggest that all our conceptions are 'reduced forms' of the totality of the system, that is why "failure to consider this larger context could lead to the identification of 'response opportunities,' which, if implemented, lead to significant unintended consequences or surprise" (Turner et al. 2003, 8076). Interestingly, during the last decade, despite the emphasis on community and its coping capacity, ever-evolving definitions of disasters in the international discourse have shown reluctance on reinforcing the agenda of 'root causes'; rather, an event- and impact-focused definition continues to prevail conventionally, as apparent from the terminology adopted later by UNISDR:

> A serious disruption of the functioning of a community or a society involving widespread human, material, economic or environmental losses and impacts, which exceeds the ability of the affected community or society to cope using its own resources. (UNISDR 2009, 9)

Consequently, event-driven and impact-focused conceptions continue to define the objectives and mechanisms of disaster policy and planning processes in India. Based on the DM Act of 2005, the National Disaster Management Authority (NDMA) was set up as an apex body in the country for laying down policy, plans, and guidelines for disaster management. NDMA also prepared a policy framework[11] by keeping in view the following national vision:

---

[10]UNISDR. ISDR Terminology: Basic Terms of Disaster Risk Reduction. http://www.unisdr.org/eng/library/libterminology-eng%20home.htm. Accessed 22 May 2007.

[11]Themes underpinning this policy framework focus on five aspects: (1) community-based disaster management, (2) capacity development, (3) consolidation of past initiatives and best practices, (4) cooperation with agencies at national, regional and international levels, and (5) compliance and coordination to generate a multi-sectoral synergy. From the national vision and aforementioned themes, the objectives guiding the policy formulation have also evolved. These objectives are: (1) Promoting a culture of prevention and preparedness—by centre-staging DM as an overriding priority at all levels and at all times, (2) Encouraging mitigation measures based on state-of-the-art technology and environmental sustainability, (3) Mainstreaming DM concerns into the development planning process, (4) Putting in place a streamlined institutional techno-legal framework in order to create and preserve the integrity of an enabling regulatory environment and a compliance regime, (5) Developing contemporary forecasting and early warning systems backed by responsive and fail-safe communications and Information Technology (IT) support, (6) Promoting a productive partnership with the Media, NGOs and the Corporate Sector in the areas of awareness generation and capacity development, (7) Ensuring efficient response and relief with a caring humane approach towards the vulnerable sections of the society, (8) Making reconstruction an opportunity to build back better and construct disaster-resilient structures and habitats (NDMA http://ndma.gov.in. Accessed on 10 December 2015).

> to build a safer and disaster resilient India by a holistic, pro-active, technology driven and sustainable development strategy that involves all stakeholders and fosters a culture of prevention, preparedness and mitigation.[12]

NDMA further states that the vision of a disaster-resilient India is to be achieved primarily by developing a multi-disaster and technology-driven strategy focused on community based approach through the collective efforts of government and Non-Government Organizations (NGOs). It is further pointed out that this policy framework is in compliance with International Strategy for Disaster Reduction (ISDR), the Rio declaration, the Millennium Development Goals (MDGs) and the Hyogo Framework, 2005–15. Interestingly, the notion of 'technology-driven strategy' (mentioned in the national vision) dominates the objectives of guiding the policy formulation as, out of eight objectives mentioned by NDMA, four focus on technocratic solutions which straightforwardly target 'physical corrective engineering' and 'post-event emergency response' as the main responses to disasters. The argument is that considering mitigation measures merely based on state-of-the-art technology would be a narrow approach by making considerations of social vulnerabilities secondary in practice to the science of nature (natural events) and technological solutions. Similarly, multi-hazard early warning as proposed must address hazards beyond environmental hazards. It must include aspects of complex emergencies and compound disasters. In other words, the multi-hazard early warning system has to broaden its horizon from collection of a few natural hazards to covering threats as a result of globalization, distress migration, global climate change, information-communication-technologies (ICT) failure, natural disaster-triggered technological (natech) disasters, creeping disasters, risks from various existing/upcoming technologies, and many more upcoming potential threats.

Based on the policy framework developed by NDMA, the National Policy on Disaster Management, 2009 came into existence followed by the development of a National Disaster Management Plan (NDMP), 2016. NDMP is said to be a 'dynamic document', which is supposed to incorporate changes through periodic reviews and updates based on 'global practices and national experiences.' NDMP further adopts following vision:

> Make India disaster resilient, achieve substantial disaster risk reduction, and significantly decrease the losses of life, livelihoods, and assets—economic, physical, social, cultural, and environmental—by maximizing the ability to cope with disasters at all levels of administration as well as among communities. (GOI 2016, 2)

Interestingly, this vision of disaster resilience and risk reduction emphasizes minimizing the loss of life, livelihood, and assets, but in practice, loss of social, cultural, and environmental assets is yet to appear prominently. This is evident from the latter part of the NDMP where the focus of Disaster Risk Reduction (DRR) related investment remains confined to structural (physical infrastructure)

---

[12]National Disaster Management Authority, GoI. http://ndma.gov.in/en/about-ndma/vision.html. Accessed 10 December 2015.

and non-structural (laws, mechanisms, and techno-legal regimes) measures driven by hazard-wise responsibility matrices for disaster mitigation. Disaster risk continues to be defined through the lenses of 'human induced' (Chemical, Biological, Radiological, and Nuclear) and 'natural' (cyclone, floods, seismic, tsunami, landslide, drought, cold wave) hazards. The real challenge before India's contemporary disaster management framework is to broaden its horizon from the focus on selected hazardous events to a more dynamic conception based on flexibility to adapt to unrealized threats and vulnerabilities.

Thus, the recognition of disaster management as a cyclic process—integrating prevention, preparedness, and mitigation with disaster management in development practices—reflects the shifting nature of the disaster discourse in India. At the same time, disasters continue to be conceptualized as 'external' as well as events limited in time and space; thus, the focus remains on attempting to mitigate the impact of a select natural hazards and technological accidents. The next section will demonstrate how the application of such a simplistic strategy in disaster management practices has failed to engage with wider socio-political and ecologically complex realities. The inability to recognize these complexities, in fact, make the practice of disaster mitigation itself a kind of disaster.

3 Vulnerability and Post-tsunami Recovery in India

After the tsunami of 26 December 2004, the Indian government undertook massive recue and relief operations in mainland India, and the Andaman and Nicobar Islands. Severe damage was reported from the Andaman and Nicobar Islands, Tamil Nadu, Kerala, and the Union Territory of Pondicherry. Moderate impact of tsunami was also reported from Andhra Pradesh. Immediate post-disaster response in the Andaman and Nicobar Islands was driven by the Central Government, while on the mainland it was taken up by the State governments. It is further pointed out that entire post-tsunami situation was managed with national resources; in fact, the Indian government even offered limited assistance to Indonesia, Sri Lanka and Maldives (Murty et al. 2006). Despite all these efforts, discrimination in disaster relief on the basis of caste and class, failure to understand the ground realities, weak enforcement of environmental regulations, and rampant corruption in the aid delivery were widely reported during the post-disaster recovery aftermath of the tsunami of 2004 (Asian Tsunami 2006).

Immediately after the tsunami, Government of India (GOI) approached the Asian Development Bank, the United Nations, and the World Bank in order to assess the socioeconomic and environmental impact of disaster. In response, all three organizations conducted a 'Preliminary Damage and Needs Assessment' through a Joint Assessment Mission (JAM) in the states of Andhra Pradesh, Kerala, Tamil Nadu and the Union Territory of Pondicherry (now Puducherry), excluding the Union Territory of the Andaman and Nicobar Islands. Maximum damage was identified in fisheries, housing, and infrastructure, while losses related to livelihoods in fisheries, agriculture and micro-enterprises, due to pre-existing vulnerability to poverty of communities involved in these sectors, were also recognized. Based on these estimates, the JAM team made short-, medium-, and long-term recommendations for reconstruction. These included multi-hazard risk assessment, community involvement, and community-based disaster risk management, multi-hazard early warning system, repairing cyclone shelters, linkage with environmental issues, strengthening institutional, techno-legal and techno-financial disaster risk management system, and so on (ADB, UN, and World Bank 2005). The most significant aspect of the JAM report was its conceptual basis: a methodology developed by the United Nations Economic Commission for Latin America and the Caribbean (ECLAC) for the damage and need assessment of disasters in these countries. This model can assess merely physical and economic vulnerabilities in specific geographical and sectoral contexts. Thus, its inability to capture wider socio-political, and other vulnerabilities, in the Indian context made the task of assessors limited in scope. The result was a proliferation of complex problems or post-disaster vulnerabilities during the recovery programme ahead. The Washington Post reported the post-tsunami tragedy as follows:

> The South Asian tragedy has ripped open centuries-old fault lines of caste in rural India's rigid social hierarchy. In the district of Nagapattinam, where more than 6,000 people died, untouchables from about 10 villages have openly protested what they call discrimination against them in the provision of relief supplies and access to shelters.[13]

This report exposes that how fishermen forced Dalits (the above-mentioned untouchables) of the same village to leave shelters, and also how fishing families claimed the losses suffered by Dalits as small as compared to them, therefore, Dalits require less aid than the fisherman. Another similar reporting states the discrimination of Dalits as following:

> The plight of villages such as Pudupalayam (pronounced poo-doo-PAH-lay-uhm) is especially challenging in a country such as India, where a decade of economic growth has spawned shopping malls and a burgeoning middle class, but still hasn't overcome a social system that remains divided largely along gender and caste lines. 'All the aid agencies and even the government are only looking after the fisher-folk,' complained Shivalingam, a young and outspoken community leader who like many here uses only one name. 'They got their boats and their nets. We don't have any work.'[14]

According to this report, Dalits were overlooked during distribution of relief material because fishing families were identified as most affected. This resulted in more and more provisions of boats and nets for fishermen while there was hardly any provision for cows and goats lost as had been reported by Dalits. Further, the

---

[13]R. Lakshmi. 'Tsunami Opens Fault Lines in Old Caste System: India's Untouchables Allege Discrimination in Allocation of Aid.' *The Washington Post*, 18 January 2005. http://www.washingtonpost.com/wp-dyn/articles/A16472-2005Jan17.html.

[14]K. Moritsugu, 'Tsunami Aid Distributed Unevenly between India's Fishing and Farming Villages,' *South Asian Journalist Association*, 7 December 2005. http://www.saja.org/resources/moritsugu/main.htm.

proliferation of boats resulted in reduced catches per boat, due to which some fishermen wouldn't have been able to recover their costs. In another article, Ken Moritsugu writes that, in the rush to rebuild, 'shelters built in low-lying areas ended up flooding during rainy season', and also that the delivery of relief as quickly as possible led to the delivery of poor-quality boats which were not seaworthy.[15] Other issues, which were later unearthed, included unsuitability of plastic tents in the hot and humid conditions of coastal areas, enforced long-term stays in temporary houses due to delay in construction of permanent houses, and unsustainability of the cash for work programme.

Thus, the caste-based, socially generated, and politically reinforced productive and allocative pattern of Indian society made Dalits more vulnerable in historical time; their assets were also assigned small value. Looking through the lens of Western discourse of developmentalism, based on geographical similarities of India with Latin American and Caribbean countries, JAM applied the UNECLAC model with the sectoral approach, and, therefore, could not assess the genotypic nature of the historically-produced vulnerability of Dalits, the causes of which are deeply rooted in the centuries-old political ecology/economy of this country. More importantly, following a constructivist approach, the argument is that it was not merely the application of Western principles (sectoral model of market economy) which made Dalits more vulnerable in the aftermath of the tsunami, but that, even though the millennia-old caste system was outlawed by the Indian Constitution, many Indians still observe the system and follow it illegally, which increased the post-tsunami vulnerability of Dalits manifold. Thus, the two-way process of Indian society's inability to overcome untouchability in practice and the failure of damage-assessing agencies to recognize Dalits as more vulnerable after the disaster suddenly accelerated their post-disaster vulnerability beyond control. Gal Frenkel has highlighted that marginalized and vulnerable groups had limited or in many cases, no access to relief assistance and benefits of rehabilitation programmes. He further pointed out that relief and rehabilitation efforts were unable to recognize all those affected by the tsunami. Frenkel summarizes this as following:

> The rapid assistance to the fisherfolk and assistance in recognizing their boats and nets has neglected many other people who have been affected by the tsunami and who depend on the sea, the crippled fishing industry, and the coastal agriculture for their livelihood. These include vendors and labourers, salt pan and lime workers, shrimp farmers, women involved in seaside lime stone production, small coastal business, petty traders, Dalit and tribal inland fisherfolk, and farmers whose land has been made uncultivable by the influx of salt water… Moreover, no compensation has been received for loss of livelihood. (Frenkel 2005, 2)

There was also immeasurable damage to the environment, fish stocks, and local adaptation patterns evolved over the years. The exact estimation of this damage is beyond the scope of the scientific community due to the highly complex nature of

[15]K. Moritsugu, 'A Rush to Rebuild Leads to Wasted Effort,' *South Asian Journalist Association*, 7 December 2005. http://www.saja.org/resources/moritsugu/side.htm.

various factors involved in time and space dyamics. Destruction of mangroves, coral reefs, sand dunes, and other coastal ecosystems along the coastline, which provided natural defences from sea waves, was as severe as the loss of fish stock. Coastlines were eroded, sediments from the land were deposited over the coral reefs, sea mud was deposited over agricultural land, and shallow soil was stripped off from some low-lying islands. Shallow and groundwater was contaminated by saline water of the sea, while faecal bacteria, from damaged or destroyed septic tanks and pit toilets, contaminated water supply systems. Rapid cleaning up of debris containing toxic and hazardous materials led to their inappropriate disposal, and finally, the impact on agriculture and local adaptation patterns was also unaccountable. According to Revathi[16], Executive Director of a Chennai-based NGO ToFarM, saline sea water contaminated fresh water over the land and the consequent reverse osmosis (RO) took out the water from the plants. Speaking at a workshop in New Delhi in 2007, she explained that though the government provided gypsum to farmers to desalinate the water, no one knew how much gypsum was to be applied and where and how much time it would take to desalinate the water. Even the scientific community was unable to provide this information either because of uncertainties, nor were they able to provide any quick solution to tackle the problem of deposition of sea mud over agricultural land. By raising the problem of destruction of the local adaptation pattern during the reconstruction programme, Revathi M. argued that, in Nagapattinam, coastal villages had natural defences of sand dunes that protected them from seawater or tsunamis. These sand dunes developed naturally over the years due to consistent sand depositions by the on-shore waves, but during the post-tsunami reconstruction programme, construction agencies started sand mining from these sand dunes, and thereby destroyed this very natural defence.

Once the fishing communities were declared to be the most affected/vulnerable, most aid agencies targeted the fisheries sector immediately. Consequent over proliferation of boats and nets as aid was beyond the absorption capacity of the coast, and resulted in the reduced fish catch per boat. In a hurry to supply relief material, poor-quality boats were supplied which were not durable. Further, the failure to recognize the geo-climatic vulnerability of the tropical coastline of India led to the supply of plastic tents as relief material, which were not suited to the hot and humid climate of the target region. The climatic vulnerability was further reflected in the rainy season, when failure to anticipate it led to flooding of shelters built in low-lying areas. The most important factors behind all these undesirable events were the responses based on an outsider's perspective, and were aggravated by the second tsunami of aid. The proliferation of a number of organizations with no prior experience in relief led to the duplication and wastage of relief. The supply of

[16]Revathi was speaking in the national workshop on 'People, Policy and Partnerships for Disaster Resilient Development', 3–4 November 2007 at India Habitat Centre, New Delhi. The workshop was organized by the National Alliance for Disaster Risk Reduction (NADRR), a joint initiative of several NGOs in India.

inappropriate aid—inappropriate clothes, expired or culturally inappropriate food, unwanted drugs, medicines, and tents—added to the poor quality of relief work.

4 Aid Politics and Re-thinking Indian 'Paradigm Shift'

The politics of disaster management is deeply rooted in the existing socio-political structure and the dynamics of class interests. The poor and marginalized are "most vulnerable to the negative consequences of a disaster and face significant challenge in recovery" (Reid 2013). Disasters also consolidate and increase social inequalities as the rich and the batter off can cope up better in emergency situations. Some people also profit from disasters by grabbing economic opportunities present in the relief and rehabilitation programmes (Bankoff 2003). Werker (2010) argues that considering "disasters as natural" and "disaster relief as apolitical" is flawed as unfortunately the laws of political economy apply to as well. Therefore, in the politics of disaster relief, "aid is driven by donor priorities" (Werker 2010). More specifically, the perception of humanitarian aid, once considered to be apolitical assistance, has also been changing over the last couple of decades. Hilhorst et al. (2010) identify three dimensions of social life of humanitarian actions that require a careful examination: One, normative[17]; two, complex processes that shape humanitarian action in everyday practice[18]; and three, the ways in which it affect societies.[19] What matters is a careful investigation of how humanitarian action is shaped in practice and how it affects societies where it is implemented (Hilhorst et al. 2010). Scholars argue that it is important to understand the consequences of 'new humanitarianism', which is driven by the growing convergence between politics and aid. Traditional humanitarian principles of neutrality, impartiality, independence, and universality are being increasingly undermined to fulfil political and strategic objectives through the integration of humanitarian assistance into Western foreign policy. It is further suggested that historically, humanitarian assistance has not been able to fill the vacuum of effective political engagement. Therefore, rather than concentrating on the strategy to contain conflict, humanitarian actions must focus on "alleviating sufferings" of the people "irrespective of colour and creed when all else has failed" (Duffield et al. 2001). Some people argue

---

[17]Different sets of principles and objectives of humanitarian actions: these actions interface with normative frames that prevail at the sites of implementation; negotiate socially between these different frames of reference to give meaning to their actions in practice, humanitarian actions must be more sensitive to the complexities of scale—how national policies are translated in the localities.

[18]Humanitarian assistance is the outcome of the political choices and social interactions of humanitarians and a large range of surrounding actors, institutions, and processes.

[19]Its impact on local conflict dynamics: where aid has large presence, humanitarian actions can be seen affecting the ordering process of society at large, changing people's outlooks, altering power of constellations, transforming institutions and leaving footprints on spatial organizations.

that the growing outcomes of research might change the public perception of causes of disasters. For instance, growing outcomes of research on climate change and other severe weather related events might reverse the public perception of "disasters not as acts of nature", rather, as responsibility of political leadership. Consequently, the accountability will shift from natural or supernatural forces to actions of individuals—political leadership in the government—for relief and prevention (Oliver and Reeves 2015, 5).

Catherine Bragg argues that in the recent past, especially after the tsunami of 2004, it is observed that many poor or less developed countries "started strengthening their national disaster management structures including central and decentralized agencies, legislative frameworks and overall governance" (Bragg 2015). In case of a disaster, many such countries may not ask for international or multilateral assistance. Hence, the traditional role of governments as either donors/providers or hosts/recipients is changing too. Rather, national governments are becoming more active and going beyond their traditional role of 'hosts' (of international aid) and entering into the domain of implementation. Bragg further points out that implementation gets often translated into action through the notion of 'civil protection' where people from civil protection or military background often head national disaster management agencies (Bragg 2015). The same is true in case of the structure of disaster governance in India.

Apart from the adoption of Disaster Management Act, 2005, and the National Policy on Disaster Management, 2009, Indian government has come out with a National Disaster Management Plan (NDMP)[20] in 2016, which is said to be in compliance with the Sendai Framework for Disaster Risk Reduction, 2015–2030. All these initiatives are flourishing under the newly adopted 'paradigm shift' approach of the government for disaster governance. Although the DM Act, the subsequent policy, and the plan, have explicit provisions for developing financial mechanisms for disaster governance, the question of disaster aid or humanitarian aid is not adequately addressed in the new institutional framework and financial mechanisms. This is despite the fact that media, researchers, experts, and other professionals have highlighted the politics of disaster-aid as a matter of careful analysis for ensuring the effective delivery of aid and post-disaster responses. The new approach of 'paradigm shift' in Indian disaster-management discourse continues to speak the language of hazard centrism where hazard-wise responsibility matrix has become the basis of disaster governance, and thus, of determining the financial arrangements accordingly. Existing provisions of the Disaster Response

[20]Based on a hazard-wise responsibility matrix that clearly defines roles and responsibilities of centre and state government agencies for each hazards, NDMP focuses largely on four aspects namely: inter-agency coordination; structural measures; non-structural measures; and capacity developments. Inter-agency coordination covers aspects of disaster governance such as response, warning, information, and data. Structural measures consider developing resilience based on securing and improving physical infrastructure. Non-structural measures rely on techno-legal regime that includes building codes, norms laws, rules, and guidelines. Finally, capacity development is defined in terms of training, curriculum development, awareness, and mock drills.

Funds and Mitigation Funds limit themselves to post-event responses, inter-agency coordination, improvements in physical infrastructure, techno-financial, techno-legal initiatives, and capacity building, primarily through training and awareness. Given the experiences of the politics of disaster aid after the tsunami of 2004, the concern for making provisions for regulating and governing appropriate disaster aid is yet to take centre stage in mainstream policy and planning. In absence of such provisions, post-event responses will continue to remain mismanaged and less productive (as was the case of tsunami in India).

5 Conclusion

Given the present lack of definitional clarity of the terms ‘disaster’ and ‘vulnerability’, these concepts are still of great significance when seen in a context-specific way. Oversimplification, essentialization, and generalization may overlook better conceptualization and may often lead to undesirable policy responses or policy formulations. The essence of vulnerability research till date is the recognition of vulnerability as a process. It assumes that the causes of disasters are not merely occurrences of natural phenomena; rather, various structural processes such as economic, social, and political contribute to the production of disaster vulnerability in the long term. Such a historically produced pattern of vulnerability needs reflections other than the conventional focus of disaster management practices on deployment of merely engineering and behavioural solutions. Although the contemporary disaster management framework in India is lauded as an exemplary initiative, there needs to be much greater coherence between its vision, goals, objectives, policies, and plans. It is important to state that the country must acknowledge the relationship between disasters and development beyond the challenge of improving physical infrastructure, warning system, and post-event responses. The real challenge before India’s contemporary disaster management framework is to broaden its horizons from a focus on selected hazardous events to a more dynamic conception of disasters, based on flexibility to adapt, and to ensure the inclusion of unrealized threats and vulnerabilities for which the nation must prepare itself if the proclaimed “paradigm shift” is to move beyond rhetoric to actual practice.

References

Bankoff, G. 2001. Rendering the World Unsafe: ‘Vulnerability’ as Western Discourse. *Disasters* 25 (1): 25–27.

Bankoff, G. 2003. *Cultures of Disasters: Society and Natural Hazards in the Phillipines*, 91–92. London: RoutledgeCurzon.

Bankoff, G. 2006. The Historical Geography of Disaster: ‘Vulnerability’ and ‘Local Knowledge’ in Western Discourse. In *Mapping Vulnerability: Disasters, Development and People*, 1st South Asian ed, ed. G. Bankoff, G. Frerks, and D. Hilhorst, 26. Earthscan: London.

Bragg, C. 2015. Disaster Management and Multilateral Humanitarian Aid: Parallelism vs. Combined Forces. In The Humanitarian Challenge: 20 Years European Network on Humanitarian Action (NOHA), ed. P. Gibbons and H. Heintze, 1–16. New York, London: Springer.

Duffield, M., J. Macrae, and D. Curtis. 2001. Politics and Humanitarian Aid. *Disasters* 25 (4): 269–274.

Editorial. 2006. Asian Tsunami: Measuring Relief. *Economic and Political Weekly*, 41 (52), 5292.

Frenkel, G. 2005. India After Tsunami—The Rights of Affected People. *Focus Asia-Pecific* 39: 2.

Gazette of India. 2005. *The Disaster Management Act, 2005*, 2. New Delhi, 26 December 2005.

Gilbert, C. 1998. Studying Disaster: Changes in the Main Conceptual Tools'. In *What is a Disaster? Perspectives on the Question*, ed. E.L. Quarantelli, 11–13. London: Routledge.

GoI. 2016. *National Disaster Management Plan*, 2. New Delhi: National Disaster management Authority.

Government of India. 2004. *Disaster Management in India: A Status Report*, 3–4. New Delhi: Ministry of Home Affairs.

Hewitt, K. 1983. The Idea of Calamity in a Technocratic Age. In *Interpretations of Calamity from the Viewpoint of Human Ecology*, ed. K. Hewitt, 3–8. London: Allen and Unwin.

Hilhorst, D. 2006. Complexity and Diversity: Unlocking Social Domains of Disaster Response. In *Mapping Vulnerability: Disasters, Development and People*, 1st South Asian ed, ed. G. Bankoff, G. Frerks, and D. Hilhorst, 53–55. London: Earthscan.

Hilhorst, D., and G. Bankoff. 2006. Introduction: Mapping Vulnerability. In *Mapping Vulnerability: Disasters, Development and People*, 1st South Asian ed, ed. G. Bankoff, G. Frerks, and D. Hilhorst, 6. London: Earthscan.

Hilhorst, D., D. Dijkzeul, and J. Herman. 2010. Editorial: Social Dynamics of Humanitarian Actions. *Disasters* 34 (S2): S127–S129.

Lindell, M.K. 2011. Disaster Studies. *Sociopedia.isa*. https://doi.org/10.1177/205684601111.

Murty, C.V.R., S.K. Jain, A.R. Sheth, A. Jaiswal, and S.R. Dash. 2006. Response and Recovery in India after the December 2004 Great Sumatra Earthquake and Indian Ocean Tsunami. *Earthquake Spectra* 22 (S3): S731–S758.

National Centre for Disaster Management. 2002. *High Power Committee on Disaster Management*, 71. New Delhi: Indian Institute of Public administration.

Oliver, A.J., and A. Reeves. 2015. The Politics of Disaster Relief. *Emerging Trends in the Social and Behavioral Sciences: An Interdisciplinary, Searchable, and Linkable Resource*, 5.

Oliver-Smith, A. 1998. Global Changes and the Definition of Disaster. In *What is a Disaster? Perspectives on the Question*, ed. E.L. Quarantelli, 187–192. London: Routledge.

Quarantelli, E.L. 1998. Epilogue: Where We Have Been and Where We Might Go. In *What is a Disaster? Perspectives on the Question*, ed. E.L. Quarantelli, 234–270. London, New York: Routledge.

Reid, M. 2013. Disasters and Social Inequalities. *Sociology Compass* 7 (11): 984–997.

Smith, K., and N. Petley (eds.). 2009. *Environmental Hazards: Assessing Risk and Reducing Disaster*, 5th ed. London: Routledge.

Turner, B.L., R.E. Kasperson, P.A. Matson, et al. 2003. A Framework for Vulnerability Analysis in Sustainability Science. *Proceedings of National Academy of Sciences* 100 (14): 8076.

UNISDR. 2009. *UNISDR Terminology on Disaster Risk reduction*, 9. Switzerland: UNISDR.

United Nations, World Bank, and Asian Development Bank. 2005. *India: Post-Tsunami Recovery Programme Preliminary Damage and Needs Assessment*. http://www.undp.org/cpr/disred/documents/tsunami/india/reports/dnassessment.pdf. Accessed 7 Feb 2007.

Warner, J., D. Hilhorst, and P. Waalewijn. 2003. Public Participation in Disaster-Prone Watersheds: Time for Multi-Stakeholder Platforms? *Disaster Studies, 6* (Disaster Studies Group, Wageningen: Wageningen University).

Werker, W. 2010. Disaster Politics: International Politics and Relief Efforts. *Harvard International Review*, August 3. http://hir.harvard.edu/disaster-politics/. Accessed 15 Sept 2016.

Wisner, B., P. Blaikie, T. Cannon, and I. Davis. 2004. *At Risk; Natural Hazards, People's Vulnerability and Disasters*, 2nd ed, 47–48. London: Routledge.

Tsunami Recovery Impact Assessment and Monitoring System (TRIAMS): Methodological Approach for Recovery

Lakshman Dissanayake

1 Introduction

The TRIAMS process recognises the importance of the development of a conceptual framework to analyse not only the impact of a disaster like a tsunami but also the headway made in the recovery efforts in each of the tsunami-affected area. In this context, the Karolinska Institute proposed a clear concept paper in collaboration with the World Health Organization (WHO) and the International Federation of Red Cross and Red Crescent Societies (IFRC). It showcased the importance of collecting qualitative information, in addition to providing quantitative indicators in four areas of recovery: vital needs, basic social services, necessary infrastructure, and livelihood.

The essential components of TRIAMS process include identifying the impact and the output indicators in the primary sectors of recovery. The process is designed to collect both quantitative and qualitative data from the perspective of the beneficiary and gather additional qualitative data, which would explain the key output findings and also the outcome indicators.

None of the countries affected by the tsunami had produced indicators which illustrated recovery interventions below the district level. Most of the data presented were focused on the output indicators and did not give qualitative inputs.

Beneficiary satisfaction and perceptions were partly measured through quantitative methods, where the indicators were listed in the matrix and provided values or numbers of things and addressed the questions of 'what', but not the 'why'. To answer the question of 'why', it was suggested that a qualitative component be included in the TRIAMS process. To understand the progress of tsunami-response and adjust plans as per the recovery data, the stakeholders needed to have an insight about why things were progressing in a particular direction; the reasons for different

L. Dissanayake (✉)
Colombo, Sri Lanka
e-mail: vc@cmb.ac.lk

S. Reddy (ed.), *The Asian Tsunami and Post-Disaster Aid*,
https://doi.org/10.1007/978-981-13-0182-7_15

responses could vary considerably across districts, or even within the same districts. In the TRIAMS process, once the quantitative data is collected, and unanticipated findings are identified—or the problems are noted—qualitative methods, such as focus group discussions with the beneficiaries (using various participatory tools), can be employed to investigate the challenges. The analyzed data and the results should be cross-checked with various stakeholders. This could then be immediately fed back to programme, for planning and implementation and bringing about the changes as per need in the project design. Based on the specific problems identified during implementation of recovery activities, the frequency and scope of the qualitative methods would vary; like the scale and the type of challenges noted and handled at the sub-national level (TRIAMS Workshop Report 2008, 77).[1]

Therefore, there is a need for inclusion of qualitative component in the TRIAMS process, focusing on the four areas of recovery proposed in the concept paper. Identifying problems in the recovery process empowers affected communities to propose how these problems can be resolved (Chambers 1989, 1995; Cerena 1991; Burkey 1993; CUCEC 2005a, b, 2006). This will provide a unique opportunity for Colombo University Community Extension Centre (CUCEC) to place qualitative analysis in a result- or outcome-oriented process, with the active participation of local communities.

However, it is imperative to mention here that the quantitative aspect of the impact assessment is very significant in the TRIAMS report. Equal emphasis has not been given to the qualitative aspect of the impact assessment. 'The People's Consultation'—as well as its second phase documents titled 'Relief to Development' and 'Identifying Vulnerable Communities'—clearly identifies how the qualitative approach can provide insight into the status of the recovery process, and propose ways of moving from the relief to the developmental stage (CUCEC 2005b). Therefore, it is not merely a descriptive exercise based on passive situation analysis, but a very result-oriented process with the active participation of communities and other relevant stakeholders involved in the recovery process.

2 Objective

The objective of the proposed study carried out by CUCEC was to gather information through a qualitative approach, with the active participation of affected communities and those involved in the recovery process. This would help to assess tsunami-recovery impact, and propose ways of accelerating the recovery process with regard to four areas of recovery: vital needs, basic social services, infrastructure, and ways of achieving sustainable livelihood development.

[1]http://www.who.int/hac/crises/international/asia_tsunami/triams/triams_report.pdf.

3 Methodology

Before proposing an appropriate method, it is essential to mention that CUCEC had substantial experience in carrying out qualitative research, particularly in disturbed environments (whether due to natural or man-made disaster). It is also important to note that the Focus Group Discussion (FGD) methodology proposed by the author, for 'The People's Consultation Project' deviates from textbook Focused Group Discussion. It is different in nature as the proposed methodology can represent the entire community drawn through focused discussion. Comparative studies can also be implemented over time, unlike in conventional FGD methodology.

It is imperative to indicate here that the first need is to ascertain if 'consultation' or 'research' has to be undertaken, as these two have different methodologies. Qualitative research methods can be utilized for community consultations. But converting the consultation process to a research study is detrimental to the consultation exercise because research ideas can impose unnecessary constraints on a consultation process. Consultation is about taking advice from people, asking their opinions, and not just recording experiences or collecting information. According to the TRIAMS Workshop report, it is quite clear that the 'qualitative component' described under 'Beneficiaries Perspective and Satisfaction' (2008, 77) envisages a consultation process implemented as stated in its objectives.

In the consultation meeting conducted in preparation for the TRIAMS in Bangkok, several stakeholders reaffirmed the need to obtain feedback, and the satisfaction with the results and recovery interventions by the beneficiaries, more regularly and reliably from beneficiaries. It was also proposed that the beneficiaries' inputs on unmet needs and future projects be solicited.

4 Selection of the Sample

To make a sample representative of the affected population, 30% of the affected Grama Niladhari (GN) divisions was determined to be the total sample size, by considering the GN division as the primary sampling unit. As the total affected GN divisions in the six selected districts (Galle, Matara, Hambantota, Trincomalee, Batticaloa, and Ampara) were 571, the total sample size was 171. This meant that 171 FGDs needed to be carried out in this exercise, and that it was also within a workable range. After obtaining the total sample size, the number of GN divisions considered for the consultation process was selected by allocating proportional distribution. Respective samples within each GN division were be selected by using simple random sampling. It is important to note that the type of the FGD (livelihood, elderly, women, disabled and children) were determined in consultation with the GNs (Table 1).

Table 1 Distribution of the sample size for the FGDs

District	Divisional Secretariat Division	No. of affected *Grama Niladari* division	Distribution of sample size
Ampara	Addalachennai	13	**4**
	Akkaraipattu	6	**2**
	Kalmunai	42	**13**
	Aladiwembu	2	–
	Lahugala	3	–
	Karaitivu	15	**5**
	Ninthavur	11	**3**
	Potuvil	20	**6**
	Sainthamarathu	9	**3**
	Thirukkovil	16	**5**
Sub-total		137	**41**
Batticaloa	Eravur Pattu	4	**1**
	Koralai Pattu (Valach)	9	**3**
	Manmunai North	18	**5**
	Kattankudy	3	**1**
	Koralai Pattu North	12	**4**
	Manmunai S. and Eruvil	29	**9**
	Manmunai P. (Araipattai)	11	**3**
Sub-total		86	**26**
Trincomalee	Town and Gravets	20	**6**
	Kuchchaweli	16	**5**
	Kinniya	12	**4**
	Muttur	8	**2**
	Eachchilampattu	4	**1**
	Seruvila	1	–
Sub-total		61	**18**
Hambantota	Ambalantota	5	**1**
	Tangalle	25	**8**
	Hambantota	6	**2**
	Tissamaharamaya	3	**1**
Sub-total		39	**12**
Matara	Devinuwara	16	**5**
	Dickwella	16	**5**
	Matara Four Gravets	18	**5**
	Weligama	22	**6**
Sub-total		72	**21**
Galle	Ambalangoda	3	–
	Galle Four Gravets	49	**15**
	Balapitiya	26	**8**
	Bentota	13	**4**
	Hikkaduwa	56	**17**
	Habaraduwa	29	**9**
Sub-total		176	**53**
Grand total		**571**	**171**

When the selection of the sample was made, we also considered the suitability of the sample for this study. Therefore, in addition to the random sampling procedure adopted in selecting the sample GN divisions, we had to examine the level of the tsunami's impact on those chosen GN divisions randomly. If the impact was relatively insignificant, we replaced them with the adjoining GN division, which had a significant impact. For this purpose, we used the People's Consultation Report as the base (CUCEC 2005b). Since the study captured the severity of the effects and the level of recovery about RRR during the first six months after the tsunami disaster in all other sectors, particular emphasis was placed on damages to housing, and interventions for recovery.

5 Methods of Consultation

5.1 Consultation of Affected Communities

The TRIAMS objectives highlight the importance of consulting women and children. They have become vulnerable because of the tsunami and have unmet needs. Our experiences suggest that the elderly and disabled people also need to be consulted, in addition to the above. The four areas of recovery indicate that consultation with women, children, the elderly, and disabled will provide a clear understanding of the recovery process. But separate FGDs are required to understand the development in the livelihood sector of affected people.

5.2 Consultation Process for Livelihood Sector

Within a sample village, 15 people whose livelihoods were affected were to be elected by the community to be part a Focus Group Discussion. The selected 15 were to comprise both men and women of different ages. The sampling of the villages was done to represent the sectoral dimensions of livelihoods.

5.3 Consultation Process for Women

Women in the selected villages were asked to select 15 women of different age categories to represent the affected women community in each village. Particular emphasis was given to select those who lost spouses during the tsunami and thus became more vulnerable. Focus Group Discussions were carried out to ascertain information from these women.

5.4 Consultation Process for Disabled People

People disabled due to the tsunami in the selected village or GN division were consulted at their place of residence to communicate on the achievements after the tsunami, their unmet needs, and their future expectations. Disabled people were consulted through representatives since some disabled people had conditions that impeded communication through deafness, blindness, or a learning disability.

5.5 Consultation Process for Children

Children who lost both or either of their parents were consulted in the selected village or GN divisions. In this particular exercise, children were categorized into two groups: 10 years of age and below, and above ten years. Children from the younger group were consulted together with their guardians, while the older group was consulted separately. Group consultation was suggested for the latter, as they may have been able to gain confidence from the presence of their friends.

5.6 Consultation Process for Elderly

Those affected by the tsunami disaster who were aged 60 years and above, were consulted in groups from the selected sample villages. Samples for the team were chosen by the elderly people in the respective village with the help of a facilitator to represent the elderly community.

6 Application of Formative Dialogue Research Methodology

The objective of the TRIAMS was to monitor ongoing tsunami-recovery efforts over the next five years and make periodic assessments of the impact of recovery activities. Having a continuous dialogue with other stakeholders was necessary to make the impact assessment more effective. In this regard, the application of Formative Dialogue Research Methodology was recommended (Rasnayagam and Amarasinghe 1999; Senanayake and Dissanayake 1999). The researcher had a duty solely to maintain scientific standards and pursue relevant and good-quality research. This was to ensure that findings were disseminated to the stakeholders in a manner which would facilitate critical reflections and dialogues with the researcher and within the programme.

Three perspectives are central to the formative process research.

- The actor perspective: It focuses on the problems of the actor. Since it is the actors who are supposed to cause change, they require a sense of ownership of the problems. The starting point is, therefore, the actors, their actions, the reasons for and the consequences of their actions. Formative process research is in concord with Max Weber's admonition to observe reality "from the actor's point of view".
- The contextual perspective: The perceptions and actions of the actors are understood regarding the context in which they occur. Different settings, situations, and sense of reality would result in flexibility for manoeuvring. Those in a position to determine which worldview would prevail would be able to influence space for manoeuvring.
- The process perspective: Interaction processes are a key to the study theme. The undertaking of an investigation is viewed as a process. Formative process research itself takes place in interaction with the process and exerts an effect on it.

In this respect, it was essential to implement consultations with the different stakeholders by organising one meeting for each group: Grama Niladharis, sub-national governments, district and provincial governments and (International) NGOs. These consultations would provide a unique opportunity for those who initiated interventions to assess contribution about relief, reconstruction, and rehabilitation activities and refine or modify strategies for future operations accordingly, rather than waiting for the final reports of the qualitative study to be released.

7 Approach

Meetings were organised initially with the GN and Divisional Secretariat (DS) in each district, to make them aware of the TRIAMS project and obtain their fullest cooperation to implement fieldwork. Second, DSs were consulted separately to organise meetings with GNs since their help was needed for fieldwork at the village level. GNs were first consulted to investigate the appropriateness of the sample at a meeting held at the respective DS Office with the GNs of all Tsunami-affected GN divisions in the selected DS division. After checking the appropriateness of the sample, GNs were met with separately to find out the dominant group in each division among the five groups—the women, children, elderly, disabled, and livelihood groups—which were proposed to be investigated. Once the selection of the appropriate group was done for each GN division, we visited each GN division with the GN to identify the respective groups and requested them to make their representation, except for the disabled and children. Approximately 15 people were selected by the respective group to be included in the FGD. However, all the children and disabled people were consulted to obtain responses for the recovery process. Consultations were held at various places such as temples, schools, community halls, and so on.

It is quite important to mention here that the research assistants who were part of the research study were selected from recently graduated social science students who were currently unemployed. Training workshops were organized before the commencement of the fieldwork to create awareness about the methodologies adopted in the consultation process. The experiences gained from the fieldwork can be regarded as a way of improving their capacity and capability for research.

8 User-Friendly Format and Maps

Another important aspect of this study was the development of a user-friendly format to summarize the qualitative information gathered, so anyone could easily understand where the problems existed and who the marginalized groups were. Also, an attempt was made to quantify the level of recovery by requesting the investigated people to distinguish three levels: low, medium, and high. This concept was introduced because this particular variable can be mapped to identify the level of recovery of recovery indicators in the respective district/DS divisions.

As we held FGDs for five groups, the ideas or opinions expressed at the FGDs are of those groups. It is also important to note that the same types of questions about four areas of recovery were put forward to the groups. But the relative importance of certain factors emerged at different magnitudes depending upon the group under investigation. For example, issues related to basic services and education, in particular, became more important when children were consulted. Therefore, the methodology adopted in this study can be regarded as appropriate since it captures all the dimensions of the four areas of recovery under the study in the five vulnerable categories.

9 Use of Statistics to Examine the Internal Consistency of Information

Gamma Statistic

Since the collected information was transformed into ordinal variables, use of the Gamma statistic is possible to find the internal consistency of the data. If the information is consistent, then certain variables need to show positive association in the Gamma statistic. Table 2 shows the relationship between food consumption and livelihood improvement; schooling and livelihood improvement; and housing and livelihood.

Table 2 Gamma statistic for selected variables, Galle District

Relationship	Gamma
Livelihood and housing	0.771
Livelihood and schooling	0.800
Livelihood and food consumption	0.786

$N = 53$

10 Median Values for Further Analysis

Since we categorised the recovery indicators to three levels (low, medium, and high), we created ordinal variables as indicated earlier. When we have ordinal variables, we can use the median or mode to obtain an understanding of the behaviour of that indicator as indicated in Table 3. Therefore, the methodology developed here is simple but provides a compelling message of how the qualitative information can be quantified and interpreted to understand the behavioural pattern without much difficulty (Table 4).

Table 3 Median values for variables, Southern Districts, Sri Lanka

Recovery indicators	Median values
Access to water from an improved source	3
Availability of sanitation facilities	3
Food consumption	2
Status of housing	3
Status of health	3
Status of schooling of the children	3
Psycho-social status	3
Status of roads	2
Status of coastal protection	1
Status of receiving loans for livelihood improvement	1
Status of social protection programmes	1
Status of receiving fishing boats and fishing gear	1
Status of SME/self-employment sector	2

$N = 86$

Table 4 Level of recovery: FGD for tsunami-affected women's sector, Kahawagammedda, GN Division, Bentota DS Division, Galle District

Areas of recovery	Recovery indicators	Opinion requested	Opinion expressed	Level of recovery
Vital needs	Access to water from an improved source	Was there a water problem with the tsunami disaster? Has that been improved? How? If not what are your opinions? Any constraints?	The entire water supply net work in the village was damaged by the tsunami The water system was renovated by 'NWS&DB'. Only 8 damaged wells were reconstructed by 'Red Cross' but only five wells can be used for drinking purposes The *Nelumpokuna Gama* tsunami resettlement has 49 houses in two units, 39 and 10 respectively. Pipe water services are not available in the resettlement. Both these units have a well each. For about a period of one month 'Red Cross' supplied water in a browser. This has stopped since the bowser overturned due to the bad condition of the road	Low
	Availability of sanitation facilities	Was there a sanitation problem with the tsunami disaster? Has that been improved? How? If not what are your opinions? Any constraints?	Out of the 26 affected households, 18 are resettled at Nelumpokunagama Eight families have not accepted houses and their toilets are not reconstructed. The drainage system in the village was affected and 70% is reconstructed by the *'Pradeshiya Sabha'* There was no proper garbage dumping area even before the tsunami In the 39 housing units of the resettlement, one septic tank serves three toilets. As the pit walls are not paved, the sides have collapsed, exposing the pit. The bad smell emanated and the capacity of the pits has become a big problem There is no proper drainage system in the resettlement. The water flows through the compounds There is no proper garbage disposal system in the resettlement area. The garbage is dumped in the marsh, polluting the whole area, environmentally and health wise	Medium

(continued)

Table 4 (continued)

Areas of recovery	Recovery indicators	Opinion requested	Opinion expressed	Level of recovery
	Food consumption	Has household food consumption deteriorated? How was it overcome? If not what problems do you have? How can it be overcome?	Household food consumption has deteriorated to a certain extent due to high cost of living. Villagers have lost their usual livelihoods after resettling in new locations	Low
	Status of housing	Is there anyone still living in temporary shelter? What were the problems associated with housing? Have any delays taken place? Who was responsible for the delay?	Four houses were fully and twenty two houses were partially damaged. For the four fully damaged houses and 14 partially damaged houses, new houses were given in the *Nelumpokuna* resettlement There are 49 houses in the resettlement. 39 houses were constructed by the '*Lankarama Buddhist Fund*' and the balance ten by the '*Bentota Social Service Foundation*'. The villagers were not satisfied with the quality of these houses The following are the common problems of houses in the resettlement: 1. Doors and windows are not properly fixed 2. The roofs leak 3. Wall cracks, floor cracks 4. Roof sheets are loose due to improper fixing 5. Walls are not plastered 8 partially damaged households declined accepting the houses. They were not compensated with a house or by cash (Rs. 100,000/-)	Medium
	Status of health	What is the status of health in this locality? Is the attendance of the midwife regular? Are you happy with the immunization campaign? If not why?	There are no 'Health Centers' or dispensaries. Mobile clinics were held, sponsored by 'Red Cross', foreigners and *Bentara* hospital The attendance of midwife is regular Clinics for children and mothers are held. The mosquito menace has increased due to the accumulation of garbage and improper drainage system. Various skin diseases exist	High

(continued)

Table 4 (continued)

Areas of recovery	Recovery indicators	Opinion requested	Opinion expressed	Level of recovery
	Land issues	Did you have any problems associated with land titles? If yes, what are the problems? Are these problems associated with people's socio-economic status?	There are two units of triple combined houses, ten units of double combined houses and 28 single houses The combined houses have a land extent of about two perches on either side. The extent allocated for the single houses is 5 perches. Though the boundaries are demarcated legal possession is not allocated. Boundary stones are fixed, but not divided Land disputes are causing unrest in the village	Low
Basic social services	Status of schooling of the children	Were school children affected by the disaster? If yes, in what way? Have they recovered? How have they recovered? Any constraints for recovery?	Schools, libraries, technical schools, vocational training centers or pre schools were not affected due to tsunami. Minor floods have reached the *Gurukanda* building, but damagers are not reported. 'Red Cross', an American priest and 'CCF' has distributed books and other items to school children	High
	Psycho-social status	Were people faced with some psycho-social issues with the disaster? What is their situation now? If recovered, who helped them to recover? If not what services do they expect?	People were affected by psycho-social problems at the initial stages of the disaster. However, today people have recovered with assistance from social networks	High
Infrastructure	Status of Roads	Were any roads damaged? What is their position now? What constraints exist for reconstruction?	The resettlement is situated in a high elevated land and about 100 m from the main road About 30 m from the main road is tarred. The access road to the resettlement by vehicle is through an alternative long route. The roads within the resettlement are eroded. Due to the unavailability of drains, water flows through the compounds. Public transport is not operating closer to the resettlement One km of the Main Galle road and the railway was repaired by 'RDA' and 'CGR' Six interior by roads were damaged and the 'Pradeshiya Sabha' had renovated about 90% of the damaged roads	Medium

(continued)

Table 4 (continued)

Areas of recovery	Recovery indicators	Opinion requested	Opinion expressed	Level of recovery
	Status of Bridges	Were any bridges damaged? What is their position now? What constraints exist for reconstruction?	Bridges were not damaged Five damaged culverts in the village were reconstructed by the 'Pradeshiya Sabha' Four culverts in the settlement are not wide enough	High
	Status of Harbours	Were any harbours/jetties damaged? What is their position now? What constraints exist for reconstruction?	There are no harbours or anchorages in the GN	–
	Status of damaged hospital/health clinics	Were any hospital/health clinic damaged? Were they rehabilitated? If not why	There are no hospitals or Health Clinics	–
	Status of damaged natural habitat	Was any natural habitat damaged? Were they restored? If not why	The *Wetakeiya* belt was destroyed by 90% and the 'Department of Coastal Protection' successfully replanted the coast with *Wetakeiya*, Mudilla and Kottan But the plants were washed away by the rough sea waves recently	Low
	Status of coastal protection	Has any coastal protection (bio-fencing, seawalls, quarry walls, breakwaters) been constructed/repaired?	The dike was damaged and now under construction	High
Livelihoods	Status of agriculture	What type of damage observed in agriculture? What is the situation now?	Agriculture is not a significant livelihood in the GN	-
	Status of receiving loans for livelihood improvement	Did tsunami-affected people receive loans to improve their livelihoods? If not, why? If yes, how did they obtain loans? For what type of livelihood activities? What are the sources of loans?	People were not aware of the '*Susahana*' scheme Villages have not received loans People were not aware of '*Susahana*" loan scheme	Low
	Status of social protection programmes	Was there any social-protection programme? If yes, describe them	'Fisheries and Farmers Pension Schemes' are in operation No Elders Society and other social protection programmes	Low

(continued)

Table 4 (continued)

Areas of recovery	Recovery indicators	Opinion requested	Opinion expressed	Level of recovery
	Status of receiving fishing boats and fishing gear	Has everybody received fishing boats and fishing gear? If not what are the problems?	4 boats and 5 canoes were damaged and replaced in excess The donation was done by a foreign company, 'Fisheries Office' and 'World Vision' Damaged nets and fishing implements were donated in excess by 'World Vision', 'Red Cross' and 'Salvation Army' 4 fish vendors lost bicycles and carriers donated by foreigners	High
	Status of SME/ self-employment sector	Were any SMEs/self employment activities disrupted? What is their position now? Any constraints for recovery? What are they?	3 groceries were damaged and have recommenced business on their own The people who were engaged in business at weekly fairs were affected and not compensated 4 carpenters lost their tools. They were not compensated Mat weavers, Pan cultivators and sewing were affected and not compensated They have problem of recommencing	Low

11 Special Note on the Four Areas of Recovery

Vital Needs: The scarcity of water in the resettlement areas is a major concern.
Basic Social Services: Educational facilities for the children and psycho-social status are satisfactory.
Infrastructure: Access roads to the resettlements should be constructed.
Livelihood: SME/self-employment activities had not recovered because necessary assistance had not been provided. Livelihood activities for men and women should have been developed.

References

Burkey, S. 1993. *People First: A Guide to Self-Reliant Participatory Rural Development*. London: Zed Books.

Cerena, M. (ed.). 1991. *Putting People First: Sociological Variables in Rural Development*, 2nd edn. (1st edn. 1985). Oxford University Press for the World Bank.

Chambers, Robert. 1989. Vulnerability Coping and Policy. *IDS Bulletin* 20.

Chambers, Robert. 1995. Paradigm Shifts and the Practice of Participatory Research and Development. In *Power and Participatory Development: Theory and Practice*, ed. Nici Nelson and Susan Wright. Intermediate Technology Publications.

CUCEC. 2005a. *How do Sri Lankan Institutions Respond to Relief and Rehabilitation After the Tsunami?* Colombo University Community Extension Center.

CUCEC. 2005b. *The Report on People's Consultations on Post-Tsunami Relief, Reconstruction and Rehabilitation in Sri Lanka*. United Nations Development Programme.

CUCEC. 2006. *Identifying Vulnerable Communities in Tsunami-Affected Areas in Sri Lanka & Proposing an Appropriate Method for Mobilizing Vulnerable Communities*. Colombo University Community Extension Centre.

Rasanayagam, Y., and P.D. Amarasinghe. 1999. *Social Mobilizing and Poverty Alleviation in the Moneragala District*. MONDEP Formative Process Research Project, Working Paper No. 2.

Senanayake, S.M.P., and Lakshman Dissanayake. 1999. *Development of Non-Farm Employment Activities in the Monaragala District*. RDSC Working Paper no. 1, Regional Development Studies Center, University of Colombo.

TRIAMS Workshop Report. 2008. Tsunami Recovery Impact Assessment and Monitoring System: Second Regional TRIAMS Workshop, Bangkok, 21–23 March 2007. World Health Organization; United Nations Development Programme; International Federation of Red Cross and Red Crescent Societies.

Tsunami Aid and Its Effectiveness

P. G. Dhar Chakrabarti

1 Introduction

The Indian Ocean tsunami of 26 December 2004 was one of the deadliest natural disasters in recorded history. It was triggered by an earthquake of the magnitude of 9.1 on the Richter scale, the second largest ever recorded on a seismograph. It had the longest duration of faulting—of 10 min—which caused the planet to vibrate by as much as 1 centimetre, and released an energy equivalent to 26.3 megatons of TNT, about 1500 times more powerful than the atom bomb dropped over Hiroshima during the Second World War. The earthquake resulted in the sudden vertical rise of the seabed by several meters which displaced massive volumes of water. The result was a wave up to 30 m in height that struck the coasts along the Indian Ocean in as many as 15 countries of Asia and Africa: 4 in South-East Asia (Indonesia, Thailand, Malaysia, and Myanmar), 4 in South Asia (Sri Lanka, India, Maldives, and Bangladesh), 1 in West Asia (Yemen) and 6 in Africa (Somalia, Tanzania, South Africa, Seychelles, Kenya, and Madagascar). In all, 230,273 people were confirmed dead, 45,752 are still missing, 125,000 were injured and 1.74 million were displaced. The largest number of deaths (167,799) took place in Indonesia followed by 35,322 in Sri Lanka, 18,045 in India and 8,212 in Thailand. The casualties included deaths of over 9,000 tourists from 31 countries. Never before had so many countries over such a wide area been affected by a natural disaster.

P. G. Dhar Chakrabarti (✉)
National Institute of Disaster Management, New Delhi, India
e-mail: dharc@nic.in

S. Reddy (ed.), *The Asian Tsunami and Post-Disaster Aid*,
https://doi.org/10.1007/978-981-13-0182-7_16

2 Unprecedented Humanitarian Assistance

Humanitarian assistance of over USD 14 billion was mobilized from the world over for the relief and recovery of the communities and countries affected by the tsunami. This is the biggest international response to a natural disaster on record (WHO 2013). 39% of this assistance, amounting to USD 5.4 billion, was contributed through voluntary private donations. No other natural or manmade disaster or civil conflicts had triggered such massive international private donations for a humanitarian cause (WHO 2013) (Fig. 1).

USD 3.2 billion of private donations was channelized through NGOs and the rest through the UN system. Private contributions to the International Federation of Red Cross and Red Crescent Societies alone amounted to USD 1.8 billion, six times more than what national governments contributed to this global humanitarian organization.

A survey found that 30% of private donations came from people who had never participated in such charities. Surveys conducted by the Tsunami Evaluation Coalition found that nearly one-third of the population in France, Spain, and the USA had donated in cash or in kind for tsunami response (Telford and Cosgrave 2006). The timing of the disaster, at a time when many people were enjoying seasonal Christmas holidays, was seen as an important factor underpinning the

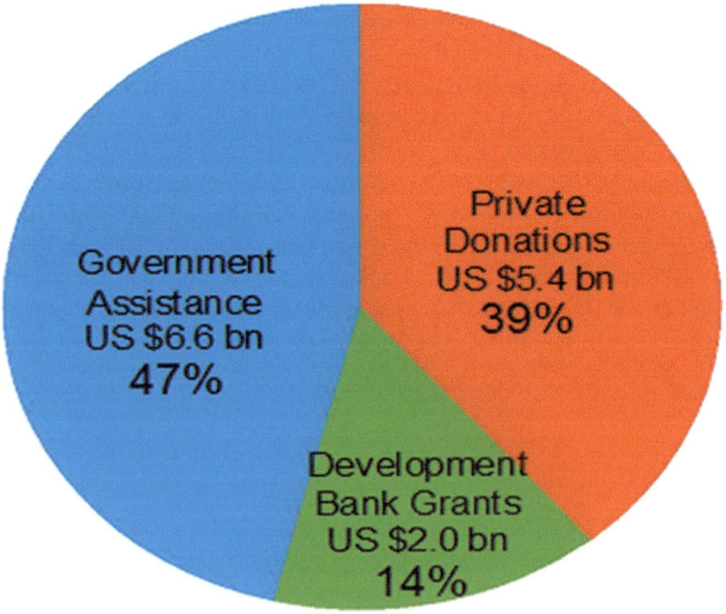

Fig. 1 International humanitarian response to Indian Ocean Tsunami. *Source* Development Initiatives (2006)

public response. Other factors were: the huge number of lives lost, including lives of many foreign tourists, for an act of nature which was nobody's fault; extensive media coverage especially amateur videos of the disaster shot by Western tourists; and a strong sense of identification with the affected people. Public awareness of and involvement in global development and humanitarian issues also contributed to this response

Box 1: Contributions from Private Donors in the United Kingdom

The Disasters Emergency Committee (DEC) of the United Kingdom is one of the largest platforms for coordinating humanitarian assistance in disasters and other conflict situations in the world.

The DEC issued an appeal for donations for the Indian Ocean tsunami on 29 December 2004. In just two months it was able to raise £392 m (USD 627 m), which was eight times the amount of donations received for the Sudan crisis appeal which had been running for eight months.

The DEC spent the funds for short-term relief as well as long-term recovery. More than 750,000 households in Sri Lanka, Indonesia, and India were directly assisted with humanitarian relief. Besides, large reconstruction programmes were carried out in these countries, supplementing the initiatives of the local government and bi-lateral and multi-lateral funding agencies. As per the report of the DEC, its member agencies, which include British Red Cross, Oxfam, Save the Children, and Pan, constructed 13,700 houses, 55 schools and 68 health centres with funds raised by the UK public. Brendan Gormley, the former DEC Chief Executive, stated in an interview with the BBC in 2009, that 80% of all households in the United Kingdom had supported the tsunami appeal, which was unprecedented.

At USD 6.6 billion, assistance from national governments for the tsunami victims was less than the USD 9.4 billion committed to Iraq in 2004 or the USD 9 billion pledged to Hurricane Mitch in 1998, but what was unprecedented is the participation of a large number of countries that came forward to offer their assistance for the relief and recovery of the affected countries and communities. 99 national governments and two intergovernmental organizations joined these efforts (Flint and Goyder 2006: 21). This included

- 77 national governments that were not traditional donors. In addition to cash, some of these countries extended their assistance in kind, and deployed military assets for humanitarian response and relief.
- 13 national governments became first-time donors and some of them continued with their donations in subsequent situations of humanitarian crisis.
- 5 donors (Australia, Germany, Japan, the USA, and the European Union) were responsible for providing over 50% of the funds

Normally, during flash appeals and the annual UN pledging conference for the Central Emergency Response Fund, countries pledge a certain amount of donations, but their actual contributions fall short of their commitments. In the case of the Indian Ocean Tsunami, due the magnitude of devastations and humanitarian crisis, many donors contributed much more than their pledges. Denmark, for example, contributed substantially higher than its normal aid for disaster relief. Australia contributed AU$1 billion for Indonesia in addition to existing assistance. Both the World Bank and the Asian Development Bank broke their ceilings on grant assistance, and contributed nearly 15% of all humanitarian assistance for the tsunami recovery.

The United Nations and its agencies were able to mobilize USD 6.1 billion from national governments and private donors, covering 89% of its targeted appeals for conducting post-tsunami response, relief, recovery, and reconstruction in the affected countries. This was the largest donation ever collected by the UN for their operations in a single natural disaster (Fig. 2).

It is interesting to observe that contributions by private individuals and organizations far exceeded the combined contributions of all countries to the UN and its agencies. The countries offered their assistance bilaterally, in cash and in kinds, to the affected countries, just as private individuals gave their donations to the Red Cross and other NGOs (Table 1).

In addition to the international response, the domestic response for humanitarian assistance in the affected countries was also sizeable. The Tsunami Evaluation Coalition estimated that national governments of the affected countries invested USD 3.4 billion and another USD 2.1 billion was raised by the corporate sector and private individuals in the affected countries (Flint and Goyder 2006: 21). This does not include the contributions of the affected population and remittances received from their non-resident friends and relatives. Further, this did not include the loans raised from multilateral financial institutions for reconstruction of damaged houses and infrastructure and for setting up tsunami early warning systems.

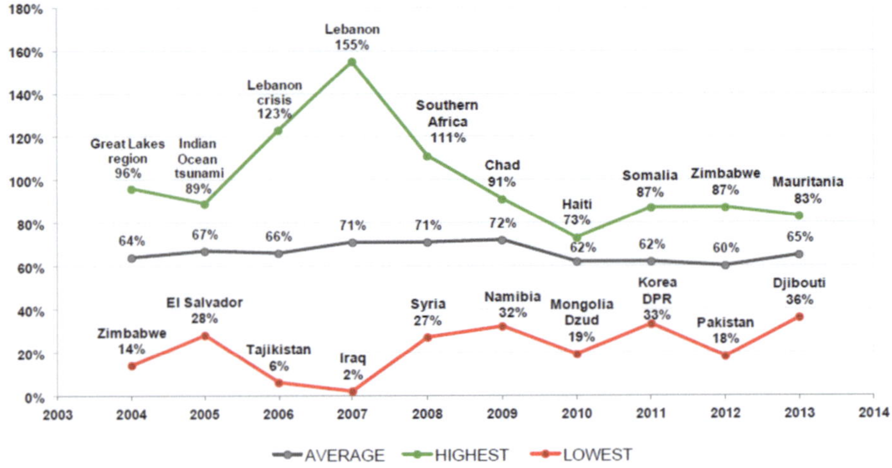

Fig. 2 Highest and lowest responses to UN coordinated appeals (2004–2013). *Source* Development Initiatives (2006)

Table 1 Contributions to the United Nations for humanitarian assistance to the victims of the Indian Ocean Tsunami (in USD)

Donors	Commitments/ Contributions	Ratio of commitments to total (%)	Uncommitted pledges
Japan	502,579,970	8.2	0
Britain	136,128,698	2.2	13,364,408
United States	134,107,148	2.2	217,776,074
Germany	129,651,389	2.1	0
Canada	116,487,566	1.9	0
European Commission	98,842,102	1.6	68,992,664
Norway	81,468,409	1.3	1,832,061
China	62,248,729	1.0	1,600,000
France	60,359,661	1.0	27,890,183
Italy	57,864,154	0.9	1,206,273
Denmark	44,119,185	0.7	0
Netherlands	41,666,548	0.7	0
UAE	41,430,427	0.7	270,000
Sweden	41,370,476	0.7	6,466,126
Australia	36,398,508	0.6	8,046,409
Finland	32,726,978	0.5	482,509
Greece	31,009,201	0.5	0
Allocation of unearmarked funds by UN	30,464,876	0.5	0
New Zealand	27,079,695	0.4	1,183,118
Switzerland	26,361,383	0.4	0
Qatar	25,000,000	0.4	0
India	23,000,000	0.4	0
Ireland	22,357,836	0.4	8,767,866
Spain	21,971,308	0.4	0
Others	196,613,478	3.2	207,728,714
Private (individuals and organizations)	4,127,818,551	67.1	53,250,000
Total	6,149,126,276	100	618,856,405

Source Data taken from the Financial Tracking Service of the UN OCHA. https://fts.unocha.org/

3 Spatial and Sectoral Distribution of Aid

It is estimated that almost twice the totally assessed damage and loss (USD 9.3 billion) in all the affected countries was raised from international and domestic sources for the relief, recovery, and reconstruction of the affected areas and communities (Flint and Goyder 2006: 21). Every study conducted so far has been unanimous that the total aids received from all sources were enough to meet the needs of the people affected by the Indian Ocean tsunami. But it cannot be stated

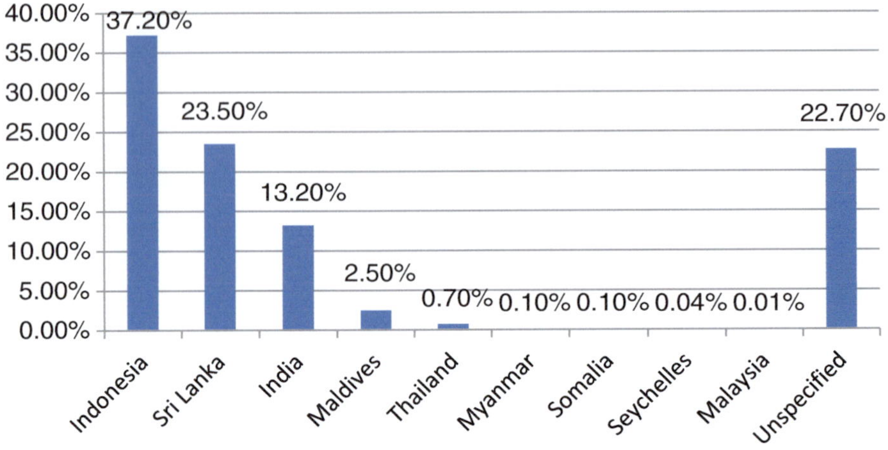

Fig. 3 Distribution of Tsunami aid to affected countries. *Source* Flint and Goyder (2006: 21)

with the same degree of confidence that distribution of aid among countries and communities were uniformly adequate. While some countries received disproportionately higher assistance than required, a few countries found that the resources were not adequate to meet the needs of affected areas and population. Within countries, there were disparities in the allocation of resources among the affected communities and sectors. A significantly high share of aid (22.7%) spent on unspecified countries (see Fig. 3) suggests that these were spent to meet cross country initiatives as well as common expenses on logistics, administration, travel, and so on.

Most of the tsunami aid had gone to the four worst affected countries: Indonesia, Sri Lanka, India, and the Maldives, but these did not match the extent of losses suffered by them. Indonesia suffered the maximum both in terms of loss of human lives and impact on economy: it accounted for 55% of the human impact (number of deaths, missing persons, and persons affected) and 48% of the total economic impact. But Indonesia received 37% of all commitments and 50% of the commitments that were allocable by country. On the contrary, Sri Lanka and Maldives received a much larger share of aid than the losses suffered by them. Comparing the impact percentages to total commitments suggests that Indonesia has been under-aided and Sri Lanka over-aided. The Government of India, like that of Thailand, did not issue any appeal for international assistance, and therefore the funds shown against India were channelized through UN agencies and NGOs.

In the initial months, most of the tsunami aid was pledged for emergency relief. As the relief phase graduated slowly, to recovery and long-term reconstruction, the sectoral preferences changed. Almost all the pledges made in January 2005 were for emergency assistance, while a quarter of the pledges/releases during 2006–2010 were for long-term reconstruction. The proportion of funds allocated to relief and reconstruction assistance varied markedly between donors. The US, Spain, Germany, and Sweden gave a clear preference to reconstruction (more than 60% of total tsunami aid from these countries were directed for long-term reconstruction. On the other end of the spectrum, more than 80% of humanitarian assistance from the Netherlands, Japan, Canada, and Ireland was meant for relief and early recovery. Food and non-food items received the largest allocation (21%), followed by health, water and sanitation, early recovery, coordination, and multisector aid (10–13% each). Shelter and measures for disaster preparedness such as early warning, training and capacity building, received low allocations (6 and 4% respectively), presumably because these were addressed through various other tracks of multi-lateral financial assistance that are not monitored by the financial tracking services of UNOCHA. Sectoral allocation of assistance from the Red Cross/Red Crescent Movement and international/national NGOs were difficult to be categorized as they did not follow a uniform system of sector classifications and some of them did not make their detailed databases available for global tracking system (Fig. 4).

The Tsunami Evaluation Coalition analyzed sectoral allocations by seven major donors: USA, Germany, Denmark, Australia, United Kingdom, Ireland, and the European Commission (see Fig. 5). The study shows an overwhelming concentration of aid in traditional humanitarian sectors such as food and non-food, and health. Shelter, multi-sectoral reconstruction, and disaster preparedness did not receive the kind of attention that was required for building resilience.

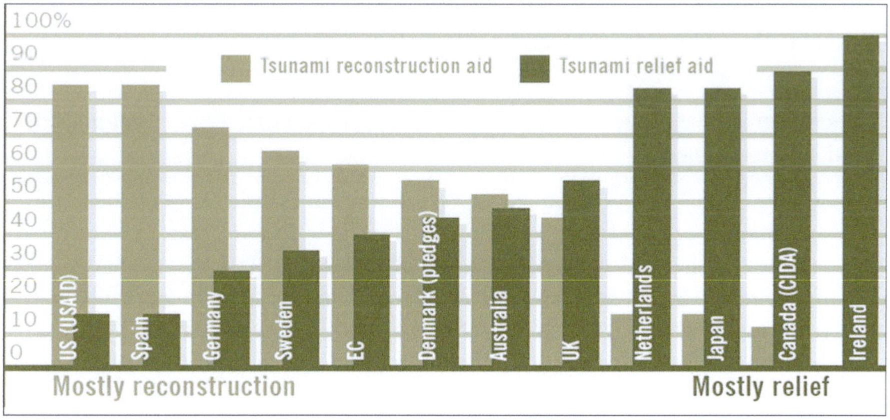

Fig. 4 Tsunami aid for relief and reconstruction. *Source* Telford and Cosgrave (2006)

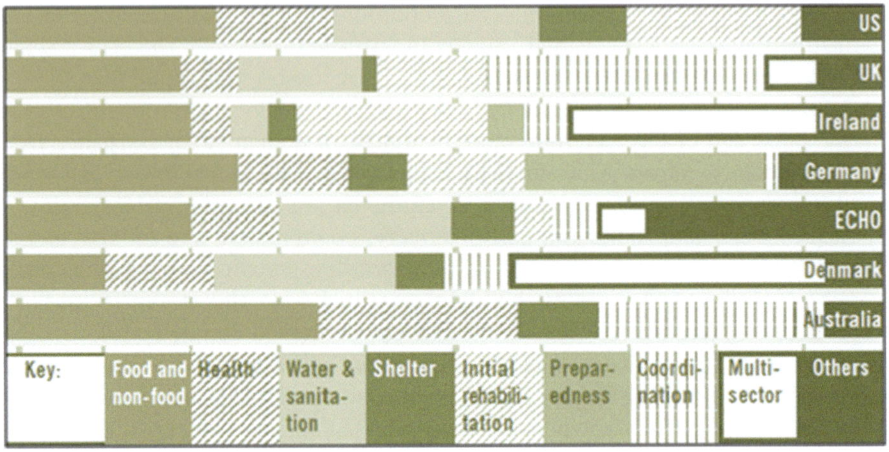

Fig. 5 Sectoral allocation of Tsunami aid by major donors. *Source* Telford and Cosgrave (2006)

Box 2: Largest Amount of Aid per Affected Person

Volumes of aid per person affected in Indian Ocean tsunami were of a completely different order of magnitude than from those of previous disasters.

- If all the tsunami commitments were shared out equally between the 2 million people directly affected, each person would receive roughly USD 5,500.
- If all the tsunami commitments were shared out between the 3.7 million people in areas with major impacts, each person would receive about USD 3,000.
- Total disbursements from DAC donors alone amounted to USD 1,000 per directly affected person and USD 550 per person in areas with major impacts in 2005.

People affected by other disasters have been allocated much lower volumes of aid. For example:

- In the South Asia earthquake (2005), 1 million people were estimated to be directly affected and 4 million to live in affected areas. Total commitments per head were between USD 250 and USD 1,000.
- In Somalia (2005) between 700,000 and 1.1 million people were affected, with an allocation of between USD 114 and USD 178 per head.
- In Eritrea (2005) 2.2 million affected people had an allocation of US$50 each.

The figures for Somalia and Eritrea show that people in less prominent emergencies were allocated much less aid per capita than people affected by more 'popular' emergencies.

Source: Flint and Goyder (2006: 21)

4 Effectiveness of Tsunami Aid

The Tsunami Evaluation Coalition (TEC)—a unique initiative of the UN and its agencies, the donors, NGOs and the IFRC to carry out joint evaluations of the international response to Tsunami—raised six questions to evaluate the effectiveness of the response (Telford and Cosgrave 2006). These are:

1. How informed was the response?
2. Was the response sufficient?
3. Was the response timely?
4. How flexible was the response?
5. How well coordinated was the response?
6. How effective was the response?

The answers TEC found to these questions reflect the strength and weaknesses of the existing architecture of international humanitarian response and, at the same time, highlight the scope and opportunities of improvement, based on the experiences gained in this and subsequent events of humanitarian emergencies arising out of natural disasters and civil conflicts.

How informed was tsunami response? TEC concluded that international response to the Indian Ocean tsunami was "needs-motivated rather than needs-informed" (Flint and Goyder 2006: 21). Strong public sympathy for the victims and competition between donors influenced pledges and decisions regarding allocation of assistance to countries and sectors more than the assessed needs. The process of pledging was largely political, and donors used "contextual opportunism" to define their priorities (Flint and Goyder 2006: 21). These were not based on any realistic assessment of needs for assistance, the volume of funds that each donor could provide, or on the logistic or operational capacity of the implementation agencies. Bilateral or regional interest, proximity, and previous presence, were defining factors in the allocation of resources Once reconstruction started, the donors' need for "visible signature infrastructure projects" significantly influenced priorities and financial flows (Flint and Goyder 2006: 21). Such ad hoc response to disasters should be replaced by objective assessments of needs of the affected persons. Assessment of emergency assistance can be done in consultation with the national governments, NGOs, and UN agencies. Detailed assessments for recovery and reconstruction can be made through multi-sectoral and multi-stakeholder Post Disaster Ned Assessment (PDNA) (World Bank 2013).

Regarding the sufficiency of tsunami response, it is pointed out that the international pledges and donations exceeded the total economic impact of the disaster in most of the countries. If national and local resources are included, then in some of the affected countries total funds available for immediate relief and long-term recovery probably exceeded the assessed needs. Problems of plenty in some places resulted in adverse social consequences like lower participation in work, higher consumption of alcohol, increased violence against women, and so on. Overall resources were sufficient to meet relief and recovery needs, even though for some areas and sectors resources were not adequate. This indicates that there is need for overall planning and coordination for the optimal allocation and utilization of available resources for better response and recovery. There were many instances when the pressure to spend a huge amount in a short time exceeded the capacity of the system to make judicious use of resources. The result has been a recovery effort that was neither as effective nor as efficient as it should have been. Aid adequacy or over-adequacy is a welcome departure from normal situations when response to flash appeal is always much less than adequate. The Indian Ocean tsunami was an unprecedented disaster which received unprecedented response from the community of donors. This provided an opportunity to utilize the donations for 'building back better' for reducing the risks of disasters. Unfortunately, there was no coordination mechanism to oversee the planning and implementation of the post-disaster response, relief, recovery, and reconstruction. If the funds were in abundance for relief these could be used for recovery, reconstruction and resilience, but this did not take place.

The timeliness of the tsunami response has been well documented (Telford and Cosgrave 2006). In the UK the world record for online donations was broken with over £10 million donated through the Disasters Emergency Committee (DEC) website in 24 h.[1] Funding timeliness was also a record for the IFRC appeal. Although there was some time lag between pledges and actual releases, particularly during the recovery and reconstruction phase, overall, international financial response was timely and that delays in funding have not been a constraint on implementation.

How flexible was the response? More than half of the funds mobilized through the UN Appeal were not earmarked for any specific country, sector, or agency. Donations from the general public to NGOs were also un-earmarked. This provided considerable flexibility in the use of funds than is normally available in most of cases of international humanitarian assistance. However, the financial pledges of most of the government donors were earmarked for specific countries and/or sectors/projects. This limited the scope and ability of national coordinators to re-direct funds to underfunded activities.

How well coordinated was the response? Too many actors and agencies were involved in humanitarian response in most of the affected countries which made it difficult for effective coordination at all levels. Globally, the UN Office for Coordination of Humanitarian Response (OCHA), which has the mandate for such coordination, found it difficult to plan and coordinate as it had no clue about what each agency was doing at the national and local levels. Government donors, on the other hand, found the UN system to be slow, which affected its legitimacy in defining priorities and resource allocation. National governments in the affected countries varied in their capacity to plan, coordinate, implement, and monitor the programmes. At the local level too many agencies working in same areas and for same purpose resulted in avoidable duplication and wastage of resources,

[1]http://www.dec.org.uk/.

while difficult and inaccessible areas were found to be lagging in required investments and initiatives for recovery and reconstruction. This definitely calls for a review of existing coordination arrangements in national and local governments, the UN, and the NGOs. Proactive national and local leadership facilitated such coordination; for example, in India NGOs and UN agencies were assigned specific tasks to supplement and not duplicate the work of government agencies.

Was the response impartial? One of the principles of Good Humanitarian Donorship (GHD) is impartiality, which means that relief, recovery, and reconstruction works should be undertaken solely on the basis of need, without discrimination between or within affected populations. Adherence to this principle can be assessed at three levels: between tsunami-affected countries, within tsunami-affected populations, and between different emergencies. As noted earlier, most of tsunami-aid was directed to the four worst-affected countries: Indonesia, Sri Lanka, India, and the Maldives. Comparing the aid flow with the impact of disaster might suggest that Indonesia was 'under-aided', and that Sri Lanka and Maldives were 'over-aided'; however, it also needs to be pointed out that the total economic loss as a percentage of GDP was almost four times larger in Sri Lanka than in Indonesia. On balance, it can be concluded that, broadly speaking, funding has been allocated to countries in proportion to their needs, even though there were some areas within countries, especially the remote inaccessible areas and communities, which were under-funded.

Overall, with so many funds available, most people received some assistance, and many received more than they ever expected. However, this does not necessarily mean that assistance has been totally impartial. Countries like Indonesia, Sri Lanka, and Maldives, which issued appeals for international assistance, received more funding than India and Thailand that did not ask for any. A lot also depended on how organized and articulated the affected population was in demanding and utilizing assistance for their recovery and reconstruction. The fishing communities, for example, tended to be more organized and therefore more successful at accessing aid in most of the countries. The needs of vulnerable groups such as women, older people, and children tended to be overlooked and had not been met at the pace or scale expected. Civil conflicts and government policies towards internally displaced persons influenced differential flow of aid to Sri Lanka and Aceh.

The GHD principle stipulates that donors should "strive to ensure that funding of humanitarian action in new crises does not adversely affect the meeting of needs in ongoing crises" (IFRC 2003: 21). TEC did not find any evidence that funding of other emergencies had declined as a result of the tsunami response. Excluding the Tsunami Flash Appeal, UN Consolidated Appeals for 2005 were 61% funded, which is within the average range of such funding. Almost every government donor asserted that the tsunami allocations were in addition to the normal allocations for humanitarian response and hence this would not affect their response to ongoing or future crises, arising out of either natural disasters or conflict situations.

Even though there is no evidence to suggest that the tsunami response had reduced the level of financial aid for other emergencies, it is also very clear that the tsunami response received substantially higher funding than any other crisis situation and this led to a very inequitable allocation of humanitarian assistance. Considering the magnitude, spread, and extent of damage and losses in tsunami this may not be treated at par with other localized disasters.

How accountable was the response? The online Financial Tracking System (FTS) of OCHA provided a good device to monitor the pledges, commitments, and disbursements by the donors, but the system did not function optimally due to a variety of reasons. The most important reasons were the voluntary nature of reporting and the absence of common terminologies for humanitarian assistance. While government agencies were more prompt and consistent in reporting disbursements, NGOs were not so. For example, only 15 of the UK NGOs (5%) involved in tsunami operations participated in the FTS. As many as 29 member countries of the Development Assistance Committee (DAC) of the OECD adopted common terminologies for humanitarian assistance, but this was not necessarily followed by the non-DAC donors and the NGOs. Accountability to beneficiaries remained another weak area for both humanitarian agencies and recipient governments.

Finally, how effective was the response? Could the relief and reconstruction efforts be more effective and efficient with more or less money, or with a different allocation of funds? These are some of the key questions that, unfortunately, the funding studies did not address in any great detail. The broad picture seems to be that the relief effort was effective, but reconstruction was less so. The size and speed of the local and international financial response made a major contribution to the effectiveness of the relief effort. Sometimes, ready availability of funds might also have reduced effectiveness as the rush to do good with so much money led to competitive behaviour among NGOs, which was not conducive to the effective deployment of funds. Uncritical and uninformed allocation of funds between agencies, without any consideration of the track record or competence of the agencies, did not contribute to the effectiveness or efficiency of humanitarian assistance. A related, and more fundamental, question was whether the agencies selected had the capacity to implement the type and scale of reconstruction activities that were undertaken. Such reconstruction works, to be effective, should always be undertaken in collaboration with the governments at the appropriate level, as governments alone can resolve critical issues such as selection of beneficiaries, selection of sites, acquisition of land, and provision of infrastructure, like connectivity, and utilities, like water and electricity. Many housing projects taken up by the NGOs did not end up being successful as these were taken up in isolation without the support of and coordination with local administration. Coordination was best in countries with a strong government, such as in Thailand, India, and the Maldives. Therefore, developing capacities of national and local governments to coordinate activities of multiple stakeholders is the key to increase the effectiveness and efficiency of international humanitarian response to disasters.

Many significant initiatives were taken in this regard. The Indonesian Province of Aceh, which was the epicentre of the disaster, saw the creation of a dedicated body, the Aceh-Nias Rehabilitation and Reconstruction Agency (BRR) with a 4-year mandate to coordinate all recovery activities as well as implement a number of government projects. In Sri Lanka, the tsunami was a catalyst for the creation of the Ministry of National Disaster Management and Human Rights in 2006. The Government of Maldives moved swiftly to set up National Disaster Management Centre (NDMC) to coordinate all the activities. In India, tsunami triggered the drafting of the federal legislation on disaster management and its enactment on the first anniversary of the disaster.

The Government of Indonesia established the Global Consortium for Tsunami Recovery and set up the Multi Donor Fund (MDF). The UN created the Office of the Recovery Coordinator for Aceh and Nias (UNORC) to facilitate coordination in a complex recovery context involving countless international and national stakeholders. As new structures, void of institutional baggage, these agencies benefited from the ability to be flexible and quickly adapt to local circumstances. The Government of India did not create any new coordinating mechanism, but devolved significant authority to local administrators, which became crucial in the recovery effort in Tamil Nadu, Puducherry and Andaman and Nicobar islands.

International and national responses to the Indian Ocean tsunami are replete with many success stories of rebuilding communities devastated by deaths and destructions. The most remarkable is the recovery story of the Banda-Aceh province of Indonesia that faced the maximum burnt of the tsunami. The huge influx of humanitarian assistance into the province made it possible to rebuild infrastructure and to revitalize institutions to much better conditions than these were ever before the disaster, thus providing better opportunities for the Acehnese to further develop and progress in all areas. These improvements were both facilitated and sustained by improved security conditions in the province after the signing of the Helsinki peace accord that ended a 30-year old conflict that had cost almost 15,000 lives (McKeon and Masyrafah 2008).

Conversely, tsunami aid failed to resolve the ethnic conflict in Sri Lanka. In fact the conflict might well have been precipitated as there were wide spread allegations of discrimination in aid dispersals in the Tamil-majority Northern Province that was worst affected by the tsunami (McGilvray and Gamburd 2013).

On the whole, tsunami aid was effective in providing immediate relief to the sufferers and in rebuilding lives, livelihoods, and infrastructure in all the 15 countries that were affected by the disaster, although the degree of aid effectiveness varied among and within countries and sectors. No comprehensive matrix of indicators has been developed to measure the aid effectiveness in countries and sectors, and therefore it is difficult to quantify the effectiveness and make *inter se* comparisons. Among the range of factors that were responsible for aid effectiveness were: the quantum of aid released in specific countries and sectors; the extent of local involvement and participation in recovery initiatives; the proactiveness of national and local leaderships; mechanisms for coordination with stakeholders and implementation agencies; and arrangements for monitoring and evaluation of programme implementation.

In the maze of studies that have been undertaken on the Indian Ocean tsunami response and recovery, the long-term impact of recovery did not engage the kind of attention that it should have. Various sociological and anthropological studies[2] undertaken in the recent years have beginning to unravel that even in countries and communities where the overflowing of aids helped the process of recovery it did have many unintended consequences such as exacerbation of existing vulnerabilities based on gender discriminations, income inequalities, and lack of social protections. Absence of proper exit strategies of most of the aid initiatives left many communities to fend for themselves and they found it difficult to sustain the standards created through the relief and recovery efforts. A decade after the tsunami many communities were facing problems of maintenance of assets created out of generous aids. Recovery and reconstruction initiatives focused on short term gains and often overlooked long-term issues like protection of environments and sustainable social and economic development. The transition from recovery and reconstruction to long-term development remained one of weakest links of the massive aid initiatives.

5 Linking Relief, Rehabilitation, and Development

Linking relief, rehabilitation and development (LRRD) means looking beyond immediate relief and recovery into long-term issues of development: reconstruction of damaged houses and infrastructure, rebuilding of livelihood and productive systems, and regenerating environmental assets. These larger issues of development are not generally addressed during humanitarian response as available funds are hardly adequate even to meet the immediate needs of relief. Unprecedented resources were mobilized after the Indian Ocean tsunami, the bulk of which was not earmarked for any specific sector or activity, providing excellent opportunities as well as flexibility to link rehabilitation with some of these long-term issues of development. This was utilized fairly well during the transition from relief to early recovery, but not so well for the next stage of recovery to reconstruction.

The humanitarian system initiated early support for livelihood rehabilitation in the form of distribution of assets, such as small boats and fishing nets, and cash-for-work, which provided the affected population with the security they needed to start rebuilding their homes and livelihoods. But the further process was not always smooth and fast.

The aid community failed to develop appropriate strategies to link rehabilitation with wider issues of development. There were risks that some rehabilitation efforts could ultimately prove ineffective and unsustainable. Some of these poorly conceived interventions actually undermined future development as they encouraged over-fishing, damaged the credibility of microfinance institutions, and created dysfunctional communities through insufficiently planned resettlement.

---

[2]A few studies worth mentioning include Hastrup (2011), Lindberg (2014), Shaw (2006), Reddy (2013), Saini (2013), Barenstein and Leemann (2016).

Most aid actors demonstrated limited understanding of what kinds of interventions may eventually prove sustainable with respect to livelihoods, community development, and resource management. Standard packages, such as small boats, did not necessarily contribute to rebuilding a fishing industry, nor did they encourage the wider private sector development that is needed to support the livelihoods of people living on the coasts (Christoplos 2006a, b).

Economic activities stimulated by the tsunami recovery created hundreds of thousands of new employment opportunities and livelihood options, but these were neither anticipated nor utilized optimally through backward and forward linkages with issues like skill development, credit facilities, or marketing arrangements.

The process of rehabilitation remained very slow in the housing sector, where an unacceptably large proportion of the affected population continued to live in deteriorating tents over two to four years after the disaster. Early promises—that tens of thousands of houses would be built in a few months—could not be fulfilled leading to lot of discontent among the affected population. Such promises demonstrated arrogance and ignorance regarding the process of development. Issues such as land rights, environmental impacts, and links to infrastructure and services inevitably take time to be effectively addressed, but this was not acknowledged in housing reconstruction plans and projects.

LRRD is best served when it supplements national and local initiatives for relief, rehabilitation, and development. This requires mapping of these initiatives, analysis of gaps in the initiatives, and designing of recovery programmes to address these gaps in a systematic manner. This required consultation and coordination with national and local agencies in planning and implementation of the programmes, which were largely missing in most of the countries, resulting in a fragmented approach. The aid agencies largely worked in silos, with guidance and directions only from their own headquarters, rather than supporting and being supported by the national and local authorities. This did not augur well for linking relief and rehabilitation with development.

Genuine LRRD requires that programmes for rehabilitation are aligned with the policies, procedures, capacities, and actions of national and local level institutions and actors, such as government, municipal authorities, local NGOs, masons, and affected people themselves. Projects were prepared and programmes drawn up without consultation with these key actors and without appreciation of government regulations and procedures that were time consuming. Houses and other facilities were designed without consideration of either local lifestyle and cultural ethos, or locally available building materials, or even locally available capacities for construction. Some of these critical and basic issues were appreciated much later when the damage had already been done and it was difficult to realign the pre-conceived projects with the realities on the ground. Studies pointed out that most agencies demonstrated ignorance of the historical trends and knowledge of how aid programming could avoid repeating past mistakes and contribute to prevailing development opportunities (Barrientos et al. 2005).

Most of the reconstruction programmes had to be carried out in coastal zones in most of the affected countries where stringent environmental regulations did not permit new constructions to be taken up. Coastal communities have been traditionally living in these zones as their livelihood was connected with fishing and other occupations related to the coasts. The tsunami provided opportunities for rebuilding settlements away from environmentally sensitive and disaster risk zones. Locating safe zones for resettlement that were acceptable to coastal communities became an extremely difficult and cumbersome process that did not always end up with success. Complex trade-offs involving government and communities had to be reached, which not only delayed the projects but also compromised with the principles of 'building back better' (Christoplos 2006a, b).

The tsunami aggravated both the chronic and transient poverty of the people by damaging their immediate productive tools and severing opportunities for sustainable employment and livelihood. Effective LRRD manifests itself in a judicious balance of efforts to tackle both these dimensions of poverty. While substantial progress was achieved in alleviating many shades of transient poverty, nothing much could be done to address the issues of existing chronic poverty that was compounded by the tsunami, or new forms of poverty that were created by the destruction of natural assets or environment. Chronic poverty was unlikely to be removed by small asset replacement initiatives; these are best addressed by economic development and/or social protection. The deluge of aid created opportunities for integrating these developmental issues into the tsunami response, but these opportunities were hardly utilized in any of the affected countries.

Another issue of LRRD was linking reconstruction with disaster risk reduction, which is highlighted as a priority of action in both the Hyogo Framework of Action and the Sendai Framework for Disaster Risk Reduction. Apart from the issues of compliance with risk-sensitive land use regulations, which was compromised at several places, there were larger issues of mainstreaming disaster risk reduction in housing, livelihood construction, and social protection programmes. It was observed that these issues were largely not addressed in recovery programming (Buchanan-Smith and Maxwell 1994). Generous international funding for early warning and disaster preparedness post-tsunami provided opportunities of developing standards and good practices, of mainstreaming DRR with recovery and reconstruction, which was not utilised optimally. It was pointed out that, instead of reducing risks of disasters, some of the reconstruction programmes might have created new risks or exacerbated existing risks of disasters (Kennedy et al. 2008). There is a need for more detailed evidence-based assessment of the impacts of aid programmes on environments and natural resources.

6 Lessons Learnt from Tsunami Aid

Many important lessons have been learnt from the successes, constraints, and shortcomings in the mobilization, management, and utilization of unprecedented international assistance to countries, communities, and people affected by the Indian Ocean Tsunami. These provided opportunities for improving the framework of global systems and processes of humanitarian assistance to natural disasters and civil conflicts.

First, the quick inflow of a large volume of humanitarian assistance, immediately after the tsunami, made a speedy response to the humanitarian needs of the affected population possible, but it also created problems of plenty for many organizations, particularly the local NGOs, who were not used to handling large funds in such a short time. Many fly-by-night operators, without proven track records of commitments and experience in humanitarian response, emerged to grab a share of this influx of funds through their own appeals, but these were not followed up with reports and evaluations of how the funds were spent and for whose benefit. This raised larger issues of transparency and accountability of such operations.

Many credible organizations anticipated these dangers and withdrew from the rising responses to their appeals for assistance. For example, Médecins sans Frontières (MSF) closed its tsunami appeal after just one week, having raised six times more from their tsunami appeal than it had raised for the Darfur crisis in two months. A week later, Oxfam also began closing flash appeals for the tsunami and urged the public to donate to other, less high-profile, emergencies.

The amount of public funds raised for the tsunami, in comparison to other large-scale humanitarian disasters, highlighted the unequal and often unfair flow of funds for emergencies, and underscored the need to make humanitarian funding more equitable. The Good Humanitarian Donorship (GHD) principles of 2003 had provided a framework of internationally recognized standards for effective donor behaviour. Learning from the experiences of past disasters, including the Indian Ocean tsunami, the European Union adopted, in 2007, the European Consensus, which reiterated its commitment to the principles of humanity, neutrality, impartiality, and independence. A study conducted in 2014 found that these principles were not consistently acted on. The Forgotten Crises Assessment (FCA) Index developed by the European Community Humanitarian Office (ECHO) was another effort to raise awareness about the world's 'forgotten crises' that fail to receive donor and media attention.

The Indian Ocean tsunami provided an opportunity to establish an institutional mechanism for mobilizing humanitarian assistance from all sources on a regular basis so that a corpus of funds is always available for immediate response to catastrophic as well as forgotten disasters. UN General Assembly established this mechanism in 2005 by creating the Central Emergency Response Fund (CERF) with an annual target of pooling donations of USD 500 million. CERF made it possible to fast track emergency responses almost instantly with the onset of a crisis situation without going through the process of flash appeal. More recently, in April 2014, international NGOs launched their own pooled fund, known as Start Fund, designed specifically to fill gaps in the emergency funding system, by providing an early response for emergencies that fail to attract sufficient funding.

The second lesson learnt from the tsunami response is that sound and effective coordination mechanisms must be in place at all levels, and across all sectors, to avoid duplication and to ensure that humanitarian aid is available to all affected communities and individuals without any discrimination, favour, or exclusion; and,

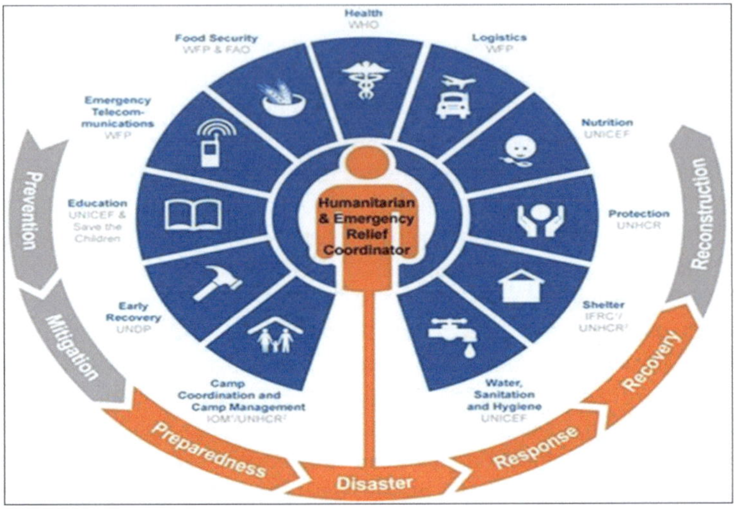

Fig. 6 The UN cluster system. *Source* UN Office for the Coordination of Humanitarian Affairs

furthermore, that such assistance is spread out to cover all basic needs. Generous international funding for the tsunami created serious challenges of coordination with multiple agencies at various levels. For example, nearly 200 international NGOs were operating in Aceh province of Indonesia alone, in addition to numerous agencies of the UN, regional organizations, and the national, provincial, and local governments. Overall coordination with the multitude agencies for disaster response, recovery, and reconstruction was by no means an easy task, even though there were arrangements for coordination within the UN and the Government. In order to address these challenges, the United Nations established a cluster approach for disaster response with one Under Secretary General responsible for overall coordination across the UN systems of disaster preparedness, response, emergency relief, and recovery as presented in Fig. 6.

An evaluation of the cluster approach, commissioned by the UN Office for the Coordination of Humanitarian Affairs (OCHA), found that, while this improved humanitarian response and relief, it has had difficulty addressing issues of pre-disaster risk reduction as well as post-disaster reconstruction, as these were beyond the mandate of the coordinating agency For disaster risk reduction the responsibility of coordination lies with the UN Office for International Strategy for Disaster Reduction (UNISDR), while there is no specific agency for reconstruction, which leaves large grey areas in inter-agency coordination. The local actors (national, provincial, and local governments, as well as the NGOs) are often excluded from the cluster approach, even though they play critical roles in disaster preparedness, response, and relief.

Another way for better coordination was to ensure that there is better compliance to common international standards of humanitarian response, relief, recovery, and reconstruction. Three of the most well-established and internationally recognized quality and accountability initiatives are the Humanitarian Accountability Partnership, People in Aid, and Sphere. All these standards were established much before the tsunami; but the valuable lessons learnt after the Indian Ocean tsunami and subsequent disasters made it imperative that the standards be reviewed and revised. Revised versions of three standards were issued, incorporating several lessons learnt during tsunami operations, particularly regarding minimum standards for life-saving activities such as provisions of food, water supply, sanitation, health, and shelter. In order to promote coherence among these standards and to make it easier for humanitarian agents to put the standards into practices, these three initiatives have been harmonized into a Core Humanitarian Standard (CHS), developed by the CHS Alliance which was launched in December 2014.[3]

The third lesson learnt in tsunami operations was that layers of exclusion and inequities remained despite the abundance of assistance. The joint evaluation report of the TEC identified several inequalities in the way that assistance was delivered. These included: national inequalities (for example, between conflict and non-conflict areas); inequalities by sector (for example, between the fishing sector and other sectors); geographical inequalities (between areas that were more and less accessible); social inequalities (between male and female, between ethnic and religious minorities and majorities, and between residents and migrants); and social inequalities (between poor and marginalized groups and better-off households). One of the factors that exacerbated these inequalities was a focus on replacing physical assets, such as houses or boats, which perpetuated existing inequalities. Households owning these assets before the tsunami had them replaced, resulting in more aid to better-off households. In addition, some groups, such as the fishing communities in India, were better organized and therefore were better able to both press for and access aid.

TEC's joint evaluation further pointed out that the needs of vulnerable women, children and older people were often overlooked. Women tended to be more vulnerable as they often had less access to resources, and emergency living conditions created higher work burdens and increased domestic and sexual violence. A survey conducted in Sri Lanka, as part of the TEC joint evaluation, found that, in general, women were less satisfied with the tsunami response than men. Another survey in Tamil Nadu in India found that ex-gratia payment of relief was made mostly to male heads of households who did not share it fairly with female members who were responsible for running the households. Consumption of alcohol among male members increased sharply and the expenditure on food and nutrition declined, affecting the health of mother and children. Many rehabilitation activities centred on male-dominated sectors, such as fishing; the livelihoods of women and other marginalized groups were largely overlooked. Many women felt that the livelihood projects geared towards them (such as mat-weaving) were not sufficient to provide a decent income. These were some of the socio-economic issues that could not be

[3]https://www.corehumanitarianstandard.org/the-standard.

addressed in the short run and required more in depth understanding and appreciation for designing appropriate policy and programme interventions for the long term.

The tsunami prompted more research into how disasters affect women differently. For example, a 2007 study in more than 140 countries covering the period from 1981 to 2002 found that natural disasters (and their impacts) kill more women than men (Neumayer and Plümper 2007). This effect is even more pronounced for women from poorer backgrounds. A study conducted in Aceh, Indonesia, after the tsunami, highlighted ways to ensure that humanitarian responses are sensitive to gender (Minza 2005). These include: providing separate toilet/latrine facilities and safe accommodation to ensure privacy and protect women from sexual harassment; providing accessible health facilities to provide care for pregnant women and babies as well as access to contraceptives; prioritizing livelihood activities for women as well as men, especially women headed households; involving women representatives in the coordination of aid distribution, and so on.

While some of these measures have been incorporated in international standards on humanitarian response, there are recurring challenges in applying these principles consistently into practice. More work is needed to ensure that the needs of marginalized and vulnerable groups are consistently and adequately taken into account in post-disaster relief, recovery, and reconstruction as much as in pre-disaster prevention, mitigation and preparedness.

The Indian Ocean tsunami brought into focus another critical area that remains to be addressed in a significant manner. This is the issue of capacity building of local communities who are always the first responders as well as the ultimate sufferers. A 2008 review of reconstruction in post-tsunami Indonesia and Sri Lanka found that community involvement is essential to building back safer and stronger communities (Kennedy et al. 2008). In the absence of effective enabling systems for participation of local communities, national and international programmes for disaster relief, recovery, and reconstruction have mostly been led, coordinated, and implemented by outside actors (Featherstone and Antequisa 2014).

After the tsunami, many new initiatives had been undertaken in several countries to strengthen community-based disaster preparedness. Most of these initiatives are limited to pilot projects that have shown encouraging results in building resilience of local communities to disasters. These success stories should be replicated in larger scale through active involvement and participation of local communities in disaster preparedness and response as well as in relief, recovery, and reconstruction. Participation of and accountability to local communities in all matters pertaining to delivery of humanitarian assistance should become integral part of national and international standards of relief. An oft quoted idiom that emerged during the tsunami relief and recovery is 'build back better'; the idea being that national governments, international organizations and NGOs should not only restore communities, but build safer, more resilient communities, able to withstand future disasters. This found express recognition in the fourth Priority of Action of the Sendai Framework for Disaster Risk Reduction that is titled 'Enhancing disaster preparedness for effective response and to "Build Back Better" in recovery,

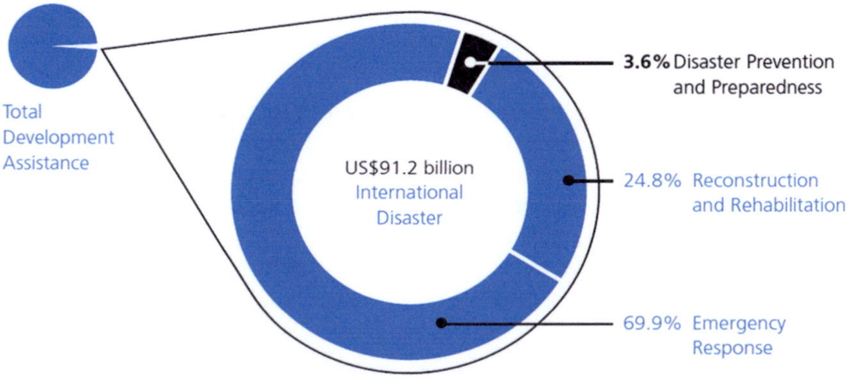

Fig. 7 Disaster-related international financing 1980–2010. *Source* Disaster Risks for a Resilient Future, GFDRR (2012)

rehabilitation, and reconstruction' (UNISDR 2015a, b). There has been some success in 'building back better' in large housing and infrastructure reconstruction programmes, particularly those funded by multi-lateral and regional banks, but building resilience in social and economic recovery programmes have been much more challenging.

Building resilience, through poverty reduction and the strengthening of public services, has the potential to reduce the financial and human costs of humanitarian emergencies and to prevent reversing important development gains. However, despite recognition of the far-reaching importance of disaster risk reduction (DRR), international investments on these activities have remained abysmally low (See Fig. 7). Global assistance for DRR represents only a small fraction of total funds allocated for disaster risk management, which in turn represents a small fraction of overall international development assistance (UNISDR 2015a, b). On average, only around 2% of the total international development assistance is allocated to disaster risk management and 75% of that budget is allocated to only a few countries (GFDRR 2012).

National and local governments have also tended to invest much more on disaster response and relief, than on disaster risk reduction. However, taken together, such investments may far outweigh international aid for disaster risk reduction (Kellet et al. 2014). According to a study conducted by Overseas Development Institute (ODI) in six countries—two each from Asia, Africa, and Latin America—national investments on DRR were several times higher than international aid. The study shows that the Philippines Government is investing 20 times more than the international community on disaster risk reduction, and the Government of Indonesia is investing almost 10 times more (Kellet et al. 2014).

Another critical area that remains to be addressed is capacity building of local communities who are always the first responders as well as the ultimate sufferers. A 2008 review of reconstruction in post-tsunami Indonesia and Sri Lanka found that community involvement is essential to building back safer, stronger communities (Kennedy et al. 2008). Capacity building and accountability to local communities are internationally recognized priorities, but various evaluations of the humanitarian system have shown that more efforts are needed to ensure that these principles are consistently put into practice. For example, a study found that, although there have been efforts to include local and national actors in the response to typhoon Haiyan (one of the strongest tropical cyclones on record which struck the Philippines in 2013), it has still been mostly led, coordinated, and implemented by international actors (Featherstone and Antequisa 2014).

7 Conclusion

The need for global humanitarian assistance is rising due to the increasing frequency and intensity of disasters, as well as the proliferation of insurgencies, conflicts, and migration of distressed communities. The resources mobilized through government, inter-government, non-government, and private sources are not adequate to meet these growing demands for humanitarian assistance, resulting in significant gaps in humanitarian funding. Overall gaps are estimated to be in the range of nearly half of the requirements. The Indian Ocean tsunami was a rare occasion when the mobilization of global humanitarian assistance almost matched, and in some places even exceeded, the needs. While this provided opportunities to go beyond immediate humanitarian relief to address some of the long-term issues of recovery and reconstruction, this also highlighted the inadequacies of the existing systems of humanitarian response in long-term planning and coordination, effectiveness in delivering assistance to the last mile, and issues related to transparency and accountability. Furthermore, this highlighted the significant inequality in the speed and level of funding for different humanitarian emergencies.

The thirteen years since the 2004 Indian Ocean tsunami have seen a number of positive steps for improving the predictability, efficiency, and sustainability of humanitarian response. New funding mechanisms for humanitarian response, such as CERF and Start Fund have been launched to significantly reduce the time lag in mobilization of resources for response. This has enabled humanitarian funding for low-profile emergencies that fail to attract media and public attention, thereby helping to address some of the issues of equity in humanitarian response. Principles of good humanitarian assistance have found wider acceptance among donors and aid agencies. Global standards for response and relief have been further improved and adopted by many stakeholders, including a growing number of national governments.

However, as yet, these measures have not been able to overcome the problems of insufficient funding and inequalities in the funding system. The following actions would be required to reduce the impact of future emergencies and to continue to improve the equity and quality of humanitarian responses.

(a) **Enhanced international funding for humanitarian response**: While funding for UN-coordinated appeals has improved over the years, there are still wide gaps, which seem to be widening due to growing needs for humanitarian assistance. Considering the size of the global economy, and the national and per capita income of OECD and other countries, these gaps are not very difficult to be bridged, provided there is political will and pressure of public opinion for a more just and humane world. Closing this gap would provide much needed relief to the millions of people affected by natural disasters and conflict every year.

(b) **Investment in disaster risk reduction and resilience**: Long-term measures for reducing the vulnerabilities to disasters—through poverty reduction, strengthening of public services, and investment in disaster risk reduction—has the potential to prevent and/or reduce the costs and overall impacts of humanitarian emergencies. However, funding for prevention and preparedness still remains low, and needs to be enhanced substantially through domestic and international assistance.

(c) **Impartial, needs-based international funding**: Much of international assistance is committed without adequate assessment of either the immediate needs for humanitarian assistance or the long-term needs of recovery and reconstruction, due to media pressures and/or for political reasons. Often aid is pledged for specific disasters or/and selected countries and communities. This has resulted in aid inequality within and among countries and sectors, and created an aid-imbalance between immediate response and relief, and long-term recovery and reconstruction.

(d) **Government donations**: Both the number of government donors and quantum of funds donated have increased over the years, but not all these donations are without conditions. Political and strategic considerations have influenced many donations that are earmarked for specific countries and for earmarked activities that are not always the priority for humanitarian response. Government donations are also not always channelled through the UN system. This has perpetuated the existing inequity and unpredictability in the funding system. Adherence to the principles of Good Humanitarian Donorship can help to ensure that donation decisions are based on humanitarian need. Increased contributions to pooled funds like CERF can also improve the equity and speed of humanitarian responses.

(e) **Private donations**: The quantum of private donations for the Indian Ocean tsunami, accounting for more than 67% of all donations, demonstrated the potential of private philanthropy for global humanitarian funding. Even if half of this amount is mobilized on a recurring basis, it may reduce the gaps in humanitarian funding. A range of factors—including location of disasters, nationality of affected persons, level of media coverage, pressure of public opinion, and so on—influence private donors. These cannot be the same for every crisis situation. Therefore channelizing private donations for multilateral- and NGO-pooled funds might help to ensure that private funding is more regular, predictable, and needs-based.

(f) **Coordination of humanitarian response**. Coordination among multiple agencies involved in humanitarian response at the global, national and local levels are crucial for the effectiveness of such response. Existing coordinating mechanisms should be further improved and new mechanisms created with standard operating procedures for institutionalizing coordination at all levels. Successfully functioning mechanisms in some regions and countries should be replicated on a larger scale, especially in countries and regions facing recurring disasters and conflict situations.
(g) **Addressing needs of marginalized and vulnerable populations**. While inclusive humanitarian responses are a global priority, as laid down in Good Humanitarian Donorship principles, Sphere standards, and Core Humanitarian Standards, more efforts are needed to ensure that this priority is consistently put into practice.
(h) **Capacity building for preparedness and response**. Capacity building of local communities for preparedness for disasters and other crisis situations are crucial for effective response. Similarly, capacity building of humanitarian actors is important for compliance of humanitarian standards. Hence, concerted efforts are required to develop and improve all round capacity in humanitarian interventions, particularly as they relate to strengthening local capacity for disaster preparedness and response to emergencies.
(i) **Addressing pre-existing conflicts**. Disasters provide opportunities for bringing together communities divided by pre-existing ethnic, religious, or political conflicts as they are similarly affected and have issues of common concern that can be addressed jointly. Disasters further provide opportunities to governments to approach communities estranged by conflicts and violence by providing them humanitarian assistance for rebuilding their lives and livelihoods. The Failure to provide timely and adequate assistance in times of need may exacerbate pre-existing conflicts. Therefore, it is extremely important that humanitarian agencies design appropriate conflict-sensitive approaches for humanitarian response that either de-escalate, or at least avoid exacerbating, tensions between different groups.
(j) **Tracking and monitoring humanitarian aid**: UNOCHA has developed an international system for tracking humanitarian assistance for emergency response. The system is able to track and monitor every government and private donation that is routed through the UN system. However, the system is not able to track such assistance outside the UN system unless these are reported voluntarily. Much of bi-lateral and private donations are not reported and hence not tracked at all through the UN system. The system is also unable to capture data on remittances during humanitarian emergencies, non-monetary donations of goods and services, and domestic humanitarian responses. The system is also unable to track where and how humanitarian aids are disbursed, and to monitor the outcome and impacts of such aid through an indicator-based monitoring system. This definitely calls for a more comprehensive system for tracking and monitoring international humanitarian response before, during, and after disasters and other emergency situations.

Most of these issues came up for discussion in the first ever World Humanitarian Summit held in Istanbul on 23–24 May 2016. The Summit resulted in about 1,500 commitments from 400 UN Member States and other organizations. These include a set of 51 commitments known as the 'Grand Bargain' to reform humanitarian financing to make emergency aid more efficient and effective.[4] However, the Summit failed to adopt a common framework or agreement on humanitarian assistance, despite the mounting number of emergency situations arising out of civil conflicts and natural and manmade disasters. It remains to be seen how the Grand Bargain translates itself into an agenda for action to make national and international humanitarian assistance more efficient and effective in the years to come.

References

Barrientos, A., D. Hulme, and A. Shepherd. 2005. 'Can Social Protection Tackle Chronic Poverty?' *The European Journal of Development Research* 17(1).

Christoplos, Ian. 2006a. *Links between Relief, Rehabilitation and Development in the Tsunami Response: A Synthesis of Initial Findings.* London: Tsunami Evaluation Coalition (TEC).

Christoplos, Ian. 2006b. "The Elusive Window of Opportunity" for Risk Reduction in Post Disaster Recovery, *UNISDR.*

Development Initiatives. 2006. *The Global Humanitarian Assistance.* London: Discript Ltd.

Buchanan-Smith, Margaret, and Simon Maxwell. 1994. Linking Relief and Development: An Introduction and Overview. *IDS Bulletin* 25(4).

Featherstone A., and C. Antequisa. 2014. Missed Again: Making Space for Partnership in the Typhoon Haiyan Response. *Action Aid, CAFOD, Christian Aid, Oxfam GB.*

Flint, Michael, and Hugh Goyder. 2006. *Funding the Tsunami Response: A Synthesis of Initial Findings.* London: Tsunami Evaluation Coalition (TEC).

GFDRR. 2012. Managing Disaster Risks for a Resilient Future: A Strategy for the Global Facility for Disaster Reduction and Recovery 2013–15. Washiington DC, USA.

Hastrup, F. 2011. *Weathering the World: Recovery in the Wake of the Tsunami in a Tamil Nadu Fishing Village.* Berghan Books.

IFRC. 2003. Principles and Good Practices of Humanitarian Donorship. Stockholm.

Kellet, J., A. Caravani, and F. Pichon. 2014. *Financing Disaster Risk Reduction—Towards a Coherent and Comprehensive Approach.* New York: UNDP.

Kennedy, Jim, Joseph Ashmore, Elizabeth Babister, and Ilan Kelman. 2008. The Meaning of 'Build Back Better': Evidence From Post Tsunami Aceh and Sri Lanka. *Journal of Contingencies and Crisis Management* 16(1).

Lindberg, M. 2014. *Post Tsunami Recovery in Thailand: Socio-Cultural Responses.* Routledge.

McGilvray, Dennis B., and Michele R. Gamburd. (2013) *Tsunami Recovery in Sri Lanka: Ethnic and regional Dimensions.* London: Routledge.

McKeon, Jock, and Harry Masyrafah. 2008. Post-tsunami Aid Effectiveness in Aceh: Proliferation and Coordination in Reconstruction. https://www.brookings.edu/research/post-tsunami-aid-effectiveness-in-aceh-proliferation-and-coordination-in-reconstruction. Accessed on 11 May 2016.

Minza, W. 2005. *Gender and Changes in Tsunami-Affected Villages in Nanggroe Aceh Darussalam Province.* Oxford: Oxfam GB.

[4]https://consultations.worldhumanitariansummit.org/bitcache/075d4c18b82e0853e3d393e90af18acf734baf29?vid=581058&disposition=inline&op=view.

Neumayer, Eric, and Thomas Plümper. 2007. The gendered nature of natural disasters: the impact of catastrophic events on the gender gap in life expectancy, 1981–2002. *Annals of the Association of American Geographers* 97 (3): 551–566.
Reddy, S. 2013. *Clash of Waves: Post Tsunami Relief and Rehabilitation in Andaman and Nicobar Islands*. Idos Books.
Saini, A. 2013. Post-Tsunami Socio-Cultural Changes among the Nccobarese: An Ethnography of the Nicobarese of the Southern Nicobarese Islands. *Tata Institute of Social Sciences*.
Shaw, R (ed.). 2006. *Recovery from the Indian Ocean Tsunami Disasters*. Emarald.
Telford, John, and John Cosgrave. 2006. *Joint Evaluation of International Response to Indian Ocean Tsunami: Synthesis Report*. London: Tsunami Evaluation Coalition (TEC).
UNISDR. 2015a. *Making Development Sustainable: The Future of Disaster Risk Management. Global Assessment Report on Disaster Risk Reduction*. Geneva, Switzerland
UNISDR (United Nations International Strategy for Disaster Reduction). 2015b. Sendai Framework for Disaster Risk Reduction 201-52030. http://www.wcdrr.org/uploads/Sendai_Framework_for_Disaster_Risk_Reduction_2015-2030.pdf. Accessed Apr 2015.
World Health Organization Regional Office for South East Asia. 2013. *Tsunami 2004: A Comprehensive Analysis*, vol. 1. New Delhi.
The World Bank. 2013. Post disaster need assessment. Washington DC, USA.

MIX
Papier aus verantwortungsvollen Quellen
Paper from responsible sources
FSC® C105338

FSC
www.fsc.org

If you have any concerns about our products,
you can contact us on
ProductSafety@springernature.com

In case Publisher is established outside the EU,
the EU authorized representative is:
Springer Nature Customer Service Center GmbH
Europaplatz 3, 69115 Heidelberg, Germany

Printed by Libri Plureos GmbH
in Hamburg, Germany